IN THE NAME OF THE GREAT WORK

The Environment in History: International Perspectives

Series Editors: Dolly Jørgensen, *University of Stavanger*; Christof Mauch, *LMU Munich*; Kieko Matteson, *University of Hawai'i at Mānoa*; Helmuth Trischler, *Deutsches Museum, Munich*

ENVIRONMENT AND SOCIETY

VOLUME 1
Civilizing Nature: National Parks in Global Historical Perspective
Edited by Bernhard Gissibl, Sabine Höhler and Patrick Kupper

VOLUME 2
Powerless Science?: Science and Politics in a Toxic World
Edited by Soraya Boudia and Natalie Jas

VOLUME 3
Managing the Unknown: Essays on Environmental Ignorance
Edited by Frank Uekötter and Uwe Lübken

VOLUME 4
Creating Wilderness: A Transnational History of the Swiss National Park
Patrick Kupper
Translated by Giselle Weiss

VOLUME 5
Rivers, Memory, and Nation-Building: A History of the Volga and Mississippi Rivers
Dorothy Zeisler-Vralsted

VOLUME 6
Fault Lines: Earthquakes and Urbanism in Modern Italy
Giacomo Parrinello

VOLUME 7
Cycling and Recycling: Histories of Sustainable Practices
Edited by Ruth Oldenziel and Helmuth Trischler

VOLUME 8
Disrupted Landscapes: State, Peasants and the Politics of Land in Postsocialist Romania
Stefan Dorondel

VOLUME 9
The Nature of German Imperialism: Conservation and the Politics of Wildlife in Colonial East Africa
Bernhard Gissibl

VOLUME 10
In the Name of the Great Work: Stalin's Plan for the Transformation of Nature and its Impact in Eastern Europe
Doubravka Olšáková

In the Name of the Great Work

Stalin's Plan for the Transformation of Nature and its Impact in Eastern Europe

Edited by
Doubravka Olšáková

First published in 2016 by

Berghahn Books

www.berghahnbooks.com

© 2016, 2019 Doubravka Olšáková
First paperback edition published in 2019

All rights reserved. Except for the quotation of short passages
for the purposes of criticism and review, no part of this book
may be reproduced in any form or by any means, electronic or
mechanical, including photocopying, recording, or any information
storage and retrieval system now known or to be invented,
without written permission of the publisher.

Library of Congress Cataloging-in-Publication Data

Names: Olsakova, Doubravka, editor of compilation.
Title: In the name of the great work : Stalin's Plan for the Transformation of Nature and its impact in Eastern Europe / edited by Doubravka >Olsakova.
Description: New York : Berghahn Books, 2016. | Series: The environment in history : international perspectives ; volume 10 | Includes bibliographical references and indexes.
Identifiers: LCCN 2016024960| ISBN 9781785332524 (hardback : alkaline paper)
| ISBN 9781785332531 (ebook)
Subjects: LCSH: Environmental policy--Europe, Eastern--History--20th century.
| Environmental policy--Soviet Union--History. | Stalin, Joseph, 1878-1953--Political and social views. | Socialism—Environmental aspects--Europe, Eastern--History--20th century. | Nature--Effect of human beings on--Europe, Eastern--History--20th century. | Environmental impact analysis--Europe, Eastern--History--20th century. | Environmental degradation--Europe, Eastern--History--20th century. | Social change--Europe, Eastern--History--20th century. | Europe, Eastern--Environmental conditions--History--20th century. | Europe, Eastern--Social conditions--20th century.
Classification: LCC GE190.E852 I5 2016 | DDC 333.70947/09041--dc23 LC record available at https://lccn.loc.gov/2016024960

British Library Cataloguing in Publication Data
A catalogue record for this book is available from the British Library

ISBN 978-1-78533-252-4 hardback
ISBN 978-1-78920-502-2 paperback
ISBN 978-1-78533-253-1 ebook

Table of Contents

List of Tables	vi
Acknowledgements	vii
Abbreviations	viii
Introduction: The Stalin Plan for the Transformation of Nature, and the East European Experience *Paul Josephson*	1
1. Kafkaesque Paradigms: The Stalinist Plan for the Transformation of Nature in Czechoslovakia *Doubravka Olšáková and Arnošt Štanzel*	43
2. Untamed Seedlings: Hungary and Stalin's Plan for the Transformation of Nature *Zsuzsanna Borvendég and Mária Palasik*	126
3. The Conspiracy of Silence: The Stalinist Plan for the Transformation of Nature in Poland *Beata Wysokińska*	226
Conclusion: Environmental History, East European Societies, and Totalitarian Regimes *Doubravka Olšáková*	290
Index	303

List of Tables

Table 1.1	Growth in collective farms in Czechoslovakia	50
Table 1.2	The prognosis of cultivating rice in Czechoslovakia	71
Table 1.3	An overview of the largest water management projects in Czechoslovakia	101

Acknowledgements

The research was supported by the Czech Science Foundation (Research Grant reg. no. 15-04902S), the National Cultural Fund of Hungary (3802/04556), the Historical Archives of the Hungarian State Security, the Institute for Contemporary History of the Academy of Sciences of the Czech Republic, the Institute for the History of Science of the Polish Academy of Sciences, and the Visegrad Fund. Our sincere thanks to all of these.

Abbreviations

AARI	Association of Agricultural Research Institutes (Svaz výzkumných ústavů zemědělských)
CAA	Czech Agricultural Academy (Česká akademie zemědělská)
CzAA	Czechoslovak Academy of Agriculture (Československá akademie zemědělská) [this comprised the CAA and the SAA]
CAAS	Czechoslovak Academy of Agricultural Sciences (Československá akademie zemědělských věd)
CAS	Czechoslovak Academy of Sciences (Československá akademie věd)
CIR	Central Agriculture Institute (Centralny Instytut Rolniczy)
COMECON	Council for Mutual Economic Assistance (Совет Экономической Взаимопомощи, *Sovet Ekonomicheskoy Vzaimopomoshchi*)
CUP	Central Planning Office (Centralny Urząd Planowania)
DISZ	Association of Working Youth (Dolgozó Ifjúsági Szövetség)
FKGP	Small Holders' Party (Független Kisgazdapárt)
FM	Ministry of Agriculture (Földművelésügyi Minisztérium)
GMO	Genetically Modified Organism
GOP	Upper Silesian Industrial Region (Górnośląski Okręg Przemysłowy)
GUS	Central Statistical Office (Główny Urząd Statystyczny)
HUF	Hungarian Forint
IBL	Forest Research Institute (Instytut Badawczy Leśnictwa)
ICSU	International Council of Scientific Unions
IHAR	Plant Breeding and Acclimatization Institute (Instytut Hodowli i Aklimatyzacji Roślin)
IMUZ	Institute for Land Reclamation and Grassland Farming (Instytut Melioracji i Użytków Zielonych)

Kčs	Czechoslovak Crown (Koruna československá)
KRN	National People's Council (Krajowa Rada Narodowa)
KSČ	Communist Party of Czechoslovakia (Komunistická strana Československa)
KSS	Communist Party of Slovakia (Komunistická strana Slovenska)
MDP	Hungarian Workers' Party (Magyar Dolgozók Pártja)
MKK	Agricultural Experiment Center (Mezőgazdasági Kísérletügyi Központ)
MKP	Hungarian Communist Party (Magyar Kommunista Párt)
MNL OL	National Archives of Hungary (Magyar Nemzeti Levéltár Országos Levéltára)
MTA	Hungarian Academy of Sciences (Magyar Tudományos Akadémia)
MTH	Labor Force Reserve Office (Munkaerőtartalékok Hivatala)
MTI	Hungarian News Agency (Magyar Távirati Iroda)
MTK	Agricultural Science Center (Mezőgazdasági Tudományos Központ)
MTT	Hungarian Science Council (Magyar Tudományos Tanács)
MW	Megawatt
NÉKOSZ	National Association of People's Colleges
NHKG	New Steelworks of Klement Gottwald (Nová huť Klementa Gottwalda)
NKVD	People's Commissariat for Internal Affairs (Народный комиссариат внутренних дел, Narodnyy Komissariat Vnutrennikh Del)
OT	National Planning Office (Országos Tervhivatal)
PAN	Polish Academy of Sciences (Polska Akademia Nauk)
PAU	Polish Academy of Arts and Sciences (Polska Akademia Umiejętności)
PFZ	State Land Fund (Państwowy Fundusz Ziemi)
PGR	State Agricultural Farms (Państwowe Gospodarstwa Rolne)
PKPG	State Commission for Economic Planning (Państwowa Komisja Planowania Gospodarczego)
PKWN	Polish Committee of National Liberation (Polski Komitet Wyzwolenia Narodowego)

PRL	People's Republic of Poland (Polska Rzeczpospolita Ludowa)
PROP	State Council for Nature Conservation (Państwowa Rada Ochrony Przyrody)
PTB	Polish Botanical Society (Polskie Towarzystwo Botaniczne)
PTL	Polish Forest Society (Polskie Towarzystwo Leśne)
PWRiL	State Publishing House for Agriculture and Forestry (Państwowe Wydawnictwo Rolnicze i Leśne)
PZPR	Polish United Workers' Party (Polska Zjednoczona Partia Robotnicza)
SAA	Slovak Agricultural Academy (Slovenská poľnohospodárska akadémia)
SWMP	State Water Management Plan (Státní vodohospodářský plán)
SZDP	Social Democratic Party (Szociáldemokrata Párt)
TRJN	Provisional Government of the Republic of Poland (Tymczasowy Rząd Jedności Narodowej)
VASKhNIL	Lenin All-Union Academy of Agricultural Sciences (Всесоюзная академия сельскохозяйственных наук имени В. И. Ленина)
VŽKG	Klement Gottwald Vítkovice Ironworks (Vítkovické železárny Klementa Gottwalda)
WSR	University of Wrocław (Uniwersytet Wrocławski)

Introduction
The Stalin Plan for the Transformation of Nature, and the East European Experience

Paul Josephson

In October 1948 the Communist Party of the USSR unanimously passed the Stalin Plan for the Transformation of Nature. According to the propaganda of the time, nature itself would be subject to the party's dictates.[1] No longer would droughts, hot, dry winds (*sukhovei*), energy shortfalls, or agricultural failures prevent Stalin from achieving superhuman targets in industry and agriculture. The same propaganda claimed that major rivers would be turned into machines, with stepped reservoirs and hydroelectric power stations. Rather than flowing "uselessly" downstream, the water would serve year-round purposes of power generation, irrigation, municipal supply, and industrial processes; they might build a total of 45,000 reservoirs and ponds. Foresters audaciously approved the task of planting 70,000 kilometers of forest shelterbelts—30 to 100 meters deep—to protect farmland from winds and to keep moisture.

Although a massive undertaking for any society, the Stalin Plan for the Transformation of Nature was in fact one program dedicated to improving agricultural performance in the European part of the country, and in particular in the steppe region of the south. It then became connected with a series of related plans to transform and remodel nature in the USSR, some of which, including the 1948 Plan, became known as "hero projects" (*velikie stroiki*) of communism.

In this way the Stalin Plan was more than the 1948 decision with its focus on the European USSR. It was folded into a larger program of economic, scientific, and cultural construction. It was a product of the era of high Stalinism, of bold determination to finish rebuilding the economy from the devastation of World War II and to advance its military to compete with the United States in the nascent Cold War. Scientists and engineers throughout the nation took advantage of the 1948 proclamation and implicit state support to join their research and development programs to the plan. They gained support to create new institutes. They expanded their foci and intentions to hydroelectricity and land reclamation. Engineers designed locks and canals to improve inland river transport with the goal of linking major seas. In the designs of planners and visionaries the Stalin Plan of 1948 was soon extended to Central Asia and to the expansion and creation of a series of research and construction institutes dedicated to irrigation and canals. The irrigation water would turn vast regions of Central Asian desert, rich in soil nutrients but low in rainfall, into productive farmland and cotton and citrus plantations. There are maps of the period that show the Aral Sea, fed by Stalin-era canals, to serve rice culture. Later Nikita Khrushchev saw the 1948 Plan as part of his Virgin Lands campaign in the mid-1950s.

The Stalin Plan quickly became the focus of myriad articles in magazines, journals, and newspapers, and of public pronouncements about hero construction projects of Communism that extended into Siberia. Literary magazines paid homage to the plan as a vehicle of state and cultural construction. I therefore refer to the "Stalin Plan" for its various scientific, economic, social, and cultural meanings, not only its initial European focus, and for its inspirational and motivational messages among the workers, peasants, Communist Youth League, and Communist Party members who rushed to embrace it. To varying degrees and intensities, with different emphases and possibilities, East European socialist nations, especially in their Stalinist periods of roughly 1948 to 1953, also engaged the Stalin Plan and referred to it as such.

Here I also use the "Stalin Plan" as shorthand for a series of government resolutions for dam, reservoir, canal, forestry, roadway and other construction projects, some of which dated to the 1930s; some required significant investments and institutional expansion even to begin in the years following the promulgation of the plan. The Council of Ministers and the Central Committee of the Communist Party passed the Stalin Plan itself, "On the Plan for Forest Defense Belts, Grass–Arable Crop Rotations, and the Construction of Ponds and Reservoirs for the Guarantee of Highly Stable Harvests in Steppe Regions and Forest-Steppe Regions of the European Part of the USSR," on 20 October 1948. The dams, canals, irrigation and other projects, many with pre-war roots, were then embedded in this effort.

For example, the government and party passed resolution No. 1339 on 10 August 1937 to build the Kuibyshev hydroelectric complex; the war ended any nascent construction at the site, and the government did not publish a new resolution to build the Kuibyshev power station until 21 August 1950. The Stalin Plan itself was a belated response to droughts in 1946 in Ukraine, Northern Caucasus, Black Earth, Volga, West Siberia, and Kazakh regions that, along with Stalin's murderous investment policies favoring industry over agriculture, led to famine in 1946 (peaking in August 1947) and caused at least one million deaths. To follow through on the forest defense belts thousands of kilometers in length, the authorities created a design institute, Agrolesproekt, in 1949, that exists in a modern incarnation, to design and carry out forestry projects along the Dniepr, Don, Volga, Ural, and other rivers. Construction on the Volga–Don Canal actually predated the "Stalin Plan" by six months, and itself dated in designs to 1944 and in the popular press to an article in *Tekhnika-Molodezhi* in 1938. All of these projects were gulag slave labor projects, and many of the lead engineers and design institutes were gulag organizations.[2] While scaled back and eventually abandoned after the death of Stalin in 1953 owing to significant costs, a more accurate understanding of the technical limitations of Soviet "geoengineering" organizations, and likely the absence of the will of Stalin, the plan indicated the great potential for reworking nature in the Soviet Union given the tremendous momentum that ministries and construction firms had acquired and the absence of any public opposition to the environmentally disruptive and fantastical plans.

In East Central Europe after World War II, the newly socialist states followed major aspects of the Stalinist program: secret police and trials; centralization of cultural and educational institutions and their control by party officials; a planned economy; the construction of such "hero" cities as Sztálinváros, Hungary, Nowa Huta, Poland, and Dmitrovgrad, Bulgaria, in line with the single-profile cities arising throughout the USSR; rapid industrialization; and the collectivization of agriculture. Also included in the effort to remold society to one degree or another in socialist Poland, Hungary, and Czechoslovakia was the transformation of nature. But because of firmly established scientific traditions in those countries, their relatively small geographical size compared with the USSR, and the death of Stalin in 1953 before nature transformation programs could be fully established, the impact of nature transformation was smaller and shorter-lived than in the USSR.

On the one hand, wherever it was pursued, the Stalinist plan to transform nature was not very different from many other large-scale nature transformation projects of the twentieth century—the construction of large dams and canals in India and Brazil (for example, the Sardar Sarovar and Tucurui, respectively); the taming of the American West by the Army Corps of Engineers and the Bureau of Reclamation; the harvesting of the rain forests of Brazil and Indonesia; the Panama and Suez canals (the former at a cost of 22,000 lives); geoengineering along the major rivers of China including the Three Gorges Dam (with 1.5 million people ousted from their traditional homes in the flood plain); and many other such expensive, extensive, and environmentally questionable projects that moved ahead with significant social displacement and loss of human life.

Yet Stalinist transformation was geoengineering and agricultural engineering run amok. Leading party officials and scientists set forth aggressive programs to subjugate rivers, streams, steppe, forest, and croplands to party dictates. Planners in central cities, many of them without agricultural knowledge, dictated the establishment of citrus or cotton cultures in regions that could not, for climate, soil and other reasons, support them. Granted, many scientists opposed the hubristic plans, or at least one or another aspect of them, if not for their scientific futility, then for their waste of resources.[3] But they opposed the plans in a hostile environment that might lead to official censure or loss of a job or worse, because Stalin and the party had spoken. Many other forestry, hydrology, and other specialists embraced a transformationist ideology steeped in Lamarckian faith in the ability and desirability of scientists to adopt and adapt crops and lands to each other, and for plants and animals to pass along their acquired characteristics to the next generation; in the socialist experience this was called Lysenkoism after the Russian peasant farmer Trofim Lysenko whose Lamarckian policies were officially endorsed by Communist leaders. These specialists benefited in the support to their institutes and research programs. They had no doubt that damming, dredging and straightening waterways was always worth the cost, and that the benefits outweighed potential ecosystem damage, including the flooding of millions of hectares of land.[4]

Whether in the USSR or Eastern Europe, they carried out extensive propaganda campaigns to convince the masses of the utility and glory of their plans, co-opt their assent, and preclude their opposition. They stressed the indisputable fact that such projects were possible only under socialism—and only under the wise gaze of Stalin. It helped that many East European Communist officials had been trained in Moscow through the Communist International, beginning in the interwar years, and had been fully converted to the belief the state must control nature no less than industry and man. And Stalin's Red Army ensured that each nation would embrace Soviet socialism to one degree or another.

Nature transformation in socialist Eastern Europe was large scale, costly, and environmentally unsound—and abandoned quickly after the death of Stalin. It diverted needed investment funds from other programs, was based on unsound scientific calculations, and would have been inappropriate in East Central Europe in any event given the relatively small size of the nations compared to the USSR, not to mention different scientific traditions and social structures. The programs in each country differed according to their level of economic development, social structure, educational system and its goals, extent of natural resources, degree of urbanization, and so on. The countries had long shared borders, and had only recently abandoned private property. This too served as an obstacle to revolutionary change. Furthermore, they all faced significant pressures and costs to rebuild from World War II—and nature transformation was quite expensive.

The East European socialist nations were not a monolithic communist "bloc,"[5] and pursued nature transformation to different ends and with more or less enthusiasm. Yet certain common features of East European socialism had an impact on environmental thinking and environmental change in East Central Europe after World War II—especially the Stalinist economic plans for the region. To a greater or lesser degree, the German Democratic Republic, Czechoslovakia, Poland, Bulgaria, Romania, and Hungary embraced one-party systems that were organized through plebiscites, subterfuge, and coercion in which Stalinist Communist parties dominated public and private life. The parties organized show trials of intellectuals and other innocent people with suspect political views, and even of their own party members, with the foregone outcome of guilty verdicts, prison terms, and executions like those in the USSR; the goal was the creation of a class of worker-peasant intellectuals loyal to socialism. The state claimed to organize the economy for the worker through central planning and three-, five-, and seven-year plans. Governments pursued rapid industrialization with a focus on heavy industry, and ignored, at least initially, housing, food, and other sectors that would have benefited the worker more immediately. The peasant was forced into collectivized agriculture, although many peasants were permitted to own private plots that contributed a small, but important share of agricultural production.

Leaders and many scientists and engineers came to see the environment, like the people in it, as a malleable object, a source of energy, ore, and water to be tamed by the glorious plan in the name of the proletariat. The nations even pursued their own versions of the Stalinist plan to transform nature. Considering the human costs of World War II and the destruction of housing stock, the decision to focus on heavy industry and geoengineering rather than human recovery reveals one of the most disturbing features of the Stalinist model—the emphasis on industrial production rather than human capital. The hubris and impact of the plans to transform nature itself quickly and without adequate study, even if never fully achieved, reveal the danger of environmental management in authoritarian regimes. All of these aspects of the Stalin Plan are clear in scientific, scientific popular, and literary works as it evolved from 1948.

The Nature of the "Environment" under Socialism

The Stalin Plan to Transform Nature had prerevolutionary roots in a series of interbasin water transfer, dams and canals, peasant settlement to the ends of the empire, and other projects. But crucial for its success was a government and Communist Party apparatus anxious to transform the economy, society, and nature itself into a socialist wonderland of plenty. The decision to pursue the plan grew out Stalin's self-proclaimed great break (*Velikii Perelom*) with past institutions and approaches that arrested Soviet cultural and economic development at the beginning of the 1930s; the seeming victory of the Bolsheviks over the economy and agriculture during the five-year plans; a determination to rebuild the economy after World War II; and the enthusiasm of significant numbers of scientists and engineers to work with the Bolsheviks on heroic projects, with welcome government support for them and their research programs.

Russian dreamers, scientists, and poets had long imagined great projects. The nineteenth century poet Nikolai Nekrasov, who grew up on his father's estate, Greshnevo, near the Volga, witnessed the hard life of Volga boatmen. But in this poetic excerpt from "On the Volga," he saw the nation's future connected with river commerce.

> I foresee the beginning
> Of the new times and different scenes
> In the fortuitous life
> Of my favorite river:
> Liberated from shackles,
> The tireless nation
> Will mature and densely settle
> The coastal deserts;
> Science will deepen the waters;
> Through their smooth plain
> Giant ships will swim
> Like an uncountable crowd,
> And the exhilarating work will be eternal
> Above the endless river . . .

From their first days in power, the Bolsheviks set out to create socialist industry, establish workers' control of major enterprises, give land to the peasants, and nationalize all forests, waters, and subsoil minerals with the goal of rational use.[6] Many of Lenin's early proclamations concerned forestry, agriculture, and irrigation. Early Bolshevik leaders believed that under socialism they could indeed transform desert into productive farmland, end poor forestry practices, and even eliminate forest fires.[7] However, the Russian Revolution and resulting civil war and anarchy put "nature" at great risk. This risk abated during the New Economic Policy of the mid-1920s. Several engineering projects commenced at this time in the Caucasus and Steppe regions near Saratov, Samara, and Astrakhan. Stalin himself wrote about several of them and stressed the importance of using irrigation to increase grain production in the Trans-Volga region.[8]

At the same time nascent environmentalism grew among professionals to establish inviolable nature protection areas and encourage conservation of resources. Ecologists and other scientists, amateur naturists, and a variety of citizens participated in an extensive nature movement in the USSR. The movement had prerevolutionary roots. Some of its organizations became the largest voluntary societies in the USSR and included the Moscow Society for the Admirers of Nature and the All-Union Society for the Preservation of Nature. These organizations actively sought to temper breakneck economic development and successfully lobbied the government to establish a series of nature preserves (called *zapovedniki*). From the 1930s specialists, however, the organizations and individual scientists were co-opted and coerced to figure out ways to use nature for the economy, and they struggled to protect the *zapovedniki* and promote conservation measures against state efforts to encroach on the nature preserves and subjugate greater and greater tracts of forest and field to development programs.[9] Given the imperatives of the Stalinist economic model and the lack of autonomy of specialists, they had limited influence on the scale or content of nature transformation plans. As we shall see, in Eastern Europe these early environmentalists managed to maintain greater autonomy and worked quietly to limit the impact of transformationist plans.

But during Stalin's great break, party officials, economic planners, and engineers joined in the effort to master the empire's extensive natural resources toward the end of economic self-sufficiency and military strength. At their order, armies of workers began the process of constructing giant dams and reservoirs on major European rivers—the Don, Dniepr, and Volga. They planned extensive irrigation systems across Central Asia. They built canals and waterworks. The workers erected massive chemical combines, metal smelters, and oil refineries in both European and Siberian parts of the country, paying little attention to the pollution they produced. They put up entire cities to house the laborers whom they exhorted to meet plans and targets, irrespective of the environmental costs and the risks to the workers' own health and safety. Scientists, party officials, and writers produced self-conscious, self-serving literature praising these hero projects, without any sense of the ideological ironies, human sufferings, or environmental degradation that ultimately accompanied this "heroism." Beginning with the White Sea–Baltic Canal, they employed gulag slave laborers in murderous large-scale projects. Indeed, many of the hero projects were in fact built by gulag organizations. The major Soviet hydro-engineering design and construction firm, Zhuk Gidroproekt, actually was born in the blood and lives of seventy thousand prisoners at the White Sea–Baltic Canal Construction. The Belbaltlag slave camp director, Sergei Zhuk, used prisoners as fodder and used them to set up other canal and dam construction organizations that moved down the Volga and built Stalin's water works. (Gidroproekt exists in the twenty-first century under Vladimir Putin as RusGidro.)

Immediately after the White Sea–Baltic Canal, many Belbaltlag prisoners were sent to build the Moscow–Volga Canal. In 1931 "Comrade Stalin proposed to build a canal and turn the Volga to the walls of the Kremlin." They built the canal in four years, with 240 major structures including locks, pumping stations, dams, and tunnels. They excavated 200 million cubic meters of rock and soil, cut the shipping distance from Moscow to Gorky by 110 kilometers, and provided water for Moscow industry and residents through the Moscow River.[10] In July 1937 the first ships traversed the canal, cutting the distance between Leningrad and Moscow by 1,110 km and between Moscow and Gorky by 110 km. They built the Ivan'kovskii (1937), Uglichskii (1939), and Shcherbakovskii (1941) waterworks along the Volga River. They created the Moscow Sea at 327 square kilometers; the Rybinskoe Reservoir at 4,100 square kilometers; and other massive seas.[11] (As ships entered the canal from the Volga, they passed by statues of Lenin and Stalin on either side; Stalin's was removed in 1961 but the pedestal remains, and Dubna city residents now use it to train for rock climbing.) The camps had their own environmental and of course human costs. In this way, the pattern for the Stalin Plan of 1948 was established: large-scale projects carried out by armies of laborers, many of them slave laborers.

Not only canals and dams, but bridges, forestry, smelters, mines and factories—and their construction organizations—spread inexorably from one finished project to another in response to the whims of Stalin's planners, who drew lines across maps and circled various locales that indicated their suitability for rapid development. The planners did so in the confidence that the small landowners endemic to the capitalist system had been eliminated through class war and so could not interfere, while nature herself could no longer resist the planner's pencil or the builder's bulldozer under socialism. While in some cases the socialist regimes of Eastern Europe the authorities resorted to slave labor in their industrial, agricultural, and forestry projects, those projects were never on the scale of Soviet ones, nor, apparently, did they create entire forced labor camps dedicated to the geoengineering and nature transformation, although they used prisoners. Yet they embraced the notion that, under socialism, they could draw lines on maps with planners' impunity.

The Stalin Plan for the Transformation of Nature combined prerevolutionary ideas with glorifications of the socialist economy in the 1930s. In *Men and Mountains* (1935), called by Maxim Gorky a "prose poem," M. Ilin (pseudonym) described many projects that dated to any early era and the determination of the Soviets to remake nature—its forests, rivers, deserts—and turn nature into an instrument of the socialist economy. Planning and science would turn the steppe into farmland; new dams and canals would open cotton plantations; the Amu Darya and Syr Darya rivers would be diverted to Central Asia for fruit growing and would enable ships to sail from the Caspian Sea, while hydroelectric power stations would furnish "light for cities and electric power for machines." Already in May 1932 Stalin and Molotov signed "a decree against the elements" for "Ending Droughts in the Volga Territory" with forest shelterbelts so that water would gush forth and flow over the fields.[12] Engineers would make the Volga deeper and faster to become the main traffic artery of the country and would link, also by railway, the center of the country to the Arctic through the Kama and Pechora rivers, and to the Baltic and White Sea.[13] Rivers would be bridled, weather controlled, there would be no floods, Arctic ice would be melted for agriculture, and so on.

Since, in the 1930s, planners focused investments on rapid industrialization and the collectivization of agriculture, large-scale nature transformation would wait until after World War II. Yet Soviet authors celebrated a number of signal achievements that presaged the hero projects of the late 1940s. In agriculture, for example, collectivization destroyed what they believed were ingrained and unscientific peasant farming techniques and replaced them with a kind of socialist agribusiness. Modern machinery required collectivization, because new tractors, combines, and other equipment could run to the horizon and back during plowing, sowing, and harvesting. There is, in fact, some evidence that modern agribusiness inspired some American farmers who visited the USSR in the 1930s and saw what machinery could do on massive plots.[14] In any event, in the USSR, the experience with big earth-moving and harvesting machines in agriculture encouraged thinking about how to alter nature itself, and led to the development of massive bulldozers, section dredges, and other equipment.

During the first five-year plans, several distant transformation projects from the Arctic to Central Asia, and from the plains of Europe to Siberia and the Far East, accompanied urbanization—so-called socialist reconstruction. Scores of new industrial towns and cities appeared, notably Magnitogorsk in the Ural region, Norilsk, Kirovsk, Molotovsk, and others. These would serve as models for the East European hero cities of metallurgy, concrete, and power production. Yet Stalin did not overlook nature itself. At the Seventeenth Party Congress in 1934 Stalin called for increases in irrigation systems in the Trans-Volga region and afforestation through the planting of forest shelterbelts to fight drought. He said:

As you know, this work is already taking place, although it cannot be said that it is being carried on with sufficient intensity. As regards the irrigation of the Trans-Volga area—the most important thing in combating drought—we must not allow this matter to be indefinitely postponed. It is true that this work has been held up somewhat by certain external circumstances which cause considerable forces and funds to be diverted to other purposes. But now there is no longer any reason why it should be further postponed. We cannot do without a large and absolutely stable grain base on the Volga, one that will be independent of the vagaries of the weather and will provide annually about 200,000,000 poods of marketable grain. This is absolutely necessary, in view of the growth of the towns on the Volga, on the one hand, and of the possibility of all sorts of complications in the sphere of international relations, on the other. The task is to set to work seriously to organize the irrigation of the Trans-Volga area.[15]

The pre-war years were a prelude to the transformation of nature. From 1927 to 1941 laborers planted more than 468,000 hectares of shelterbelts at collective farms – forty-three times more than was planted in the entire Russian Empire from 1817 to 1917. By the third five-year plan the amount of irrigated land had increased by 75 percent since the revolution.[16] In all cases, it is difficult to verify Soviet data, which was frequently exaggerated for the consumption of domestic and foreign audiences. But there is no doubt that the Soviet planners truly increased the production of industrial goods, electricity, irrigation, and so on many fold.

Electricity would power this transformation; Soviet molecule counters had enumerated 108,000 rivers in the USSR with 15 percent of the world's hydroelectric potential, many of them in Siberia. They referred to rivers as "white oil." In terms of the USSR, 80 percent of the nation's hydroelectric potential was on the Siberian Ob, Lena, Enisei, Amur and Angara rivers.[17] Between 1928 and 1953, hydroelectricity capacity grew forty-seven fold. Soviet engineers moved forward unabashedly in a variety of climatic and geological conditions along rivers from the Arctic Circle to the steppe, to arid Central Asia, and to Siberia, along the Dniepr, Svir, Kura, Syr Darya, Dnestr, Narva, Rioni, Kovda, Kama, Don, Niva, and Razda rivers.[18] No river would escape the search for ways to power and water nature transformation.

Planners focused on large-scale integrated geophysical technologies to achieve the goal of the transformation of nature. The so-called hero projects—dams, weirs, hydroelectric power stations, irrigation and transport canals, sluices, forest defense belts, and the like—were the epitome of modern technology; yet, paradoxically, they were built with poorly equipped armies of men who had fewer steam shovels and bulldozers, fewer trucks, fewer horses(!), and fewer wheelbarrows than they needed. Capital was expensive, while labor inputs were easier to requisition, even if it meant using gulag slave labor.

Stalin Begins to Rebuild in Earnest

In the midst of a difficult recovery from World War II, with millions of people still living amongst the rubble or in dug-out earthen huts called *zemlianki*, and with factories, power stations, and infrastructure in ruins, the tireless Stalin set the nation on a course to transform nature, with geoengineering projects from the White Sea to the Ural Mountains and to Central Asia, with hydroelectric power stations rising on the Volga and other major rivers, with Lysenkoist agriculture spreading into formerly barren or underperforming fields, and with harvests of grain, barley, rye, cotton, and fruits increasing up to fivefold. In six to seven years, irrigation would enable the production of sufficient grain, sugar beets, vegetables, fruits, and livestock to feed a hundred million people.[19] Even more fantastic, the Soviets promised that within twenty-five to thirty years they would turn the sand of Turkmenistan and Uzbekistan into forested lands capable of producing thousands of cubic meters of wood annually, and that the Central Asian republics, each now with its own Academy of Sciences staffed with eager engineers, would engage water transfer projects to create gardens of cotton and fruits. Rivers would be tamed and rapids removed so that barge and steam ship travel would grow significantly. The White, Baltic, Azov, Black and Caspian seas would be tied together into one transport nexus. The major focus was stepped reservoirs and power stations on the Volga, including the largest in the world, the Kuibyshevskaia, the Stalingrad, and others that would permit the irrigation of millions of hectares of the Sarpinskaia and Nogaiskaia Steppe.[20] A propagandist, Kasimovskii, wrote, "In the country of Soviets this utopia became a reality."[21]

In order to reach the level of communist plenty, not only industry, but agriculture would have to be revolutionized in the postwar years, with ever more ambitious hero projects and with increased sowing and harvests, new crops, and better animal husbandry. To achieve these agricultural targets they needed forest defense belts, irrigation systems and hydroelectricity to transform the landscape. One-fifth of land in the European part of the country was steppe and forest steppe regions of the European USSR that suffered from droughts and dust storms. Soviet scholars were determined to overcome the constant droughts that revealed the ineptitude of the Tsarist regime in handling crises—for example, the famine of 1891. They and political leaders were also aware that civil war (1918–1920) and postwar crises (1945–46) had resulted in a total of five to six million deaths by starvation and disease.

Stalin's plan primarily required vast stretches of land to be afforested in the south of the country in order to prevent the dry winds, coming from the steppe of Kazakhstan and Central Asian deserts, from penetrating the fields and causing damage.[22] Planners also intended to plant up to 80,000 miles of forest defense belts, built from trees planted in up to three bands 30 to 50 meters in width; they succeeded at great cost in planting only 5,000 kilometers.[23] With Bolshevism, adherents of ambitious afforestation plans came to power. World War II had halted afforestation, but in 1947 the Ministry of Forestry Management advanced a program for 1.5 million hectares of forest, followed in 1948 with another 5.7 million hectares—in part in response to the famine and grain failure of 1946. As part of Stalin's plan, the "world's largest waterworks along the Volga, Dniepr, Don, and Amu Darya, the canals and reservoirs" would revolutionize travel and tie the entire country together, region to region, countryside to city.[24]

Soviet authors saw the hero projects as crucial for an agricultural revolution including crops, trees, and animal husbandry, and equally to create modern, socialist relations between burgeoning cities and the countryside. The massive waterworks on the Volga, Dniepr, Don, and Amu Darya rivers, the construction of gigantic canals, the creation of irrigation systems, and the mechanization of labor would move along seamlessly. The water works would magically create *smychka*, the legendary and anticipated bond between urban and rural life, while the proletariat would lead the peasant into the twenty-first century. Yet, even though Stalin had used collectivization to break the peasantry and to extract capital for industry, he ultimately realized that the countryside was starved of investment. He turned to hero projects as the most efficient way to achieve all ends, including subjecting nature to man's control and serving as work sites where the proletariat could be indoctrinated about the glories of socialism. Stalinist hero projects were a powerful lever to end contradictions between city and countryside once and for all.²⁵

As a prelude to the Stalin Plan, in July 1947, after decades of underinvestment in the countryside that had left collective farms impoverished in terms of modern equipment and power, government officials determined to provide farms with electricity. The countryside really was dark: many collective farms had no electricity whatsoever, not even small generators, and those with generators had difficulties getting parts and gasoline. Lucky farms had a few light bulbs, but labor continued to be nineteenth century: on the hands, shoulders and backs of peasants, with horses, and perhaps the occasional rusty tractor. Now the government passed a decision to irrigate 575,000 hectares in collective farms of the central black-earth regions with pumping terminals powered by 590 new small hydroelectric power stations, generators, and wind power, with the work to be completed within six years. The next year the government approved the construction of shelter belts in collective farms that would cover 1,350 hectares by 1951.²⁶ But have no doubts: no significant or successful rural electrification program resulted under Soviet power, in spite of the utopian belief that "Communism equals Soviet power plus electrification of the entire country" (a 1920s slogan).

The fascination with electricity produced fantastical forecasts nonetheless. What did Comrade Stalin have in mind that East European Communists wished to emulate? Perhaps drawing on Lenin's inspiration, he saw electric tractors, combines, and other machines to mechanize all agriculture and revolutionize production. In 1949, at Stalin's insistence, apparently only as a limited experiment, several Machine Tractor Stations established electrical tractor stations with thirty tractors that operated nearly 40 percent cheaper than gas-powered tractors. The feeding and watering of livestock, shearing of sheep, and other activities would also be electrified. The capitalist farmer no doubt would be shocked to learn that the Soviet collective farmer had never heard of electric sheep shears.²⁷

Stalin's plan used military rhetoric. It was "a battle with drought and salinization of soil" to overcome the technological lag of Tsarist Russia and increasingly frequent droughts. Between 1917 and 1951, irrigation systems spread south to Uzbekistan, Kazakhstan, Tadzhikistan, Azerbaijan, Armenia, the Trans-Volga, and Southern Siberia. One of the pre-war hero projects that opened in 1939 was the Stalin Great Fergana Canal; at 270 kilometers in length, it was built by 160,000 Uzbek and Tajik collective farm laborers—and no doubt slave laborers as was

Introduction: The East European Experience   15

the case in other hero projects. At the beginning of the 1930s, Central Asian camps of the secret police were established to build water works and finish some fifty-two projects, according to one source, including the Fergana Canal. But "amazingly" it was built in only forty-five days, with workers relying on shovels and picks, and toiling directly under the hot sun.[28] The canal was crucial to Central Asian agriculture with its multitudinous sunny days but arid soul. Harvests of cotton, sugar beets, oats, rice, and corn all grew substantially. Roughly 25 percent of the Amu Darya river was intended for the canal so that the Aral Sea would lower its level—catastrophically as they must have known—but they were thrilled that they had reclaimed land for agriculture, even if the saltiness of the water had a negative impact on productivity.[29] The goal was great agricultural nature transformation in Karakalpakiia and in the Karakum, and inland water shipping from the Volga to the Amu Darya.

Shelter belts, with their own prerevolutionary roots, were another important feature of the plan.[30] They had a long history in world forestry and were tried experimentally in the nineteenth century. The US government supported the planting of shelterbelts in the 1930s in the "Plains States" to try to ameliorate the devastating impact of the Dust Bowl and to put laborers back to work through the Works Progress Administration.[31] In the USSR in the 1930s, specialists were gearing up for their planting as they studied the influence of shelter belts on microclimate and, in turn, on harvest of various cultures. They considered the geometry of plots and planting, of height and density. They worried about planting along roadways where winds might create significant drifts, and so called for planting denser, shorter bushes in addition to trees. They propagandized their efforts in such journals as *Na Lesokul'turnom Fronte!* They founded such research centers as the Institute of Agroforest Melioration and Forestry.[32]

The transformation effort yielded, in addition, Cold War rhetoric. The heroic old Bolshevik who directed the State Plan for Electrification of Russia (GOELRO, set in motion in 1918), Gleb Krzhizhanovskii, a friend of Lenin and one-time head of the State Planning Administration, Gosplan, kowtowed before the "geniuses of the proletarian revolution—Lenin and Stalin" as a 70-year-old. He celebrated GOELRO for upsetting capitalist doubters that the USSR had surpassed Europe and would soon surpass the United States in electrical energy production. He reminded the postwar Soviet public that while such projects as Dnieprstroi had been met skeptically by the bourgeois press, "the enemies of our motherland were convinced finally by the reality of our plans." The bourgeois press realized that the "Leninist-Stalinist teachings about electrification" were the "unshakable foundation of our energetics." The planned system had given the nation the Svirskaia and Rybinskaia stations, the Chirchinskii cascade in Uzbekistan, projects on the Kama River, reworking of rivers in the Caucasus, and the Main Turkmen Canal. The latter would solve the problem of 300 million hectares of desert in the USSR, of which the Kara Kum alone covers 250 million hectares—27 percent of the area of Soviet Central Asia. Electrification and hero projects would subjugate the Amu Darya River to the desert. Krzhizhanovskii pointed out that while many big projects unfolded under capitalism, they did not benefit the worker—for example, the Suez Canal built by poorly paid laborers, the Panama Canal with large numbers of laborer lives lost, and other capitalist projects.[33] Krzhizhanovskii did not refer to the hundreds of thousands of slave laborers who perished building Stalin's waterworks.

The Bolsheviks considered electrification a panacea and referred to it as a "child of the revolution." They debated whether to rely on peat or coal, and whether to build centralized generating stations or decentralized. But hydroelectricity captured their imagination, as it would under Stalin when the USSR and the United States engaged in kind of a "hydropower race" to erect the biggest, most powerful stations in the world on the Volga and Columbia rivers respectively. The first projects commenced near Leningrad on the Neva, Volkhov, and Svir rivers, and at Imatra near the Finnish border, in the 1920s. The reconstruction of the Volga began in the 1930s in its upper reaches, with dams at Ivan'kovskaia, Uglich, and Rybinsk. After the war as part of the Stalinist plan, these projects moved ahead with the construction of such massive hydroelectric power stations as the Kiubyshevskaia, at its completion the largest in the world—and perhaps the largest gulag project ever with as many as two hundred thousand prisoners, the Stalingradskaia, and others. Soviet engineers saw power stations as "factories of electricity" that work the ore of water—its mechanical energy. This all gave way to the "Big Volga" project to transform the river from its source to the Caspian Sea.[34] For this project—as for the ones before and after it—since it was designated an all-union project of national significance, resources and laborers were requisitioned from around the country. Construction crews moved up and down the Volga River, trucks came from Moscow, tractors and bulldozers from Kharkov, dump trucks from Minsk, specialty steels from Petrozavodsk, cranes and steam shovels from the

Urals, turbines from Leningrad, and so on.[35] Centralized requisitioning did not overcome bottlenecks and shortages, but also created them in other sectors and regions of the country that had to devote efforts to Stalin's determination to change rivers.

A significant aspect of the Stalin Plan to Transform Nature was its focus on creating an agricultural wonderland of expansive farms, plentiful harvests, and new crops grown on irrigated and protected steppe and desert according to Lysenkoist and Michurinist ideas. What precisely did they mean by "Michurinist" and "Lysenkoist"? Ivan Michurin, a plant selection specialist, created a massive collection of plants and pursued genetics research, cytology, and hybridization. He created all sorts of fruit plants. Because of his practical contributions to agriculture, and because of his somewhat Lamarckian view that the selectioner should work to promote natural selection, the Lysenkoists embraced him as a hero of proletarian biology versus "fruitless" genetics.[36] They often quoted his saying: "We cannot wait for favors from Nature. Our task is to take them from her."

As the authors of these chapters write, Trofim D. Lysenko was a quack scientist who falsified data, misled funding agencies, and used extra-scientific channels to destroy his opponents, real and imagined. His class origin as a peasant helped his career, as did his promise to achieve significant results in the short term to help suffering Soviet agriculture. Lysenko also based some of his work on real science, for example, vernalization, the treatment of plants with cold and moisture to get them to flower in the spring. He was attracted to the work of the Michurin, who used a kind of Lamarckism to change species among plants through hybridization and grafting; but these were non-genetic, Lamarckian techniques that claimed the inheritance of acquired characteristics. No matter the Lysenkoists' claims, they could not deliver plants or animals whose new characteristics were heritable. But they promised the transformation of nature.

Lysenko rose rapidly to dominate agrobiology through extra-scientific channels. It helped Lysenko that Isaak Prezent, a Stalinist ideologue, propagandized the simple peasant Lysenko as a genius with radical new agricultural techniques. Soviet leaders, including ultimately Stalin, embraced Lysenko as a hero—a new Soviet hero. All of this enabled Lysenko to dominate the Soviet biology and agriculture establishments from the 1940s, especially after a 1948 conference that declared Lysenkoism as the only form of Soviet biology, while attacking genetics as a bourgeois science.[37]

In 1948 at the Lenin All-Union Academy of the Agricultural Sciences, Lysenko and his followers took control of genetics—in fact, rejected genetics—and with the approval of Stalin carried out a purge of geneticists from the academy, and genetics from textbooks and libraries. Lysenkoism and the idea that nature can be transformed and improved were mutually reinforcing. Lysenko's ideas were devastating to agriculture and wherever they were applied. For example, in the pursuit of shelter belts, Lysenko insisted that trees be planted in clumps or nests where their inherent "collectivism" would lead them to thrive, as opposed to being planted individually at some distance from others, like some kind of exploited capitalist laborer. In this way, the grandiose Stalin Plan for the Transformation of Nature was Lamarckian and Lysenkoist.

Taking a different view, Stephen Brain argues that the plan was an attempt "to reverse" human-induced climate change with its goals to create six million hectares of new forest to cool the air, moisten the soil and raise humidity. Scientists and forest specialists, whom Brain calls "technocrats," had Stalin's ear and convinced him of the need to fight drought through the measured application of knowledge. But the death of Stalin ended the ascendance of their "technocratic ecology." Brain notes that other utopian visionaries believed that communism would end all natural limitations. They believed that forests could be made to conform to human will, while the technocrats relied on science to take into account local variation and natural limits in their plans. It helped that Cold War rivalries and pressures egged visionaries onward, for they believed that only the progressive socialist USSR could do with nature what was needed for the people, while in capitalism droughts and dust bowls would remain. In this view, the Stalin Plan reflected larger trends toward conservatism in Soviet society in culture, literature, family policy, and education. Hence the plan meant a focus more on stability than on radical restructuring.[38]

According to Brain, Lysenko kept Promethianism alive from his positions in the Agriculture Academy and in his visions of transforming crops and nature. He jumped on the afforestation bandwagon; he did not jump-start it. And he applied his crazy Lamarckian-socialist ideas to it to suggest that trees were collectivists. He called for planting of trees in nests—especially based on his work on the dandelion-rubber plant Kok-saghyz. Collective nest planting was a complete failure. In fact, for a variety of reasons the eight great shelterbelts were only 46 percent completed, less than half of the area planned for sowing was afforested, and more than half of the seedlings planted between 1949 and 1953 died.[39]

Shockingly, transformationist plans did not adequately take into consideration "science" when it interfered with grandiose visions. How could one rework nature and then ignore soil, topography, and climate? The effort to rebuild and recast agriculture in the USSR and in postwar Eastern Europe required extensive agronomical knowledge, from soil to vegetation to techniques. Much of this knowledge had not only pre-revolutionary roots but also originated in the Russian empire. In the nineteenth century, Vasily Dokuchaev, soon to be revered by Soviet patriots, demonstrated that soil types differed because of geological, climatic, vegetative, and topographic factors, and developed important soil classifications that he introduced in *Russian Chernozem* (1883).[40] In other words, it was important to consider climate, vegetation, country, relief, and age in a modern soil science. But many Lysenkoists thought it more important to engage agriculture practically, and thought that too much research, including that of Dokuchaev, was too theoretical or lacked the imprimatur of "practice," especially socialist practice, which would prove them right in the end.

Hence from modest beginnings at Volkhovsk and Svirsk, and then Dnieprostroi, the Soviets turned to gulag-based projects on the White Sea–Baltic Canal, the Moscow Canal, and other Stalinist waterworks near the capital, to the Great Fergana Canal. The immense costs and devastation of the Nazi invasion interrupted further projects and destroyed many that had been built. After the war, the party focused on rebuilding industry. By 1947, recognizing that agriculture lagged significantly (and had never recovered from Stalin's own war against the countryside in collectivization), party leaders decided that only a large-scale nationwide campaign to rebuild nature itself would increase agricultural production, build a unified transportation network, conquer drought, and grow fruits, cotton, and other crops according to Michurinist–Lysenkoist fashion in previously arid regions. What was the impact of these plans and ideas on the East European socialist nations of Czechoslovakia, Hungary and Poland?

Red Army Nature Transformation
In 1951 in a replica-translation of a Soviet publication, Klement Gottwald, a Stalinist and early leader of Socialist Czechoslovakia, who pursued collectivization and industrialization with delight and vigor, and carried out murderous purges of Czech communist officials,

published *For the Happiness of the People: The Transformation of Nature in the Stalinist Era* that carried a frontispiece of a painting that depicted Stalin drawing channels and rivers across a map of the Soviet Union to indicate the way that nature should be. Behind him stood Politburo members and Army marshals who clearly approve of his battle map for a war on nature.

In East Central Europe the experiences with "nature transformation" shared many of the features of this dramatic painting. In Czechoslovakia, Poland, and Hungary, Communist regimes pursued rapid industrialization, including the construction of socialist cities; they aggressively sought to tame waterways for municipal and industrial water supply, and for transport and irrigation purposes; to a greater or lesser degree they embraced a revolution in agriculture, with new crops and approaches through collectivization and through flirtation with Lysenkoism; and they explored afforestation programs. They proselytized these programs with scientists and among the public in a variety of forums and through a variety of media.

The governments moved rapidly to adopt Stalinist transformation projects as part of their overall programs. In Czechoslovakia the Communists took over the government in a coup in 1948; two months later, after the nationalization of land, industry, minerals, and limiting holdings to fifty hectares, they undertook violent collectivization of agriculture along with its mechanization. The Academy of Agricultural Sciences provided socialist agronomy; it was subjugated to the Ministry of Agriculture and the Agricultural Department of the Central Committee of the Communist Party. Ultimately, because of the tradition of intensive agriculture and the relatively small size of the country, the Stalin Plan for the Transformation of Nature had relatively little impact in Czechoslovakia. But the socialists were determined to achieve fantastic results against all odds. Toward those ends, the authorities promoted the spread of Michurinist agrobiology, applied Stakhanovite methods[41] in agriculture, and established massive construction trusts to build reservoirs in pursuit of the Soviet vision of turning infertile land into fertile fields, orchards, and gardens, with forests to regulate wind erosion.

A lexicon of transformation unfolded in the media. Officials discussed a concrete plan to follow the Stalin nature plan early in 1949 with a ten-year horizon that included some aspects of all of its features, from tree-planting shelterbelts to Michurinist agriculture. In the Czechoslovak Academy of Agricultural Sciences, a cult of Lysenko developed and specialists undertook struggle against "Cosmopolitanism and Objectivism in Science." Czech specialists joined pilgrimages to see the god, Lysenko, in Moscow, and to learn how to turn theory into practice. Under socialist power the Czechoslovak Agricultural Academy, which dated to the 1920s, followed the Soviet economic development model to introduce Michurinism. The academy funded applied research at the expense of basic science when it was subordinated to the Ministry of Agriculture. Similarly, through coercion and co-optation, engineers pursued proletarianization of the engineering sciences.

In practice, what happened? Shelterbelts were already standard measures for protecting the landscape in Czechoslovakia—as they had become throughout the world. During high Stalinism the forestry specialists and party officials intensified the program in the hopes of changing the hydrology of the entire country. From Czechoslovak plans to popular brochures about them, officials and specialists emphasized the importance of forests for transforming nature. Forests were no longer a resource, nor the interest of conservationists alone, but a means through which to control nature and to follow the Soviet designs, although not in Soviet grandeur.

Experimentation with and introduction of crops had a major place in Czechoslovakian transformation plans. Just as the Siberian plains were to be planted with a special pear tree bred by Michurin, Czechoslovakia was to be planted with rice, which, according to Soviet experts, would work well. Specialists hoped to have an indigenous variety within ten years that required less water and a shorter growing season, and of course replete with higher yields. The effort to spread "Michurinist methods" in Czechoslovakia involved attempts to attract local groups of teachers and agronomists who worked in local clubs and groups that turned out to be like an agricultural extension service. "People's research" also made its way into schools. In 1954, seventy thousand primary and secondary school students were involved in organized clubs. New Stakhanov-like campaigns helped to raise sugar beet production too. Of course, focusing tremendous labor resources in any sector should increase production.

As in the USSR, the nation required an agricultural revolution and the tying of the peasant to the collective farm to feed burgeoning cities and industry. Cities were rebuilt, renamed, and established next to deposits of the ore they were intended to work; for example in Vitkovice in the Ostrava region, the "steel heart of the republic," with its Gottwald New Steelworks and Gottwald Steel Rolling Mill—like Nowa Huta near

Krakow, Poland. Grandiose plans to develop the region required, in 1950, almost one-fifth of all state investments (17.5 percent). But party officials justified the expenses because the mills were crucial to gather large numbers of agricultural workers in one place to transform them, too—in this case into an industrial proletariat. During the first five-year plan alone, Czechoslovakia invested more than 558 billion crowns in large-scale construction projects: first new steel mills, then reservoirs and hydroelectric power plants. Construction required infrastructure of roadways, railways, power lines, the excavation of 2.5 million cubic meters of earth, and the pouring of 700,000 m^3 of concrete. Stalinist transformation always meant more excavation—and more concrete.

Like Soviet "all-union" and Communist Youth League construction sites, Czechoslovak officials used a publicity campaign to attract young men and woman through slogans that were military metaphors: To the front! To the battle! These signified a war on nature. However, the campaign approach did not guarantee rational use of resources. Instead, construction sites were pools of mud and disorder; the lack of experienced designers, construction engineers, and workers slowed efforts, as did show trials, like those of engineers in the USSR, to affix blame for failure to reach targets.

Not having learned from Soviet practices that submerged hundreds of thousands of hectares of fertile land in flood plains—or having determined these were acceptable costs—officials planned a systems of reservoirs and dams on the Orava River in Slovakia that would be the largest in all of Central Europe, in part to achieve a fivefold increase in the production of electricity. They planned reservoirs on the Vltava, the longest Czech river, on the Váh River in Slovakia, and on the Danube where the Gabčíkovo-Nagymaros Reservoir would rival in size the Dnieperstroi Dam, one of the first Stalinist hero projects of the 1930s. Czechoslovakia pursued this project only in the 1980s, and it was killed by legal and environmental concerns and an international legal dispute between the Czech Republic and Hungary.[42]

However, Czech projects moved ahead more carefully than Soviet ones, applying more the effort to stretch water resources among municipal, industrial, and agricultural demands, not "controlling nature." Even in this environment, an impoundment boom resulted: before the war there were thirty-seven reservoirs and a total volume of 173 million m^3, but between 1945 and 1962 twenty-three reservoirs were built, with a total volume of 1,147.9 million m^3. The Orava Reservoir was the first great hero project, with a surface area of 35 km^2 and a volume of 350 million m³. As with many of the projects set forth under socialism in the USSR and in Eastern Europe, the Orava Reservoir had pre-socialist roots. Fittingly for a hero project, the reservoir was touted in the media, attracted thousands of workers, and served as the basis for a socialist realist production novel, Dušan Kodaj's *Oravská priehrada* (1953). Such socialist realist novels had been a standard in the USSR since the 1930s. A second hero project was a series of dams on the Váh River, the "Váh Cascade" for hydroelectricity, and it had a similar sociocultural response. Other geoengineering focused on the Vltava Cascade in Bohemia (with 3 major dams, the first of which was the Slapy hydroelectric plant with an output of 144 MW, the Lipno, and the Orlík).

The Soviet influence in nature transformation persisted after Stalin's death—in corn under Nikita Khrushchev; Khrushchev grew convinced that a crash campaign to plant corn would solve a growing shortage of fodder, yet pursued it inconsistently and without careful planning, with costly agricultural and environmental failure the result.[43] Still, in his "Secret Speech" at the Twentieth Congress of the Communist Party of the Soviet Union in Moscow in January 1956, which condemned Stalin's "cult of personality" and murderous acts, Khrushchev opened Stalinism to criticism, including in East Central Europe, leading to liberalization of regimes generally, and to rejection of the Stalinist Plan for the Transformation of Nature, or at least large parts of it, including in Czechoslovakia.

From Sztálinváros to Hortobágy
In Hungary, Communists repeated the pattern. Between 1947 and 1949 they abolished the multiparty system, nationalized private property, and centralized control over the economy, education, and culture. Power was assumed by a small group of the Hungarian Workers' Party led by Mátyás Rákosi, who had his own cult of personality. By August, Hungary was a Soviet-style "republic" with arrests, show trials, exiles, and executions of enemies, including within the party. Nationalization of the economy proceeded rapidly: first coal mines and major banks were taken over, then factories employing more than one hundred individuals, next those with more than ten, and so on. Class war against the bourgeoisie proceeded with workers, peasants, and only then intellectuals to reap the benefits of society; new intellectuals from working-class backgrounds and faithful to the system were advanced into positions of power, mirroring the process in the USSR in the late 1920s and 1930s called *vydvizhenie*. Stalinist cities contributed to this process: peasants

left villages in droves to gargantuan industrial development projects that relied heavily on Soviet resources—Sztálinváros, Kazincbarcika, and Komló.[44]

Cultural revolution also proceeded in Hungary. It included an assault on the Academy of Sciences to bring it in concert with the political and ideological directives of the party. Reductions in staff by retirements, illnesses, and stays abroad were followed by purges and demotions, while communists were advanced to full membership. This meant that scientific R & D was now controlled by the Communist Party. The academy was prepared to consider, and approve, nature transformation policies that affected water-supply management, soil cultivation and forestry, and introduced new crops completely foreign to Hungary's climate. Cultural revolution involved the establishment of new secular schools with 8-grade compulsory education. The authorities succeeded in eliminating illiteracy rapidly, but schools were overcrowded, programs put emphasis on vocational skills, and re-attestation of professors based on their political beliefs created havoc. Churches and the Church were attacked, including show trials. Socialist realism replaced all other forms in the arts, literature and music, and imitating the Soviets, the system introduced *chastushki* or made-up agitation verses. Granted, there were such achievements as social security, housing, and health insurance, if very low wages, but housing would remain a sore spot as the population grew twice as fast as housing, women were forced into work by the men's low wages, and the state could not provide adequate nurseries or kindergartens.

Following Stalinist directives, the Hungarian Workers' Party under Party Secretary Mátyás Rákosi began to build "a country of iron and steel," despite the fact that there were no iron ore deposits in Hungary, nor enough coking coal for their purposes, yet, perhaps to become a modern industrial power in Europe, invested in these energy-intensive and expensive projects that created bottlenecks for all other recovery efforts. On top of this, pushed by the Korean War and prodded by the Cold War, Hungary entered the Warsaw Treaty Organization and agreed to maintain 150,000 troops, a fantastic number given there were only 40,000 at the time.

In word, deed, and failure, Stalinism pervaded the economy. Not content with underfulfillment of the original five-year plan, the party raised targets so that, compared to 1949, as noted in chapter two, production by 1954 would go up by 200% instead of the original 86%, heavy industry would improve by 280% instead of 104%, and the light industry would go up by 145% instead of 73%, on top of which collectivization had begun. Hungary, a country with inadequate housing and poor infrastructure, instead prioritized mining, metallurgy, and heavy manufacture, and the establishment of new Socialist cities—company towns—that were vulnerable to sectoral failure.

The socialist cities established in Hungary at the beginning of the 1950s were Sztálinváros (today Dunaújváros), Kazincbarcika (1954, and its Borsod Chemical Combine), Komló (1951, coal mining), and Tatabánya (1947, also coal, but later machine tools, textiles, and telecommunications); the surrounding areas were environmentally degraded by the construction. Sztálinváros had no industrial roots and was built completely from scratch on an agricultural site near the border of the village of Dunapentele to replace and augment outdated factories in Ózd, Salgótarján, Diósgyőr, and Csepel. If steel temporarily did well—on coke and ore from thousands of kilometers away—then housing, schools, kindergartens, nurseries, hospitals, cinemas, bakeries and slaughterhouses all lagged, as did public utilities, roads, and parks.

Sztalinvaros was for party loyalists a beacon of a socialist industrial future, but for peasants it symbolized an attack on their way of life. One thousand construction workers arrived in May 1950. Those arriving sought to escape the poverty, high taxes, compulsory grain deliveries, and dislocation in the countryside like the Russian peasants who fled to Magnitogorsk to avoid de-kulakization (the process of dispossessing allegedly wealthier peasants from their land and property, often forcing them into exile). The locals deeply resented the new arrivals. Still, by Christmas, 5,860 workers had joined the construction site, and by January 1952 over fourteen thousand laborers were at work. Like at Magnitogorsk, Sztalinvaros, as Pittaway writes, was "a workshop of chaos, low wages, despotic management, and poor working conditions." Facing constant exhortation to meet impossible targets, the workers felt not in the least a part of the glorious communist future. And like throughout Eastern Europe, the Communist Party alienated them in new city designs that excluded a church from the city center; these were deeply religious peasants being forced to adopt a new worldview, suddenly if not violently. They resented the recruiters who had destroyed their way of life.[45]

In this environment, Lysenkoism had a great influence. The August–September 1948 issue of *Társadalmi Szemle*, the scientific journal of the Hungarian Workers' Party, published the Soviet Communist Party's Central Committee–approved version of Lysenko's August

1948 victory speech (over "hostile bourgeois genetics") at the Lenin Academy of Agricultural Sciences. The new Hungarian agronomy journal was "Lysenkoized" from its first number in February 1949 in which it published: "A szovjet agrobiológiáról" [On Soviet Agrobiology] and was followed by the publication of selected papers by Lysenko and Michurin. Lysenkoism was introduced in colleges and universities as early as 1949 then into elementary and middle-school curricula by 1950. In scientific research institutes, Soviet Lysenkoists proselytized the new doctrine that rejected genetics. As in the USSR, Lysenkoists assumed all positions of power in ministries, research institutes, and scientific associations, almost without exception. Plant biology suffered greatly, if stockbreeders managed to survive.

With the rise of Lysenkoism, the Agricultural and Cooperative Committee of the Hungarian Workers' Party helped to set agricultural research policies and further centralized agricultural research in 1950. It turned all ten Hungarian agricultural schools into experimental farms for Lysenkoist study of soil, crops, and technology. The Soviet model led to such new crops as cotton, rivet wheat, kenaf, tea, peanuts, and citrus fruit. Crop production would grow by 35 percent and stockbreeding by 50 percent during the first five-year plan. The introduction of non-native plants—no doubt well-adapted to the Hungarian environment—would follow to meet the expected demand of textile and food industries. Those scholars who were not convinced of vernalization and other practices were accused of lacking appropriate class consciousness or doing "ivory-tower" work not of benefit to the proletariat.

The experience with cotton gives a sense of how Stalinist transformation, Lysenkoism, and the geopolitics of subjugation to the Warsaw Pact (signed in 1955 to keep East European troops under Soviet direction as a military balance to NATO) determined the extent of Hungary's cotton culture. Officials admitted that soils ill-suited to cotton and cold weather might prevent cotton culture from taking root. The absence of warm weather and sunshine in the Hungarian project was like Brezhnev's efforts in Central Asia to grow cotton, which ignored the absence of water and other crucial climatic conditions.[46] But politics determined that planting begin experimentally on 22 hectares of land in 1948. Later that year the National Cotton Production Company was established in order to implement large-scale cotton production on over 50,000 hectares of land by the end of the plan. Nothing went right: costs of production were six times higher than value, while frosts and rain destroyed yields. Cotton required manual labor, and the small labor forces at most farms required the recruitment of additional workers who were instead being siphoned off to Socialist cities. The decision to require labor on Sundays faltered because religious peasants refused to work. Students were recruited, and then children. But their transportation to the region was poorly organized, and their food and barracks were miserable. In the fields, not surprisingly, weeds spread. Some collective farms actively began to resist cotton. On top of this, some of the cotton fields were near the border; this required special permits for workers to approach border zones. Shockingly, weather forecasts were classified as a military secret, so farms had to guess when to plant and tend. In spite of this, in spring 1951 cotton production was expanded to four northern counties. Socialist Hungary went through all of this turmoil because of the desire for autarky and because the "fraternal" East European COMECON (Council of Mutual Economic Assistance) nations were required to produce cotton for their big brothers in Moscow; a patriarchal relationship rather than one of camaraderie seemed to prevail.

Yet Hungarian Communists were not to be deterred by soil, climate, or outmoded peasant traditions. All would have to be transformed. Another failed product was the yellow dandelion, a distant cousin of the "rubber" dandelion of Soviet Central Asia, which had been studied before World War II. Scientists established experimental farms for the rubber dandelion. The crop became public through a publicity campaign about how Soviet research would allow it to grow in Hungarian soil. Optimism was not justified; as with cotton, climate, inexperience, the necessity for labor-intensive cultivation, and overconfidence destroyed the crop. Nor could a National Rubber Crop Production Company, established in 1951, and assisted by Soviet advisors, turn the thing around. Kenaf, an industrial fiber to replace jute, and citrus also withered in the fields.

Rice should have turned out differently; rice culture dates back centuries to the Turkish occupation. By the end of the three-year plan, rice was being produced on over 20,000 hectares of land. To expand production, water works (irrigation) would have to follow, and planners hoped to use land that was unsuitable for the cultivation of more delicate crops for rice production instead. Planners set their sights on parts of the Great Hungarian Plain, including fields of Hortobágy, a national park of steppe and grassy plain. But irrigation fell far behind targets, the rice required far more effort to prepare soil, tend, and cultivate than officials contended, and environmental degradation was extensive. The state determined to force class enemies, exiles, and kulaks to work the rice plantations. Work such as weeding required women to work barefoot in cold, shallow water. Eventually rice was eradicated from the fields, but the Hungarian landscape is still suffering.

At the beginning of the 1950s the East European socialist nations joined the USSR in devising plans to change the course of rivers in the name of "scientific management" of soil and water. Hungarian specialists turned in particular to the eastern territories, including the Tisza River basin. Rákosi referred in speeches to Soviet hero projects on the Volga, Dnieper, and Amu Darya rivers. He announced plans for the construction of a dam at Tiszalök on the Tisza River, and for irrigation systems, saying cotton might grow there with the application of Stalinist "science." A number of aspects of the Socialist projects had pre-war roots. Under the new government, they moved ahead in fits and starts owing to ministerial debates on responsibility and authority. As in the USSR, a huge army of workers was required to carry out the Tiszalök project since the town there only had 4,500 inhabitants. The government brought in prisoners including "kulaks" who lived in labor camps and who provided most of the work force from 1951 to 1953. The dam was finished in 1954—on schedule—but produced electricity only in 1959. Other projects, for example the Danube–Tisza Canal, were left incomplete. Similar projects involved the Körös and Berettyó rivers, and mimicked the Soviet model of impossible goals and the refusal to admit defeat. Without sufficient labor or mechanization, these plans lurched forward.

An effort to build shelterbelts also began. The Hungarian Forestry Council set plans to nearly double forests, and included investment for saplings, public roads and railways, pastures, farmlands, farm centers, and settlements. Afforestation was a success story for a change, comprising 248,387 hectares of land between 1950 and 1960. Stalin's hands therefore sit at the roots of Hungarian trees to this day.

Toward the end of the nature transformation, the Hungarian Party apparatus undertook a mass media campaign to ensure public support and the proper ideological messages. As in the USSR, party and scientific journals directed toward intellectuals published essays and Soviet propaganda on Lysenkoist and other Stalinist artifacts. The regime proselytized in good Stalinist fashion. Scientific journals published Hungarian and Soviet articles in translation that touted potential achievements and glorified Lysenkoism. Radio and newspapers were devoted to bringing the Stalinist gospel to the masses, especially those in the villages, although Radio Free Europe and Voice of America on shortwave frequencies enabled the masses to learn from the West. As in the USSR, local officials installed speaker systems on the streets to reach citizens; many villages did not have these systems for a few years, so local public service announcements were publicized in the traditional way by drumming. The party organized militant, scientific lectures at Houses of Culture, which tens of thousands of people attended. Associations of scientists set up exhibitions at cinemas and railway stations, while short newsreels popularized agricultural innovations and Soviet technologies, including experiments with new non-native industrial crops in the spirit of Lysenko and Michurin's teachings, such as the claimed "success" of cotton. In addition to the constant coverage of cotton from planting to harvest, including Stakhanovite pickers, Sztálinváros and the Danube Iron Factory also became a constant focus in newsreels, as did how many bricks and how much concrete went into socialist industrialization. Heroic stories about the planting of tea, figs, and watermelon also indicated that man was in charge of weather. Changing the course of rivers was a Stalinist "monument to peace," while the imperialist states sought war. In all reports, the Soviet example served supreme—for example, the transformation of the Volga for the Tisza River that runs across the Great Hungarian Plain.

With the death of Stalin, the Stalinist Plan for the Transformation of Nature was mostly abandoned in Hungary, Poland, and Czechoslovakia. In Hungary the Central Committee of the Hungarian Workers' Party finally recognized the failure of Soviet programs in their country, and criticized industrial crop production for decreasing the farmland used for grains and for generally inadequate yields, although the committee also suggested one source of these problems was inadequate attention to Soviet scientific achievements and superior production knowledge. The Minister of State Farms and Forestry also recognized mistakes in agriculture, although did not refer extensively to nature transformation projects, which still held great promise among leaders. Still, cotton survived, perhaps because the Soviet Union made cotton production obligatory in all COMECON member states. At least the hands of Michurin or Lysenko were no longer needed to till Hungarian soil, and after 1956 Hungarian scholars essentially dropped reference to them and returned to genetics. When Lysenko visited Budapest in January 1960 he delivered a lecture to a packed hall at the Academy of Sciences, but refused to answer the two hundred questions addressed to him and lost all support among any remaining disciples. Lysenko's followers in Hungary gradually disappeared into their offices.

Stalin in Polish Nature

The Soviet dominance of politics, technology, and culture found full expression in the Stalin Palace of Culture and Science, built in Warsaw in the style of Moscow's eight postwar Stalinist skyscrapers. This was Moscow's skyline reproduced by Soviet architects under Lev Rudnev, not a "Polish" building. In fact, Polish Communists did not want the palace even as a press campaign praised it and the contribution of four thousand Russian workers, "brigades of enthusiasts," and "Soviet friends" who worked day and night using automated technology (from the civilized USSR, of course). The palace occupied a sixty-acre site that required the razing of a hundred houses and the displacement of four thousand people at a time of housing shortage in the Polish capital.[47] How did Stalinism extend to the polity, and from there to nature?

For Poland, too, the Red Army, Secret Police, and Polish Communists in exile in Moscow offices enabled the takeover orchestrated from the Kremlin that involved the gradual strangling of other parties and falsified elections. It helped at three summits between Stalin, Churchill, and the FDR that the Allies gave Stalin carte blanche to disassemble the country as he saw fit. Poland lost 20 percent of her territory. The country already faced huge problems: loss of wealth and infrastructure, and ten million people dead or in migration. The population grew rapidly over the next decades—but this put high demands on agriculture. Reconstruction was challenging due to shortages of investment income. They received little help through COMECON, which was Stalin's response to the Marshall Plan and a tool of Stalinization—it provided assistance for heavy industry, including armaments.

As elsewhere, the authorities imposed socialism on the economy. In Poland, small peasants received land taken from wealthier landowners, kulaks and "traitors," but in small enough plots to facilitate later collectivization. Eighty-five percent of forests were nationalized. Industry and other sectors of the economy followed. Elimination of opposition parties followed under Bolesław Bierut who headed the Polish United Workers' Party, the Communist Party of Poland, which ruled from 1948 to 1989 through a Soviet-style Politbureau. The party centralized all power, used terror and political police, and relied on a large number of Soviet advisors working in Polish institutions.

The major task beyond the organization of socialism was reconstruction through a 1947–49 three-year plan that essentially succeeded in part with Western help. The next six-year plan was typically Stalinist: it offered no breathing space in pursuit of heavy industry with such hero projects as the Lenin Steelworks near Krakow, a massive chemical factory in Oświęcim, a mining and steel plant in Bolesław, automobile factories in Warsaw (Żerań) and in Lublin, and a power plant in Jaworzno. The party instituted collectivization with agricultural output to increase by 35–45 percent, with more cereals, potatoes, white rye, barley and oats, vegetables and fruit, and livestock and fodder.

The Poles had to rebuild science in terms of personnel and institutes, too, especially since they lost researchers and institutes in Vilnius (to Lithuania) and Lviv (to Ukraine). An assault against the autonomy of researchers and the alleged underestimation of the achievements of Soviet science under the great leadership of Stalin accompanied this rebuilding. Structurally, Polish education and science mirrored the Soviet model of universities, Academy of Science, and branch institutes. The Polish Academy of Sciences, with its 150-year tradition, was reborn in 1951 with a declaration of the need to resist "cosmopolitanism" (Western influences) and an acknowledgement of the glories of Soviet science. A series of agricultural institutes were reformed, restructured, or established at this time, as well including the Institute of Soil Science and Plant Cultivation, the Plant Breeding and Acclimatization Institute, and the Institute for Land Reclamation and Grassland Farming all under the Ministry of Agriculture, and forestry research centers in the Ministry of Forestry. Together with the Academy of Sciences, these institutes pursued Stalin's "inspirational" plan for the transformation of nature. Given the destruction of Polish science under the Nazis, and then its isolation from European and North American science under socialism, it is not surprising that "Stalinist" science found broad support in Poland, without the repression that buffeted Soviet genetics.

The Stalin Plan found avid followers in such Polish party publications as *Nowe Drogi* as well as among scientists, including their Michurinist–Lysenkoist ideas. But many scholars found it better only to write about Lysenkoism, but not to apply the results of this work in practice so as to retain scientific independence; this made his influence fleeting. Other specialists, including biologists, botanists, foresters, and agronomists, promoted the new biology in support of nature transformation that was proselytized at various congresses and meetings.

In Poland the afforestation of fallow lands turned out well, while geoengineering projects (canals, weirs and the like) failed. One reason for the success of the former was that forest shelter belts already had a long history in Poland, and several scholars considered the Stalinist plan in a positive light—but urged caution against a full embrace of forest belts on any grand scale—while others noted their benefit in the struggle against draught and wind, yet only after detailed studies on climate, field conditions, soils, and so on. One of the causes of the limited reception of Stalin's plan in Poland may have been the country's own widespread tradition of nature conservation, which dated from the beginning of the twentieth century and continued under socialism, as can be seen in the creation or re-establishment of a series of national parks.

No less than in Czechoslovakia and Hungary, the Polish masses learned a great deal about the glories of Stalinist nature transformation through the press, popular science publications, radio, newsreels, numerous books and brochures, lecture series and the like. Socialist realist literature touted the glories of Stalinist geoengineering. But what of actual geoengineering in Poland—the construction of canals, dams, water power plants; land amelioration projects; and irrigation works? As in the other Socialist countries, including the USSR, many of the projects dated to an earlier era and had adherents among engineers from the "bourgeois era." These engineers welcomed the interest of the government in their projects, and especially in the government's largesse, even if it was socialist. But water resources in Poland were relatively scarce, so large-scale projects could not be pursued even if engineers were interested in them. The Czorsztyn, Rożnów, and Goczałkowice reservoirs, and the hugely expensive Wieprz-Krzna Canal, all of which had pre-war antecedents, were built under socialist rule and all of them had significant negative environmental impacts.

If Lysenkoism did not take complete hold, then Polish specialists still embraced the prospect of revolutionizing agriculture. The utopian scientific thoughts of Stalinism and fascination with Soviet achievements led Poles to introduce exotic plants to large-scale farming. Such plants as rice, citruses, corn, special varieties of wheat, soya, oilseed crops (castor oil plant and sunflowers) were supposed to be acclimated. Plant acclimation was also taken up by amateur clubs in Poland, usually at schools that were set as imitations of young Michurinist clubs in the Soviet Union. The aim was to recreate the new biology achievements on a small scale where young people could learn new theories, pursue experimental cultivation of rice, other exotic plants, and medical plants, undertake wheat improvement procedures, and master grafting techniques.

Stalin's death, and that of Bierut, led to reforms (like in the USSR under Khrushchev, the "Gomułka Thaw") when matters of science, industry, and agriculture took on a decidedly less ideological tone, although Poland remained firmly in the socialist world. Khrushchev, however, remained in the thrall of Lysenko and of corn (a remarkably costly and failed effort), and he insisted that the Poles follow these approaches; he proselytized the glories of corn, insisting on its cultivation in Poland. But corn planting was never substantial, likely because it was inappropriate for the climate.

Ultimately, Stalinist transformation of nature never found full geo-engineering flower in Poland. But grotesquely costly projects with significant environmental costs that were typically "socialist"—large-scale industrial settings that were designed with inadequate concern for public health and safety and that polluted significantly—were pursued as they were elsewhere in socialist Eastern Europe, especially in such hero industrial cities as Krakow. If socialist Poland was not immune to efforts to turn small-scale agriculture and dilapidated industry into modern powerhouses of production, Poland's natural environment was not directly subjugated to transformationist plans: there was no need for forest belts; economic recovery of cereal crops was more important than transformation; and collectivization did not go as far as in the USSR.

Building a Socialist Future

Major nature transformation projects were not unique to the USSR and Eastern Europe. They have many counterparts both in the socialist world and in the capitalist world. Among notable ones were a series of projects in the American West stretching from the Mississippi River basin to the Columbia and Colorado rivers, and to the Central Valley Project in California. Building on a series of earlier canals, river basin transfers, and irrigation projects in the eastern states in the early 1800s, specialists from the Army Corps of Engineers and later the Bureau of Reclamation moved westward, embracing more and more extensive and hubristic projects, and seeking to turn the plains and deserts into sites of urban development, electrical power generation, and massive farms of fruit, vegetables, and grain stretching to the horizon.

In the late nineteenth century, the Army Corps engaged in a series of grandiose projects to tame the Mississippi River, reclaim wetlands along its flood plain, and channel the river through levees in a futile attempt to increase farmland while preventing floods. In the attempt, having destroyed wetlands that served as "sponges" for floodwaters and created a human-built environment right up to the river banks, they turned 100-year floods into 50-year events, and 50-year floods into 10-year events, each year with increasing losses and property damage. The Great Flood of 1927 covered 70,000 km² of land when levees failed, and it killed hundreds of people. Issues of race played out, too, as the corps used dynamite to blow up levees to direct water away from the wealthier neighborhoods and into poorer ones, including many of African Americans, and the National Guard use violence to force blacks to work in dangerous conditions to shore up some levees. Continued efforts to control nature with billions of dollars of geoengineering did not solve the problem, but always led to another flood or disaster. In 2005 Hurricane Katrina flooded New Orleans and other low-lying areas, killing roughly two thousand people and causing $81 billion in damage.

Nature transformation was also prominent in the western states. Along the Columbia (and Snake) River, engineers and workers erected thirteen major hydroelectric power stations and built major irrigation systems to transform arid eastern Washington and Oregon into fruit and vegetable regions now dominated by agribusinesses. But the culmination of projects of geoengineering grandeur was likely the Central Valley Project in California, which had begun in the 1930s. It runs roughly north to south from Sacramento to Los Angeles, resulting in 1.2 million hectares of irrigated land; it includes a series of major dams and hydroelectric power stations, and has had significant environmental impacts, while at the same time creating tens of thousands of low-paying jobs in agriculture.[48]

An environmental catastrophe that resulted from the overconfident determination to mold nature into a machine—and profit from it—was the Dust Bowl in the 1930s in which 400,000 km² of land in the plains states, in particular Texas and Oklahoma, but also New Mexico, Colorado, Kansas, and elsewhere, was blown away in terrible wind storms brought about in part by drought, but also by extensive deep plowing of grasslands with newly ubiquitous tractors and combine harvesters that were intended instead to create monocultures of wheat and grain for businesses.[49]

China has also pursued, for thousands of years, various nature transformation projects, but especially magnificent and heinous ones in the early Communist period under Mao Tse-Tung.[50] One of the most audacious and costly was the Three Gorges Dam, built late in the twentieth century, operational at 22,000 MW in the twenty-first century, with roots dating to the Nationalist period at the beginning of the nineteenth century, involving Bureau of Reclamation officials in the 1940s and Soviet officials in the 1950s and 1960s, with a reservoir of 1,045 square kilometers. Intended to prevent flooding and generate electricity, the dam resulted in the ousting of 1.5 million residents from their homes in the inundated Yangtze River basin. But water does not make it to many regions in China, setting up a situation like that in the United States in the 1930s and even worse—with the threat of desertification because of geoengineering. In northern China at the beginning of the twenty-first century a dust bowl also spreads across grasslands that have been hurriedly transformed into cattle farms; twenty-four thousand villages have already been abandoned as dust storms increase in frequency.

Given the fertile environment for the transformation of nature in the USSR, the heroic nature of plans, the unbridled enthusiasm of their Soviet promoters, and the determination of the Stalinists to see a copy of Soviet political and ideological precepts, economic plans, and transformationist visions imposed on the East European socialist landscape, why, then did transformation fail to play out on the grandiose Stalinist scale?

First, even with the presence of Soviet and Warsaw Pact troops after World War II, the nations preserved a great deal of autonomy in domestic affairs that reflected diverse social, political, and economic conditions. Several countries had more developed industries; others had a larger percentage of students with higher education; others' leaders were more openly "Stalinist"; and so on.

Second, while Soviet planners might with great latitude look across a great landmass on a map and see no borders—only natural barriers that they could dynamite, excavate, or ignore—and they certainly did not fear the opposition of local residents or indigenous peoples to their plans, East European officials had to consider their neighbors in Europe, not to mention the fact that private ownership still played a role in these economies to a certain extent. Surely, however, in all of these nations, the loss of the sanctity of private property enabled planners to draw lines across maps with impunity and never fear legal obstacles— and false barriers—to their plans. In the USSR, where state ownership dated back to 1917, with one-half of the world's forests and one-sixth of the world's landmass, it was a simpler matter to imagine industrial forest shelter belts or a 6,000 square kilometer reservoir.

Third, only recently had the scientific R & D and educational apparatuses been coercively converted to Stalinist models. While many leading specialists embraced "Michurinist" biology and welcomed the opportunity to pursue transformationist projects, many others recognized that "proletarian" science in many ways was false science or "hurried" science, and that the pressure on them to achieve applications "impossible" under capitalism was based on false hopes and risky propositions. How could they simply ignore ecology and agronomy, push soil, plant trees, change the course of rivers, and make cotton and citrus grow where peasants knew it was impossible? It may also be that interagency and intergovernmental disputes slowed Stalinist transformation in Eastern Europe. After all, how could the ministries of fisheries tolerate dams and weirs going up willy-nilly on rivers? How could ministries of agriculture be happy with reservoirs that inundated farmland? And how could any one ministry compete for funds for its investment projects if there were so many hulking large projects to pursue to take funds toward nature transformation?

Nature transformation did not slow in the USSR after the death of Stalin. Khrushchev pursued the costly Virgin Lands campaign to plow under 20 million hectares of land, a project that attracted three hundred thousand Communist Youth league enthusiasts. He pushed the planting of corn. If progress had been slow to transform Siberia, then it accelerated under Khrushchev and Brezhnev in the enormous effort to locate new industry far from seemingly permeable western borders. Massive industrial combines to extract mineral resources and produce steel and chemicals appeared, linked by railways: Kuznetsk on the Tom River, Kemerovo, Anzherodzhensk, Prokop'evsk, and others to harvest coal, iron, zinc, tin, copper, aluminum, beryllium, and molybdenum. Forests fell to lumberjacks although they were only armed with rudimentary equipment. Hydroelectric power stations were built on the Ob, Irtysh, Angara, and Amur rivers.

In the socialist world, economic development was the raison d'être of regimes. Planners believed they could ignore geography, climate, and nature, and make them buckle to the dictates of the plan. Thus they ignored the social displacement and environmental costs of their programs from industry to agriculture to nature transformation projects. But without Stalin and a Stalinist polity, and with the abandonment of the threats of a coercive secret police and labor camps, transformation was scaled back by the late 1950s.

Paul Josephson teaches Russian and Soviet history at Colby College in Waterville, Maine, and is also a professor at Tomsk State University in Russia. He is a specialist in the history of big science and technology in the twentieth century, combining his primary fields of interest with environmental history. He is the author of *Industrialized Nature* (2002), *Would Trotsky Wear a Bluetooth?: Technological Utopianism under Socialism, 1917–1989* (2010), and *The Conquest of the Russian Arctic* (2014).

Notes

1. On the role of propaganda, see W. deJong-Lambert, *The Cold War Politics of Genetic Research: An Introduction to the Lysenko Affair*, London, 2012, 92–94; D. Weiner, *A Little Corner of Freedom: Russian Nature Protection from Stalin to Gorbachëv*, Berkeley, 2002, 89–90; S. Brain, *Song of the Forest: Russian Forestry and Stalinist Environmentalism, 1905–1953*, Pittsburgh, 2011, 1–2.
2. On the role of innocent men and women—prisoners of Stalinism—in the construction of the mighty symbols of the Stalinist Plan, see O.V. Lavinskaia and Iu. G. Orlova (eds), *Zakliuchennye na Stroikakh Kommunizma. Gulag i Ob"ekty Energetiki v SSSR. Sobranie Dokumentov i Fotografii*, Moscow, 2008.
3. S. Brain, *Song of the Forest*, 140–67. See L. Graham, *The Ghost of the Executed Engineer*, Cambridge, 1993, for a discussion of the 'disincentives' to scientific dispute and dissent in the Stalinist USSR.
4. Dennis Shaw has worked on the scientific underpinnings of the Stalinist Plan. He notes that "proponents of the plan, Party propagandists and academic commentators . . . claimed that the scientific basis of the plan lay in the research of nineteenth- and early twentieth-century scholars like V.V. Dokuchaev, P.A. Kostychev, G.N. Vysotskii and V.R. Vil'yams." They did this to give the impression that the plan was not "introduced in conditions of post-war crisis, [but] had a solid scientific basis." The evidence indicates a hurried, enthusiastic embrace of hubristic ideas, not an understanding of the natural environment. See D. Shaw, "The Science behind the Great Stalin Plan (1948–1953): Nineteenth- and Early Twentieth-Century Precedents," 24 July 2013. http://www.ichstm2013.com/programme/guide/p/0721.html (accessed 14 March 2015).
5. Z. Brezhnzhinski, *The Soviet Bloc*, Cambridge, 1960.
6. D. Weiner, *Models of Nature*, Pittsburgh, 2000, 20–26.
7. F.P. Koshelev, *Velichestvennye Stalinskie Stroiki Kommunizma i Ikh Narodnoxhoziistvennoe Znachenie*, Moscow, 1952, 68–69.
8. Ibid., 72–73.
9. D. Weiner, *A Little Corner of Freedom*; and P.R. Josephson et al., *An Environmental History of Russia*, Cambridge, 2013.
10. "Kak Sozdavalsia Kanal im. Moskvy." http://www.webpark.ru/comment/45260 (accessed 12 January 2009) and "Stroitel'stvo Kanala imeni Moskvy," http://nashram.ru/?page_id=66.
11. E. Kasimovski, *Velikie Stroiki Kommunizma*, Moscow, 1951, 52–53.

12 M. Ilin, *Men and Mountains*, Philadelphia, 1935, 114–15.
13 Ibid., 197–98.
14 In her *Every Farm a Factory: The Industrial Ideal in American Agriculture* (New Haven, 2002), Deborah Fitzgerald explores this point on the basis of archival material from Iowa state university libraries and other sources.
15 J. Stalin, 'Report to the Seventeeth Party Congress on the Work of the Central Committee of the SPSU (B)', 26 January 1934. http://www.marxists.org/reference/archive/stalin/works/1934/01/26.htm (accessed 5 April 2016).
16 Kasimovski, *Velikie Stroiki*, 35–42.
17 M. Tsunts, *Velikie Stroiki na Rekakh Sibiri*, Moscow, 1956, 14–15.
18 Ibid., 16–20.
19 Kasimovski, *Velikie Stroiki*, 44–45.
20 On the roots of some of these projects and for an environmental history of the steppe, see D. Moon, *The Plough That Broke the Steppes: Agriculture and Environment on Russia's Grasslands, 1700-1914*, Oxford, 2013.
21 Kasimovski, *Velikie Stroiki*, 46, 54–59.
22 P. Köhler, "Zarys historii łysenkizmu w ZSRR," in *Studia nad łysenkizmem w polskiej biologii*, ed. P. Köhler, Kraków, 2013, 16.
23 A. Czermiński, *Dzieje wielkiego planu*, Warsaw, 1951, 6.
24 Koshelev, *Velichestvennye Stalinskie Stroiki*, 104–24.
25 Ibid., 66–67.
26 Ibid., 74.
27 Ibid., 98–99.
28 V.A. Kovda, *Velikii Plan Preobrazovaniia Prirody*, Moscow, 1952, 25–27.
29 Iu. Leksin, 'Po Reke', in *Vokrug Sveta*, no. 8 (August 1972). http://www.vokrugsveta.ru/vs/article/4787/, (accessed October 6, 2014).
30 On the prerevolutionary roots of shelterbelts, see Moon, *The Plough that Broke the Steppes*, 173–205.
31 T. Croker, *The Great Plains Shelterbelt*, Greenville, 1991; W.H. Droze, *Trees, Prairies, and People: A History of Tree Planting in the Plains States*, Denton, 1977; and J.B. Lang, "The Shelterbelt Project in the Southern Great Plains, 1934–1970: A Geographic Appraisal," MA thesis, University of Oklahoma, 1970.
32 N.I. Sus, ed., *Polezashchitnye Polosy*, Moscow, 1936.
33 *Velikie Sooruzheniia Stalinskoi Epokhi*, eds L. Ognev, P. Serebriantsikov, Moscow, 1951.
34 M. Davydov and M. Tsunts, *Ot Volkhova do Amura*, Moscow, 1958.
35 Ibid.
36 D. Weiner, "The Roots of 'Michurinism': Transformationist Biology and Acclimatization as Currents in the Russian Life Sciences," *Annals of Science* 42, no. 3 (1985), 243–60.
37 For discussion of Lysenkoism, see D. Joravsky, *The Lysenko Affair*, Chicago, 1970; and Valery N. Soyfer, *Lysenko and the Tragedy of Soviet Science*, New Brunswick, 1994.
38 Brain, *Song of the Forest*, 140–67.

39 Ibid.
40 On the importance of vegetation to soils and soil science, see Moon, *The Plough That Broke the Steppes*, 53–56, 61–62.
41 Stakhanov, a Ukrainian coal miner, had established fantastical norms in his shifts in the 1930s; Stakhanovism was immediately introduced in all sectors of the economy.
42 International Court of Justice, "Gabčíkovo-Nagymaros Project (Hungary/Slovakia)," 25 September 1997. http://www.icj-cij.org/docket/index.php?sum=483&code=hs&p1=3&p2=3&case=%2092&k=8d&p3=5 (accessed 17 April 2016); J. Fitzmaurice, *Damming the Danube: Gabčíkovo and Post-Communist Politics in Europe*, Boulder, 1996.
43 Josephson et al., *An Environmental History of Russia*, 153–54.
44 For complete references on the Hungary case, see Zsuzsanna Borvendég and Mária Palasik, "Untamed Seedlings: Hungary and Stalin's Plan for the Transformation of Nature," in this volume.
45 M. Pittaway, "Creating and Domesticating Hungary's Socialist Industrial Landscape: From Dunapentele to Sztalinvaros, 1950–1958," *Historical Arceology* 39, no. 3 (2005), 83–85.
46 T. Gustafson, *Reform in Soviet Politics: Lessons of Recent Policies on Land and Water*, Cambridge, 1981.
47 D. Crowley, *Warsaw*, London, 2003, 38–45.
48 For various takes on nature transformation projects in the United States, see R. White, *The Organic Machine: The Remaking of the Columbia River*, New York, 1995; P. Daniel, *Deep'n as It Come*, University of Arkansas Press, 1996; M. Reisner, *Cadillac Desert*, New York, 1993; and P. Josephson, *Industrialized Nature*, Washington, 2002.
49 On the environmental devastation of the plains states in the 1930s, see D. Worster, *Dust Bowl*, New York, 1979.
50 J. Shapiro, *Mao's War on Nature*, New York, 2001.

Bibliography

Barrow, C., "The Impact of Hydroelectric Development on the Amazonian Environment: With Particular Reference to the Tucurui Project," *Journal of Biogeography*, Vol. 15, No. 1, *Biogeography and Development in the Humid Tropics* (Jan. 1988), 67–78.

Brain, S., *Song of the Forest: Russian Forestry and Stalinist Environmentalism, 1905–1953*, Pittsburgh, 2011.

Brezhnzhinski, Z., *The Soviet Bloc*, Cambridge, 1960.

Croker, T., *The Great Plains Shelterbelt*, Greenville, 1991.

Crowley, D., *Warsaw*, London, 2003.

Cummings, B., "Dam the Rivers; Damn the People: Hydroelectric Development and Resistance in Amazonian Brazil," *GeoJournal*, Vol. 35, No. 2 (February 1995), 151–60.

Czermiński, A., *Dzieje wielkiego planu*, Warsaw, 1951.
Daniel, P., *Deep'n as It Come*, University of Arkansas Press, 1996.
Davydov, M., and M. Tsunts, *Ot Volkhova do Amura*, Moscow, 1958.
Droze, W.H., *Trees, Prairies, and People: A History of Tree Planting in the Plains States*, Denton, 1977.
Fitzgerald, D., *Every Farm a Factory: The Industrial Ideal in American Agriculture*, New Haven, 2002.
Fitzmaurice, J., *Damming the Danube: Gabčikovo and Post-Communist Politics in Europe*, Boulder, 1996.
Graham, L., *The Ghost of the Executed Engineer*, Cambridge, 1993.
Gustafson, T., *Reform in Soviet Politics: Lessons of Recent Policies on Land and Water*, Cambridge, 1981.
Ilin, M., *Men and Mountains*, Philadelphia, 1935.
deJong-Lambert, W., *The Cold War Politics of Genetic Research: An Introduction to the Lysenko Affair*, London, 2012.
Joravsky, D., *The Lysenko Affair*, Chicago, 1970.
Josephson, P.R., *Industrialized Nature*, Washington, 2002.
Josephson, P.R., et al., *An Environmental History of Russia*, Cambridge, 2013.
Kasimovski, E., *Velikie Stroiki Kommunizma*, Moscow, 1951.
Köhler, P., *Studia nad łysenkizmem w polskiej biologii*, ed. P. Köhler, Kraków, 2013.
Koshelev, F.P., *Velichestvennye Stalinskie Stroiki Kommunizma i Ikh Narodnokhoziistvennoe Znachenie*, Moscow, 1952.
Kovda, V.A., *Velikii Plan Preobrazovaniia Prirody*, Moscow, 1952.
Lang, J.B., "The Shelterbelt Project in the Southern Great Plains, 1934–1970: A Geographic Appraisal," MA thesis, University of Oklahoma, 1970.
Lavinskaia, O.V., and Iu. G. Orlova (eds), *Zakliuchennye na Stroikakh Kommunizma. Gulag i Ob"ekty Energetiki v SSSR. Sobranie Dokumentov i Fotografii*, Moscow, 2008.
LeCain, T., *Mass Destruction*, New Brunswick, NJ, 2009.
Lilienthal, D., *TVA—Democracy on the March*, New York, 1944.
Liu, C., and L.J.C. Ma, "Interbasin Water Transfer in China," *Geographical Review*, Vol. 73, No. 3 (July 1983), 253–70.
Moon, D., *The Plough That Broke the Steppes: Agriculture and Environment on Russia's Grasslands, 1700–1914*, Oxford, 2013.
Ognev, L., and Serebriantsikov, eds, *Velikie Sooruzheniia Stalinskoi Epokhi*, Moscow, 1951.
Patel, C.C., "Surging Ahead: The Sardar Sarovar Project, Hope of Millions," *Harvard International Review*, Vol. 15, No. 1 (Fall 1992), 24–27.
Pittaway, M., "Creating and Domesticating Hungary's Socialist Industrial Landscape: From Dunapentele to Sztalinvaros, 1950–1958," *Historical Archeology*, Vol. 39, No. 3 (2005), 83–85.
Pritchard, S., "From Hydroimperialism to Hydrocapitalism: 'French' Hydraulics in France, North Africa, and Beyond," *Social Studies of Science*, Vol. 42, No. 4 (2012), 591–615.

———, "Reconstructing the Rhone: The Cultural Politics of Nature and Nation in Contemporary France, 1945–1997," *French Historical Studies*, Vol. 27, No. 4 (Fall 2004), 765–99.

Reisner, M., *Cadillac Desert*, New York, 1993.

Shapiro, J., *Mao's War on Nature*, New York, 2001.

Shaw, D., "The Science behind the Great Stalinist Plan (1948–1953): Nineteenth- and Early Twentieth-Century Precedents," 24 July 2013. http://www.ichstm2013.com/programme/guide/p/0721.html (accessed 14 March 2015).

Soyfer, V.N., *Lysenko and the Tragedy of Soviet Science*, New Brunswick, 1994.

Sus, N.I., ed., *Polezashchitnye Polosy*, Moscow, 1936.

Tsunts, M., *Velikie Stroiki na Rekakh Sibiri*, Moscow, 1956.

Weiner, D., *A Little Corner of Freedom: Russian Nature Protection from Stalin to Gorbachëv*, Berkeley, 2002.

———, *Models of Nature*, Pittsburgh, 2000.

———, "The Roots of 'Michurinism': Transformationist Biology and Acclimatization as Currents in the Russian Life Sciences," *Annals of Science*, Vol. 42, No. 3 (1985), 243–60.

White, R., *The Organic Machine: The Remaking of the Columbia River*, New York, 1995.

Worster, D., *Dust Bowl*, New York, 1979.

———, "Hydraulic Society in California: An Ecological Interpretation," *Agricultural History*, Vol. 56, No. 3 (July 1982), 503–15.

 CHAPTER 1

Kafkaesque Paradigms
The Stalinist Plan for the Transformation of Nature in Czechoslovakia

Doubravka Olšáková and Arnošt Štanzel

In January 1953, at the founding meeting of the new Československá akademie zemědělských věd (ČSAZV) [Czechoslovak Academy of Agricultural Sciences – CAAS], its president, Antonín Klečka, announced the establishment of the Committee for the Transformation of Nature. This committee had the grand task of working up a project that was to result in the total transformation of nature in Czechoslovakia. The new scientific institution, the CAAS, stood at the head of this project. The existence of the academy would not have been possible without the contribution and support of the USSR, as Antonín Klečka duly emphasized in his speech. This was an institution that was founded with the aid of Soviet advisors from the Ministry of Agriculture; these advisors had an important position during Stalinization. Antonín Klečka's introductory speech was "spontaneously" concluded by the audience chanting "Long live Comrade Stalin! Long live Comrade Stalin!" Clapping along to the rhythm, here in a historical hall at Charles University, the oldest university in Central Europe, were a majority of the Czech intelligentsia.

Bright Tomorrows

Once the position of the Communist regime had been solidified, which was preceded at the very beginning of the 1950s by a period of heavy Stalinization in Czechoslovakia, another step was taken that would make Czechoslovakia more dependent on the Soviet Union. The Stalin Plan for the Transformation of Nature was announced in the

USSR in 1948, and became an official part of Czechoslovak affairs five years later.

At that time, essential conditions for the success of the plan being fully implemented already existed: in 1948, the Communist government took over power and commanded all security forces; it also had a monopoly on education, cultural policy, and, above all, the mass media. The research landscape was also restructured so that academies of sciences, established following the Soviet model, would play the main role as the scientific institutions at the top of the hierarchy in the country. The goal of these academies was to transfer and promote Soviet knowledge, technology, and scientific methods. In the countryside, new agricultural policies resulted in land consolidation, violent agricultural collectivization, and the creation of collective farms and state tractor stations, which were also established based on the Soviet kolkhoz model.

Thus, in the early 1950s, the basic conditions for building a new society on new foundations were established. This new society, however, was supposed to live in a new environment, an environment that fitted the demands placed on the new, socialist man. In the great process of the transformation of Eastern Europe, the Stalinist Plan for the Transformation of Nature became for nature what the Sovietization of politics and culture was for society. Both of these projects were the expression of a great plan to create a completely new society in which nature and society were subjected to strict scientific development and whose laws would be defined within Marxism–Leninism. Just as a new society was supposed to be formed, based on people possessing Soviet qualities, so too was a new environment to surround this society to be created.

In the early 1950s, Czechoslovakia followed the example of the USSR in planning and implementing this landscape transformation. As part of the plan, large fields were created that replaced the small fields that had been farmed by individuals; livestock production was concentrated on collective farms. The number of "great works of socialism," such as reservoirs and hydroelectric power stations, grew massively.[1] There was a boom in land improvement measures and related landscape modifications, and following the Soviet model, shelterbelts were also planned.

It would be naïve to assume that all of these things were new in Czechoslovakia. Just as Soviet science was to a certain extent based on previous research, a similar amount of continuity can be observed in Czechoslovakia. However, what truly differentiated the Great Plan for the Transformation of Nature from the past was that it represented a comprehensive approach to the transformation of the environment that affected all elements of society. This comprehensive approach can be best defined by three principal characteristics. One defining element was Communist ideology, which placed this transformation into the system of Soviet symbols, and the interactions between them. Another characteristic of this plan was its aggressiveness, which was due to an absence of political opposition. The last element was faith, which not only gave scientists symbolic wings in the early 1950s but also, thanks to homogeneous political support, gave them a very dangerous weapon: nearly unlimited political power in implementing new plans.

The Historical Context of the Rise of the Great Plan

The Last Occupied Country in Europe

Whereas Paris had been liberated in August 1944, Prague was the last occupied European capital to be liberated. And whereas the Battle of Berlin came to an end on 2 May 1945 and the German Instrument of Surrender was signed by Germany, the USSR, and the Western Allies on 8 May in Berlin, Prague was only liberated the following day—9 May 1945—and fighting on the territory of Czechoslovakia continued for two more days, until 11 May.

Despite the fact that Czechoslovakia was not as devastated as, for example, Poland, and that there were no significant changes to its borders, postwar recovery began here much later than in neighboring countries.

The Košice Government Program of the Czechoslovak government formed the backbone of recovery; it was signed on 5 April 1945 in Košice, Slovakia. Just like other European countries occupied by Germany, a Czechoslovak government-in-exile was established in London, where mainly only right and center-right politicians were based; the Social Democrats were the only left-wing party represented. After the Soviet Union declared war on Germany, the leaders of the Komunistická strana Československa (KSČ) [Communist Party of Czechoslovakia – CPC] left for Moscow where they formed their own government, with Klement Gottwald at its head. In contrast to the Polish government, Edvard Beneš, the Czechoslovak president-in-exile, entered into negotiations with exiled Communists in order to organize the activities of the resistance movement and troops in exile. The first negotiations between the London government-in-exile and the Moscow Communists were successfully held in 1943.

The return home of the Czechoslovak government was a geographically complicated affair. It involved a journey from London to Prague via Moscow and Košice. The reasons for this were mainly strategic: Berlin and Prague had yet to be liberated. Therefore, once the main leaders of the Czechoslovak government had moved from London to Moscow, a unified National Front was established in March 1945, and a unified Czechoslovak government was formed on 4 April and their program was signed the same month in Košice, a district capital located in liberated eastern Slovakia. This meant that representatives of all the political parties entered Czechoslovakia together and not separately, as was the case in Poland.

The Postwar Recovery Plan

The postwar recovery plan of Czechoslovakia, as defined by the Košice Government Program, called for the immediate recovery of the Czechoslovak economy as a priority. The actual text of the program, which was the result of discussions held among all political parties, including the KSČ, contained no overt mention of nationalizing businesses. It did, however, state that companies must be built up "under general state management and for services of the renewed construction of the national economy." This could have been perceived to be the first step towards nationalization.[2] Under the influence of the unions and an increasing mood of social radicalization, these companies "under general state management" were actually nationalized not long after 1945.[3]

This fact significantly strengthened the position of the left.[4] Besides the denazification of public offices, the democratization of the school system and cultural life, and the establishment of social benefits (such as pensions and disability benefits), the Košice Program also contained important passages about the transfer of Germans and Hungarians from the territory of Czechoslovakia and about the confiscation of their land without compensation.[5] Agriculture and the supply of agricultural products became a key issue due to the large-scale looting of the country at the end of the war by German troops. Therefore, one part of this program for rebuilding the Czechoslovak state was dedicated to the standing of peasants and farmers. The Košice Government Program emphasized the alliance between liberated Czechoslovakia and the USSR, its main liberator, as a reaction to the Munich "betrayal."[6]

The Communist Party of Czechoslovakia won the first free elections to be held in the country in May 1946 with 40 percent of the votes.

The Czechoslovak National Socialist Party was second with 24 percent, the Czechoslovak People's Party third with 20 percent, and the Social Democrats fourth with 16 percent.[7]

The democratically elected Communist Party of Czechoslovakia put together a government in 1946 with Klement Gottwald at its head. Gottwald would later be connected not only to the hard-line Stalinization but also to the implementation of the Stalinist Plan for the Transformation of Nature. The Communists stayed in power for nearly two years until, in February 1948, they took over power in the country (the circumstances of which the reader can learn about elsewhere) and began to build a totalitarian system based on the rule of one party—the Communist Party of Czechoslovakia. This party however did not serve the needs of Czechoslovakia; instead, as a member of the international Communist movement, it served the needs of Moscow. The next time that free and democratic elections would be held in Czechoslovakia was fifty-four years later in 1990.

A Timeline of the Communist Government

The period from 1948 to 1953 is characterized by the mechanical transfer of Soviet expertise—in all ways, all manners and in all parts of life—to the individual, society, and the state.[8] This era can rightly be considered the harshest era of Stalinism and was peppered with manipulated political trials of leading figures from opposition parties and movements (such as the Milada Horáková trial and the trials of Catholic church leaders), as well as with trials of members of the KSČ (such as the trial of First Secretary of the Central Committee of the KSČ, Rudolf Slánský) whose Jewish origins cannot be overlooked. After Stalin's death on 5 March 1953, there was an initial thaw. A mere nine days later, on 14 March, Stalin's Czechoslovak counterpart and faithful Stalinist, Klement Gottwald, died; he fell seriously ill at Stalin's funeral in Moscow, and after returning to Prague he passed away quietly.

Besides the deaths of Stalin and Gottwald, another milestone in the history of the Czechoslovak Communist regime came in 1956, when Nikita Khrushchev made his secret speech on the cult of Stalin and set off de-Stalinization.[9] A period of loosening up followed in the 1960s, when Czechoslovakia began to work free from the territorial sphere of influence of the USSR and began forming its own value system based on more democratic rule. This short period however came to an end in 1968, when Czechoslovakia was occupied by the armies of the Warsaw

Pact on the night of 20/21 August 1968 for straying too far from Soviet policy. Then a period of attrition, known as "normalization," followed in the 1970s. This era saw the birth of the Charter 77 opposition movement. True change would only be brought in by the change of course in Soviet policy introduced by Mikhail Gorbachev known as *perestroika*. The definitive fall of Communism arrived in Czechoslovakia on 17 November 1989.

Policy

The Agricultural Policy of the KSČ

The stabilization of the political power that took place in Czechoslovak agriculture between 1945 and 1951, and the significant restructuring of relations in the Czech and Slovak countrysides, opened the doors not only to political power, but also to Communist ideology and indoctrination. Agriculture became a strategic field in postwar Czechoslovakia; nearly all political parties fully recognized its political potential. This is why during the war the Moscow-based leadership of the KSČ had wrested control of what they considered to be two key ministries: the Ministry of the Interior and the Ministry of Agriculture.[10]

Despite the fact that today the postwar Ministry of Agriculture is not viewed as a key ministry, one of its component parts played a very important role in the history of the Czechoslovak countryside. The IX Department of the Ministry of Agriculture, which was named the Department of Land Reform, administered the National Land Fund until 1948. Due to the nature of this department it worked very closely with the Resettlement Office in Prague. These institutes were delegated the task of implementing two presidential decrees signed by Edvard Beneš covering the new situation in Czechoslovakia after World War II. These decrees were the *Decree of the President of the Republic No. 12/1945 Coll., of 21 June 1945, concerning the confiscation and expedited allotment of agricultural property of Germans, Hungarians, as well as traitors and enemies of the Czech and Slovak nation* (also known as the Confiscation Decree) and the *Decree of the President of the Republic No. 28/1945 Coll., of 20 July 1945, concerning the settlement of Czech, Slovak or other Slavic farmers on the agricultural land of Germans, Hungarians and other enemies of the state* (also known as the Settlement Decree).[11]

In February 1948, the Communists took over the government in a coup. Subsequently, the Hradec Program of the KSČ was adopted and implemented, which resulted in a new land reform.[12] Due to the KSČ's pre-war interest in the Ministry of Agriculture, continuity was ensured. The figure of Julius Ďuriš represents this continuity. He was the first postwar minister of agriculture for KSČ, from 1945 to 1951. During his six years in this post, he laid the foundation for a new era in agriculture, as well as for a new era in the institutionalization of the agricultural sciences in Czechoslovakia, which in the 1950s was inextricably linked to the indoctrination of society. In the history of the Ministry of Agriculture, he is one figure who was able to stay in this position for a few years. He was followed by a slew of ministers who were generally in office for just one or two years. In later years, the Ministry of Agriculture was essentially subjugated to the Agricultural Department of the Central Committee of the KSČ, which monitored the ministry and in many respects used it only to consult its decisions, which once made would be passed on the Central Committee of the KSČ for approval.

Despite the fact that in 1946 and 1947 the Communist prime minister Klement Gottwald had assured farmers that kolkhozes would not be created in Czechoslovakia, things changed after 1948. The private ownership of agricultural land was assured by the new constitution of 9 May, which was ratified in 1948, two months after the Communists took power. Article XII, however, reads: "The economic system of the Czechoslovak Republic is based on: the nationalization of mineral resources, industry, the wholesale industry, and banking; land ownership according to the principle that 'the land belongs to those who work it'; the protection of small and mid-sized businesses and on the inviolability of public property." Land holdings were limited to fifty hectares.

Soon afterwards the violent collectivization of agriculture began, which resulted in one of the most significant social and cultural changes in the postwar history of Czechoslovakia. Traditional rural relationships were radically severed; rural society was fundamentally changed. The consolidation of fields, which started after 1948, along with mechanization led to the abandonment of small-scale production in favor of state-directed large-scale agriculture. The implementation of part of the plan for the transformation of nature coincided with the creation of collective farms, which took place between 1949 and 1960. This era was followed by a period of consolidation of large-scale agriculture (1960–1975), and then by a period seeing the stabilization of new socialist relationships and an increase in the specialization of agriculture (1975–1989).[13]

Table 1.1: Growth in collective farms in Czechoslovakia[14]

Year	Number of collectives	Average size of collective in ha
1950	1,873	317
1960	10,816	420
1970	6,200	638
1980	1,722	2,786
1989	1,660	2,592
1990	1,749	2,454
1992	2,158	1,975

A Comprehensive Approach

All fertile land was gradually collectivized and was managed by either collective farms or by state farms. The small area of Czechoslovakia, along with the fact that there was a tradition of intensive agriculture in all parts of the country, were the main reasons that the Stalin Plan for the Transformation of Nature had relatively little impact in Czechoslovakia. As opposed to the expansive Soviet steppes and deserts, this small country in the center of Europe did not have enough room to implement fantastic plans. Thus, the regime expressed a certain amount of common sense when, instead of feverishly implementing the plan, it began gradually planning. Even here though, there were attempts to push through at least some elements of the Stalin Plan for the Transformation of Nature into practice.

During the most intensive phase of the Sovietization of Czechoslovak agriculture, fertile land came to the center of attention of promoters of Michurinist agrobiology, particularly as Stakhanovite methods were applied in agriculture. Infertile land became the subject of interest for technocrats, who planned shelterbelts and implemented new and great "works of socialism", primarily reservoirs.

In this respect, the Czechoslovak vision of transforming nature copied the Soviet vision, in which "Stalinist works" to transform large areas of infertile land into fertile fields, orchards, and gardens were at the forefront. Right behind them, in second place, were forests, whose hydrological functions, both in the land and for the hydrological cycle in nature in general, were no less important. Forests also play an important role in regulating wind erosion. In the words of brochures published on the occasion of Forest and Tree Week in 1951: "Stalinist works are man's first great entry into transforming nature.... The forest becomes an important link, without which the transformation of nature is unthinkable."[15]

An overall proposal of the concept of the transformation of nature—but not the actual plan itself, as it never saw the light of day as a holistic concept or in the form of a law in Czechoslovakia—first appeared in early 1949. In March of that year, progress in creating an inventory of infertile land was discussed at the Ministry of Agriculture. This inventory was prompted by the Act on Afforestation and Establishment of Protective Shelterbelts, which was approved in Czechoslovakia in September 1948. During the discussion on the progress of collectivizing fertile land and inventorying infertile land, the main points of the Great Plan for the Transformation of Nature were proposed, which were supposed to include specific numbers and calculations.

This plan, for example, was supposed to include "historical-economic justifications for the necessity of technical interventions and forestry and hydrological interventions in the regulation of the climate and hydrological cycle of Czechoslovakia." The Ministry of Agriculture also requested the creation of a long-term perspective plan (covering ten years), as well as an action plan for 1949 and 1950, as part of the comprehensive task of transforming Czechoslovak nature. The transformation was supposed to be approached scientifically; the plan was supposed to include scientific proposals for the best choice of tree species to plant, for the best technology, and so on. At the same time, it was intended to use the plan for propaganda, so that Communist ideals could be brought to all of society. The plan was coordinated by one central body that was to monitor both the creation of shelterbelts, as well as the construction of reservoirs, and at the same time it had enough authority to implement the plan in coordination with state offices. Besides these more or less general phrases, the proposed structure of the plan contains the important point that "a necessary number of people must be urgently sent to study these issues in the USSR."[16]

In February 1950, Rudolf Slánský, at the time the general secretary of the Central Committee of the KSČ but who, in December 1952, would be executed after a political trial, announced on the development of Czechoslovak agriculture: "We have one great advantage. We have the expertise of the Soviet Union and All-Union Communist Party (bolsheviks), who have shown us how to work on this issue, how to help small- and mid-scale farmers rid themselves of hundreds of years of drudgery, poverty, and dependence on the exploiters of the village, and how to bring progress and prosperity to agriculture."[17] Not long before this proclamation, a delegation of approximately three hundred peasants had left for the USSR; they were supposed to have been convinced by their own eyes that Soviet agriculture was the "most progressive and most mechanized agriculture in the world."[18] Therefore, nothing should have been standing in the way of the great plan.

Soviet Influences

The hard-line Sovietization of Central and Eastern Europe took place primarily in the early phases of the building of Soviet-styled regimes in the 1950s. In these countries, not only did politicians find themselves under pressure, but so too did scientific and cultural figures.[19] In this early phase of Sovietization this was manifested not only in the selection of institutional models, but also in the sciences, particularly in the indoctrination of science and leading figures in various fields of science.

There was only one case where two-way influence can be detected, namely the creation of the Czechoslovak Academy of Agricultural Sciences, which is dealt with later in its own chapter. In reality, throughout its entire existence the CAAS was under strong pressure due to the Sovietization of science and applied agricultural research, and in the end it was also under pressure to create collective farms and to Sovietize rural agricultural areas. The Lenin All-Union Academy of Agricultural Sciences (VASKhNIL) served as a model for its establishment, and remained a model during the entire existence of the CAAS (1952–1962).

In the 1940s and 1950s, the existence and fame of VASKhNIL was inextricably linked to its director, Trofim Denisovich Lysenko. The cult of Lysenko in Czechoslovakia derived from a fascination with the new possibilities that this scientist and the "findings" of his research offered. The fact that he was ushered to the head of the scientific community with the political support of Stalin, only gave his teachings and revolutionary methods more power in the eyes of society. Biologists and agronomists had their own opinions, but Lysenko's theories had resonance thanks to political support from throughout the Soviet Bloc.

The agenda for the indoctrination of Czechoslovak science was created at the very important—from today's perspective—First Ideological Conference of University Scientific Workers, held in late February to early March of 1952 in Brno. At the time when decisions were being made about the institutional form and the definition of Czechoslovak science, an agenda was also defined for it. The title of the conference—"Against

Cosmopolitanism and Objectivism in Science"—more or less defined, quite aggressively, the future program, and there was not one field of science that escaped the focus of Communist ideologues.[20]

The keynote lecture on the biological sciences, "Following the Model of Soviet Biological Sciences," was made by the director of the Central Biological Institute in Prague, Ivan Málek. His paper was a radical break from how the field had developed up to that point, and it uncompromisingly subscribed to the Soviet model. Existing findings and opportunities for publishing were severely criticized, which made following the Soviet model even more imperative. He criticized nearly all academic journals, as they were too objectivist, and T.D. Lysenko and Soviet agrobiology was presented as the only model to follow.

It is no surprise then that in the early 1950s foreign delegations streamed into the Lenin All-Union Academy of Agricultural Sciences. They not only discussed biology and agrobiology with Lysenko, but also wanted to cover the issues of how agricultural research should be structured and how an academy of agricultural sciences should work; but mainly they wanted to know how research should be applied in practice.

Top scientific figures from other fields and working in other branches of the economy were sent there on study exchanges, ranging from farmers, to doctors, to engineers.[21] The Czechoslovak delegation visited Lysenko in November 1952, right before the new Czechoslovak Academy of Agricultural Sciences was opened. The discussion was one of those discussions that often took place at the time when the USSR was an out-of-reach model. They discussed plowing methods, vernalization, and also how VASKhNIL worked.[22] A large majority of the scientists who were present at this monumental discussion with Lysenko became founding members of the CAAS or members of the various biological institutes of the Czechoslovak Academy of Sciences.

Whereas the ideological conference held in early spring 1952 in Brno focused on scientific disciplines, technology and technical fields did not remain out of the sights of these new scientific cadres. Thus, another conference was held in the autumn of that year, whose title left no one doubting whether a similar approach would be applied in the technical sciences. The conference bore the name "Against Cosmopolitanism and Objectivism in Civil Engineering" and was the basis for the new agenda for building "great works of socialism."[23]

During the first few years of the existence of these new scientific institutions, Soviet influence was simply dominant. From once it started up, the CAAS based its work on the assumptions of the Stalin Plan for the Transformation of Nature, which was declared by Decree of the Council of Ministers of the USSR and the Central Committee of the All-Union Communist Party (bolsheviks) on 20 October 1948. Besides the introduction of kolkhozes, an idea that Klement Gottwald denied between 1946 and 1947 with the statement "We won't have kolkhozes here", not only were Soviet organizational patterns adopted, but so too were their methods and procedures. For example, on the basis of Act No. 206/1948 Coll., shelterbelts were established. Furthermore, the propagation of seedlings was centralized and was done by state institutes or on kolkhozes, Viliam's grass-crop rotation system method was applied in agriculture, and so on. The CAAS was responsible for the introduction of many of these regulations and also coordinated their implementation.

When the CAAS started up in the spring of 1953, its first general assembly was held alongside a "conference of agricultural workers from the highest institutes of agricultural science of the USSR and other people's republics, which contributed to defining the specific tasks of the newly created academy."[24] A year later, in 1954, problems dealt with at the general assembly were still more organizational in nature, and focused especially on collective work, which was introduced based on the Soviet model. Once again the model work schedule of the Lenin All-Union Academy of Agricultural Sciences was at the center of attention.[25]

The creation of the CAAS in the early 1950s was clearly a manifestation of the Sovietization of scientific institutions in Eastern Europe. Besides just imitating Soviet models, Sovietization can also be seen in the application of heavily ideologized Soviet scientific methods such as Michurinist agrobiology, Lysenkoism, the Viliam's grass-crop system, and such like. Antonín Klečka, the president of the CAAS, played a significant role in the Sovietization process. His leadership of the CAAS was noted by Radio Free Europe, broadcasted from Munich, as being the same as Lysenko's leadership of his academy in the USSR.[26]

Events in the USSR were followed carefully in Czechoslovakia, so when, in 1954 at the plenary meeting of the Central Committee of the Communist Party of the Soviet Union, Nikita Khrushchev presented his report on increasing agricultural yields and on cultivating uncultivated land, these issues immediately began to be talked about in Czechoslovakia as well. Relative satisfaction reigned as the plan for the transformation of the land was successfully progressing. Once

Khrushchev became the first secretary of the Central Committee of the Communist Party of the Soviet Union, his word became a sacred mantra that the powers-that-be in Eastern Europe swore by. It is no wonder that, thanks to Nikita "the Corn Man" (Nikita Kukuruznik), as Khrushchev was nicknamed in the USSR due to his obsession with introducing corn everywhere, the cultivation of corn spread massively throughout Czechoslovakia. After Khrushchev came to power, growing corn became a political priority not just in Czechoslovakia, but like always, in all Communist countries.[27]

There were of course limits to the influence of the Soviets. In 1954, on one of many visits to Czechoslovakia, Khrushchev brought corn seeds along with the normal rhetoric, specifically of the corn variety Sterling and a hybrid variety Krasnodarska No. 3. Almost unbelievable Russian fairytales were told about their great yields. Researchers however were more cautious, and decided to test the adaptability of these varieties to the new environment. Three years later experimental cultivation of these and other Soviet varieties demonstrated that Soviet varieties grown in Czechoslovakia had lower yields by about 25 percent.[28] Therefore the idea of growing them was abandoned.

In this situation it was clear that Khrushchev's speech made at the Twentieth Congress of the Communist Party of the Soviet Union would only open a floodgate of doubts that had been collecting in the scientific community for a long time since they first got word of the creation of living matter from non-living matter, or the discovery of "branched wheat." It was not long before reactions were heard; as early as March 1956, the first critical voices against the mechanical adoption of Soviet methods were heard in academia; in agriculture, for example, the grass-crop system had become a thorn in the side of scientists. Besides these complex problems, individual scientists, who had been pushed aside during the reign of Stalinism, made themselves heard. Some even sought rehabilitation—and surprisingly their voices were heard not only within the Czechoslovak Academy of Agricultural Sciences but also in the Office of the President of the Republic.[29]

Two more significant changes were made in 1956. In that year, agricultural planners from the Ministry of Agriculture who had been on a study exchange in the USSR announced that a new planning method based on decentralization had been introduced there.[30] And thus, while Czechoslovakia had reached the final stage of agricultural centralization in 1956 following the Soviet model introduced in 1948, the USSR had started up a new method and had decentralized the entire system.

The entire nature of study exchanges also began to slowly change. In 1957, another Soviet agricultural delegation visited Czechoslovakia, which included P.I. Titov, the deputy minister of sovkhozes of the USSR, V.I. Rudakov, the deputy director of the Main Directorate of Mechanization and Electrification of the Ministry of Sovkhozes of the USSR, P.I. Chaikin, the deputy director of the agricultural department of the Central Committee of the Communist Party of the Soviet Union, and representatives from kolkhozes from various parts of the USSR.[31] Meeting reports discuss the positives and drawbacks of Czechoslovak agriculture; the openness and willingness to exchange information from both sides found in the minutes are surprising. When discussing Czechoslovak imperfections, the Soviets showed them how such issues are dealt with in the USSR. Equally, when discussing positive aspects of Czechoslovak agriculture, the Soviets claimed that Soviet organizations would consider adopting some of the innovative procedures used in Czechoslovakia, and that some of them would be proposed for official adoption in the USSR.[32]

However, in the 1960s, the Soviets completely lost interest in exchanging delegations. Or rather, although both Czech and Soviet scientists and agriculturists were still interested in exchanges, and when preparing various plans counted on consulting with each other, the Soviets extended, postponed, or completely broke off negotiations.[33] Was this a mere communications breakdown or was it the first tremor of a coming crisis? Or did it mean the end of hard-line Sovietization in Eastern Europe?

The Czechoslovak Academy of Agricultural Sciences[34]

In the original concept of implementing the plan for the transformation of nature the Czechoslovak Academy of Agricultural Sciences (1952–1962) was determined to play a key role as the main coordinator for the scientific implementation of the plan. It was created under the guidance of Soviet advisors at the Ministry of Agriculture and was modeled after the Lenin All-Union Academy of Agricultural Sciences (VASKhNIL).[35] The crumbling of the political elite's self-assuredness after 1953 and the definitive end of Stalinist methods after 1956 led to this institute taking a more cautious approach to implementing the plan and to promoting its more megalomaniac aspects.

It must be stated that the CAAS was not an entirely new type of institution. The Československá akademie zemědělská (ČsAZ) [Czechoslovak

Agricultural Academy – CzAA] existed during the interwar period; it was established in November 1924 as a central institute to deal with agricultural sciences in Czechoslovakia. Its activities focused on scientific research, the practical applications of such research, and finally on awareness-raising. Just one year later, in 1925, the Svaz výzkumných ústavů zemědělských (SVÚZ) [Association of Agricultural Research Institutes – AARI] was founded, which included twenty-two research institutes, sixty-seven research stations, and twenty-two experimental buildings and farms. These two institutes worked in close conjunction.

The Česká akademie zemědělská (ČAZ) [Czech Agricultural Academy – CAA] was restored in October 1945, and its activities were supposed to follow up on the activities of the Czechoslovak Agricultural Academy during the interwar period. The Communist regime, however, was inspired by the scientific policy of the USSR to create a centrally directed hierarchical system of science. The idea of creating a central scientific institute like the Československá akademie věd (ČSAV) [Czechoslovak Academy of Sciences – CAS], which besides just conducting research would also serve as a coordinating agency, naturally arose at the height of the Stalinist period for one of the key sectors involved in the development of the Czechoslovak economy: agriculture.

Right after the February coup, major changes in the leadership of the ČAZ were made, which indicated that sooner or later the focus of its activities would change as well. At the annual general assembly of the ČAZ held on 23 February 1949, Professor Antonín Klečka (1899–1986), who had been a member of the ČAZ since 1934, was elected its head. He was a respected expert on forage and agrobiology; however, after 1948, he did not resist ideological pressure and became the most significant promoter of Michurinist agrobiology in Czechoslovakia.[36]

Already at this meeting in 1949, new tasks for the ČAZ were established under the leadership of Klečka, which was a signal that ideology was being introduced into practice. The main goals of the academy were: "(1) to coordinate the scientific program to meet the needs of the five-year plan and in accordance with the 'ten points' of President Gottwald, (2) to unite agricultural science and practice, [and] (3) to further develop the new findings of Soviet science."[37] There was no question about which way this institute was heading.

Despite the fact that the ČAZ held a very important and critical position among agricultural institutes, it did not fit the notions of the political elite. After a period of searching for a new conception for the agricultural sciences,[38] the decision was made to establish a new, central scientific institute. In contrast to the concept behind the CAS, which was developed in the interwar period and in many ways represented a continuation of certain streams of thought and effort at structuring science in Czechoslovakia in this period, the idea of a central institute focusing on the agricultural sciences developed in relative separation from these grand scientific debates. The Ministry of Agriculture was behind the idea, which inter alia contributed to meeting its ideological goals.

When studying period documents, however, one cannot avoid feeling that the words of Soviet advisors at the Ministry of Agriculture were taken very seriously. The main task in the first stage of creating a coordination and research center for the agricultural sciences was introducing Soviet expertise in Czechoslovak agriculture.

In the summer of 1952, the Political Secretariat of the Central Committee of the KSČ discussed the Government Resolution on the New Organization of Agricultural Promotion, Science, and Research. It already tended to favor the application of agricultural sciences to practical tasks, which was to be helped along by the introduction of Soviet expertise into practice. Besides introducing a centralized model of planning, another result of this tendency was the implementation of the grass-crop system. However, there were some negative experiences resulting from the mechanical adoption of the Soviet model. The authors of the government resolution complained about the mechanical adoption of the Soviet model when they stated: "Plant breeding and seed and seedling production has been conceived incorrectly, leading to the creation of a centralistic, commercial distribution organization (Oseva), which has slowed the implementation of the fundamentals of Michurinist agrobiology. In particular, seed is not being bred and produced to meet the conditions of individual production areas, which is a fundamental."[39]

Another key point for the future development of research in the agricultural sciences was the unification of scientific research work in the agricultural sciences. This was to be done "in the spirit of Michurinist agrobiology" and "by developing and broadly implementing Soviet findings in our agriculture."[40] The politburo resolved that the Ministry of Agriculture should propose a law on the creation of the CAAS by 1 November 1952. The CAAS was supposed to have the same function as the CAS, which was also being planned at the time: to act as the highest agricultural research institute in the country. Besides conducting its own scientific activities, its other main tasks included cooperating in the creation of research science plans, and educating cadres. The CAAS was also supposed to direct newly established or reorganized research and science institutes. A fixed system of these institutes had been created with full geographic regional coverage, including centers in Prague and Bratislava; and, in suitable areas, specialized institutes focusing on viticulture, aquaculture, and beekeeping were to be established or reorganized. Research was supposed to be promoted and popularized. For example, it was proposed that "Michurinist agrobiology be included in the curriculum of national [elementary] schools and middle schools, and that experimental Michurinist fields be established at these schools."[41] Zdeněk Nejedlý, the minister of education, was in charge of this.

The comprehensive application of Michurinist fundamentals in practice was supposed to be the main focus of further development in the field of breeding. All types of institutes on all levels and in all areas were assigned this task.[42]

Archival documents from the Central Committee of the KSČ from 1952 directly mention the act establishing the CAAS: "It was drafted following the model of the Lenin All-Union Academy of Agricultural Sciences in Moscow and a proposal by the Czechoslovak Academy of Sciences. Soviet experts at the Ministry of Agriculture contributed to developing the proposal."[43] Besides the Lenin All-Union Academy of Agricultural Sciences, another model for the newly conceived CAAS was the German Academy of Agricultural Science in Berlin,[44] which was the first institute that attempted to apply Soviet agricultural methods in a people's democracy.[45] The establishment of an academy of agricultural sciences independent and separate from more general scientific institutes was also planned in Hungary, where, while the CAAS was being established, talks about founding a similar institute were starting.

The CAAS was directly subordinate to the Ministry of Agriculture. This was supposed to guarantee the direct and free development of the agricultural sciences, yet at the same time it also reflected the necessity of linking the activities of the CAAS with practical agricultural activities and with the practical tasks of postwar agriculture. Due to the fact that very specific tasks were carved out for the CAAS, following the model of VASKhNIL, it was proposed that the academy's president be personally responsible for operations, unlike at the CAS where the board had collective responsibility. This raises the question: To what extent did this reflect the transfer of the "cult of personality" to the scientific community? In any case, there is no doubt that the Communist regime was hoping that a figure similar to Trofim D. Lysenko in the USSR would lead the CAAS. This proposal, however, is not reflected in the resulting bylaws of the CAAS, although they emphasize the role of the president in directing the academy to an extent that was greater than normal.

The CAAS was finally established as the leading agricultural science institute in the country as of 1 January 1953 by Act No. 90/1952 Coll., and its bylaws did not differ much from those written by Soviet advisors at the Ministry of Agriculture. In comparison to the previous activities of the Czechoslovak Agricultural Academy, there were several fundamental differences, which reflected the new regime's desire to build a new type of institution. Paragraph 5 of the bylaws, for example, clearly states that the CAAS would conduct work in its own research institutes,[46] to which other non-academic institutions would be methodically subordinate. The CAAS was also given the authority to grant academic titles, which at that time was a right that only universities had. Just like at the CAS, members of the CAAS were either full members, referred to as academicians, or corresponding members. These corresponding members could be Czechoslovak citizens as well as "progressively minded foreigners."[47] The CAS was also allocated its own line in the Ministry of Agriculture's budget (§6).

International Cooperation
The CAAS was very closely linked to the activities of similar institutions in Central and Eastern Europe. The idea of working in close conjunction with these other institutes was present from the early days of the CAAS. On the initiative of the CAAS, the First Conference on Coordinating Scientific Research Work in the Agriculture and Forestry Sectors was held in Prague in 1954. Prague passed the baton to Berlin, where in October 1956 a second conference on this topic was held and where the very important "Agreement on Direct Cooperation between Central Institutes of Agricultural and Forestry Sciences of Affiliated Countries" was made.[48] This agreement explicitly focused on defining the fundamentals, as well as forms, of cooperation between each institute (i.e., the division of labor), and the main goals each institute was supposed to focus on were also defined. The CAAS was, for example, assigned the task of coordinating and leading research on sugar beet production and on improving light and less fertile soil, and on light soils threatened by wind erosion. Furthermore, the CAAS was assigned the task of leading research on the mechanization of sugar beet harvesting and the

mechanization of the harvesting and drying of fodder crops. It was also supposed to focus on research on agricultural machine systems.[49]

These conferences continued to be held in the second half of the 1950s. The third coordination conference was organized by the Lenin All-Union Academy of Agricultural Sciences in Moscow, and it was held in October 1958.

The CAAS worked in especially close cooperation with its Soviet counterparts. This can be seen in the number of foreign corresponding members and honorary members in the CAAS. Foreign members included, for example, Soviet scientists P.P. Lobanov, the president of VASKhNIL, and Academician K.I. Skryabin, vice-president of VASKhNIL. Academician T.D. Lysenko was also a member. In the late 1950s, academicians Igor Alexandrovich Budzko, Fedor Grigoryevich Kirichenko, and Avedikt Lukyanovich Mazlumov became honorary members. It was symptomatic of all proposals discussed and approved by the Central Committee of the KSČ that they all contained information about the fact that the Central Committee of the Communist Party of the USSR and the Lenin All-Union Academy of Agricultural Sciences in Moscow approved of them too.[50]

1960: A Critical Year and Demise
In the late 1950s, complaints about inadequacies in the activities of the CAAS in relation to connecting theory and practice, coordinating the research activities of the CAS and the CAAS, and so on, became leitmotivs. In all reports it is clear that the activities of the CAAS were not meeting the expectations of the political elite, who saw before them the "exceptional" successes of Michurinist biology in the USSR, at least as presented in propaganda.

There was also a change in the concept of how science and research was planned. Starting in 1957, there were discussions about requesting a change in how the research programs of scientific institutes were created, especially for large academies of sciences. The ideas of Pyotr Kapitsa, who in the USSR was calling for greater flexibility for large institutes and proposed the creation of ad hoc institutes and ad hoc work groups (which in today's words would be referred to as being interdisciplinary in nature), made their way to Eastern Europe. The aim of these ad hoc formations was to solve specific problems. They were supposed to be disbanded once they met their goal. This idea first appeared in discussions about the activities of the CAAS in 1959.

Another important contributing factor was the introduction of a centralized planning model in all COMECON countries. This model was first proposed during Moscow negotiations in 1959–1960. They were first applied in Poland, Czechoslovakia, and the GDR in 1960 where fifteen-year prognoses were created; in 1961, these prognoses were increased to cover twenty years.

In 1960, a new constitution was signed into law which proclaimed that Socialism had been achieved. The Czechoslovak People's Republic changed its name to the Czechoslovak Socialist Republic. State administration was reorganized into three levels of administration: the district level, the regional level, and the central level. New powers, delegated from the center, were supposed to be used to improve the activities of the CAAS, as in the 1960s criticism against it reached its zenith.

Its main critic was Lubomír Štrougal, who was minister of agriculture, forestry, and water management from 1959 to 1961, and later the prime minister of Czechoslovakia from 1970 to 1988. The previously criticized facts that theory had been lagging behind practical application and that the CAAS was not able to react to current needs continued to be criticized, however in the early 1960s these criticisms were overshadowed by the fact that agricultural production was stagnating in comparison with industrial production. Thus, a date was set by which the insufficiencies of agriculture were to be overcome and agriculture was supposed to catch up with industry: 1970.

After receiving this crushing criticism, it was no surprise that on 10 October 1961 the politburo of the Central Committee of the KSČ resolved to incorporate the CAAS into the CAS and that agricultural research would be directly directed by the Ministry of Agriculture. The final wording of the resolution clearly indicates that the activity of the CAAS had been too narrowly focused on applied research and on the direct application of scientific findings in practice, while theoretical research was lagging. For this reason, the decision was made to conduct theoretical research within the CAS, whereas the Ministry of Agriculture would be directly responsible for applied research.

This elimination of the CAAS was also reflected in what happened to its members. Not all automatically became members of the CAS. Of the total seventy-two members of the CAAS, the politburo proposed that only ten to twelve become members of the CAS; the same measures were to be applied to the thirteen honorary foreign members, although no exact number was specified. Documents only vaguely state that "some of the foreign members shall be elected."[51] The proposal also followed the existing practice of politically pre-approving candidates for membership in the CAS and determining who was to be elected a full member and who a corresponding member. Only one member of the CAAS was to be elected a full member: Antonín Klečka, the president of the CAAS. The politburo recommended ten members of the CAAS for corresponding members.[52]

The CAAS was established as a central scientific institute working on the agricultural sciences in response to political needs and ideological pressure. The reasons it was created were the same reasons it was eliminated ten years later. The fact that, as of 1 January 1969, the Československá akademie zemědělská (ČSAZ) [Czechoslovak Agricultural Academy – CzAA] had been reestablished—which consisted of the Česká akademie zemědělská (ČAZ) [Czech Agricultural Academy – CAA] along with the Slovenská poľnohospodárska akadémie (SPA) [Slovak Agricultural Academy – SAA]—demonstrated that the existence of an academy of this type was justified and necessary. That this reincarnated form of the CAAS was once again eliminated in 1974 demonstrates that the oft-neglected agricultural sciences were caught up in and involved in political events and priorities more than we today realize. In the same year, Antonín Klečka, the former president of the eliminated CAAS and proponent of Michurinist agrobiology and its application in practice, received an honorary J.G. Mendel Plaque.

Practice

Forests and Shelterbelts

Just as in Poland, shelterbelts were more or less included in standard measures for protecting the landscape in Czechoslovakia, which was handed down from generation to generation.[53] In today's Czech landscape this can be seen in tree-lined roads, where the trees serve primarily to retain water, to limit drifting snow from fields in winter, and in general to limit the effects of wind in agricultural landscapes. In the late 1940s and early 1950s, the Communist authorities referred to them as *větrolamy* (literally "windbreaks"), a term that can be found in internal documents of the Ministry of Agriculture.[54]

During high Stalinism, however, the establishment of shelterbelts was intensified *in extenso* to such an extent that the board of the Ministry of Agriculture proposed creating a shelterbelt establishment plan that would be closely coordinated with the work of hydrologists, as it was assumed the extensive establishment of shelterbelts would have a fundamental impact on the hydrology of the entire country.

Forests as a Means of Transformation
Large plans soon appeared: in 1948, a "Week of Forests and Trees" was announced, the goal of which was to present Soviet forestry methods to the people of Czechoslovakia.[55] Popular brochures were also published which were aimed at sharing Soviet forestry expertise, the principles and reasons for establishing shelterbelts, and the like. The main goal of these publications was to emphasize the importance of forests for transforming nature and the overall place of forests in undertaking this comprehensive task.

An enormous propaganda campaign had a large share in increasing afforestation, which began in coincidence with the first two-year plan (an economic plan based on the Soviet model, and in effect from 1947 to 1948) and continued during the subsequent five-year plan period (1949–1954). At the time, a large five-part lecture series on the function of forests in the national economy was printed in the newspapers, the inspiration for which was primarily Soviet. Publications appeared, such as *Ochranné zalesňování v SSSR* [Protective Afforestation in the USSR] (1950) from the pen of N.I. Sus, and *Za sovětskými zkušenostmi v lesnictví* [Following Soviet Experience with Forestry] (1951) by J. Hofman and J. Jindra.

These publications clearly indicate that the postwar perception of forests and the functions they serve had radically changed: forests were not just one resource in the national economy, nor were they just of interest to conservationists, but in Communist thinking they were mainly a means through which to control nature. This was of course all to be done in the spirit of the Stalin Plan for the Transformation of Nature, "the greatest document of our time" as Czech scientists called it in the late 1940s.[56] The Czechoslovaks admired this plan for its scope, as according to calculations, during the first three five-year plans land was supposed to be transformed covering an area of 1,200,000 km², the size of England, France, Italy, Belgium, and Denmark combined.[57] According to plans, in the south and southeast of the European part of the USSR, six million hectares of new forests were supposed to be created by 1965; in total twelve million hectares of land were supposed to be afforested.[58] In the USSR, establishing shelterbelts, which could mitigate the effects of wind in dry regions, was emphasized. At the same time in these dry areas shelterbelts were used in combination with the grass-crop system; Soviet scientists expected this combination to retain water and nutrients in the soil.

The Act on Afforestation, Creating Protective Shelterbelts, and Establishing (Restoring) Ponds (1948)

In 1948, the same year that the first "Week of Forests and Trees" was announced, an act whose official name was Act No. 206/1948 Coll., on Afforestation, Creating Protective Shelterbelts, and Establishing (Restoring) Ponds was signed into law in Czechoslovakia. It is interesting to note that this law was on the books until 1961. It was inspired by the Soviet Plan for Shelterbelt Plantings, Grass Crop Rotation, and the Construction of Ponds and Reservoirs to Secure High Yields and Stable Harvests in Steppe and Forest-Steppe Regions of the European Part of the USSR, which was passed on 20 October 1948. But as the names of the two regulations show, the Czech act lacked the baroque grandeur of the Soviet one.

The act came into force on 5 September 1948, six months after the Communists officially took over power in February 1948. The act even came into being a month earlier than the Soviet decree on the transformation of nature of 20 October 1948. The expedition with which this act was passed can be attributed to two factors: Julius Ďuriš, a Communist politician, had been the minister of agriculture since April 1945 and kept his post in Gottwald's government. Therefore, agricultural policy had remained continuously pro-Communist in nature, and so new laws could be proposed and promoted much more quickly than by someone who was unfamiliar with how the ministry worked. The second reason the act was quickly passed was related to Ďuriš's overall concept of transforming the agricultural landscape and agriculture itself in Czechoslovakia. The act on afforestation was just one piece of the mosaic (alongside collectivization, the construction of water management projects, etc.) of the transformation of the countryside in Czechoslovakia. Its implementation belonged to the overall plan of the Sovietization of Czechoslovakia, a process in which Soviet advisors played a large role.[59]

When comparing the Soviet and Czechoslovak laws, it becomes quite apparent that, because the Czech act was passed before its Soviet model, it lacks much of the force and aggressiveness of the Soviet law, and instead to a large extent is based on continuity of the ideas of Czechoslovak interwar nature conservation.

The Czechoslovak law, as opposed to the Soviet one, focused on decentralization: individual measures were to be proposed on the district and regional level, which were subsequently supposed to be coordinated from the center. There was no central agency that determined through directives which regions were supposed to be changed, or by how much. Therefore, in reality this act—despite its promising name—did nothing more than merely initiate the inventorying of infertile land that was to be afforested or on which ponds were to be built. Moreover, it explicitly stated that, when selecting such land, the preservation of "natural beauty and natural monuments" (a reference to nature reserves) must not be neglected, but that at the same time nor must the rights of the owners of affected land. Therefore, all plans had to be made public by national committee offices for at least fifteen days. Section 5 (1) of the act stated:

> Land that produces no or only relatively low yields (such as dryland, wasteland, clearcuts, river banks, slopes, valley slopes, abandoned sand quarries, pits, gullies, artificial land, swamps, etc.) will be, in accordance with the provisions of this act, afforested or planted with shrubs (hereinafter "afforested"), or on such lands shelterbelts shall be created or ponds shall be established (or restored) in accordance with the water management plan. In doing so, natural beauty and natural monuments must be preserved.

In this context, it is important to realize that the Czechoslovak law was passed before collectivization was completed, meaning that all costs associated with afforestation fell on landowners. Therefore, if a national committee determined that a piece of land was infertile and included it on the list, its owners would have to make a financial obligation to "fecundate" it in accordance with the national committee's proposal. It could have been this economic aspect of the act that forced Soviet experts to approve of the passing of the act even before the Soviet law had been passed, although this may not have been the case either, as explicit proof of this is missing. However, afforesting the land with the money of private landowners rather than of the state was economically extremely advantageous.[60]

Another difference is that the act somewhat paradoxically contains provisions regulating not the transformation of nature, but the conservation of nature; for example, the act forbade the burning of grass and weeds, and the cutting of hedges and brush, between 16 March and 30 September annually.

However, similar to the Soviet law, the Czechoslovak one counted on the implementation of all proposed measures within ten years of being passed. In practice this meant that in the late 1950s all infertile land in Czechoslovakia should have been covered in forests or flooded. Landowners had five years to afforest land up to fifty hectares in area, whereas they had ten years to do the same to areas greater than fifty hectares.

Therefore, whereas the act and its use in propaganda were based on inspiration from Stalinist Russia, the act itself did not break from prewar development to the massive extent it did in the USSR. In contrast, it appears that this act and its more moderate contents were—somewhat paradoxically—the last manifestation of the dominant influence of the interwar generation of landscape conservationists on the transformation of nature in Czechoslovakia.

People's Forest Researchers

Just as in agriculture, a "people's researcher" movement developed in forestry in the 1950s following the Soviet model. Scientific institutes satisfied the needs of these aspiring "scientific foresters" by publishing various manuals and reference materials to help them to do their work. One such example is the book *Příručka pro lidové výzkumníky v pěstění lesů* [Handbook of Forestry for People's Researchers], collectively penned by scientists from the Forest Management Research Institute.[61] This was not, however, the first publication to focus on this topic, as in the early 1950s Czechoslovak State Forests published a brochure entitled *Náměty pro lidové výzkumníky* [Ideas for People's Researchers], which was officially only distributed to foresters and never found its way to the hands of common people.[62] Four years later though, the idea of utilizing people's researchers in forest management was revived.

Anyone expecting an ideological manual aiming to indoctrinate the people would have been disappointed. More than anything else, this book is a carefully written handbook full of technical details, methodologies, and detailed data about various procedures. The publication does, though, state that in 1953 every hectare of nursery could produce 400,000 seedlings, but that between 1954 and 1958 the same area was expected to yield 750,000 seedlings.[63] With this information, the authors wanted to indicate to readers the unlimited possibilities of the new regime, although otherwise we can only search in vain for any ideological indoctrination on the pages of this publication.

The biggest difference between original forestry literature written in Czech and translated literature is the amount of ideology they contain. Whereas Soviet literature, which was as a rule translated by the Soviet Ministry of Forests and Wood Production, served more to indoctrinate and convey detailed information about the ideology of the Great Plan for the Transformation of Nature,[64] Czech works refer people's researchers to interwar Czech forestry journals and to traditional Czech forestry methods and tree species.

Chatushkas about Afforestation

In order to depict the overall social mood in 1950s Czechoslovakia, there is one more prime example related to forestry and shelterbelts that we will examine: young, aspiring poets and writers wrote odes to establishing shelterbelts throughout the countryside. Among these writers though were some who would go on to later literary fame and would also become famous for their opposition to the regime. Jan Trefulka (1929–2012), the author of *Častušky o zalesňování* [Chatushkas about Afforestation], was one of them. In his defense, it should be noted that although his dedication to Communism started with *chatushkas* about forests (see the transcriptions below), it also ended symbolically in a literary forest as well.

His first critique of the bleakness of building communism dates to 1973 and is titled *Veliká stavba* [The Great Work]. In the depths of the forests stands a "great work," not dissimilar to Kafka's *Castle*, which hides the secret of how to build a communist society.

> *A Chastushka about Afforestation*
> In the desert, in the sand, in the dunes, we shall plant an apple tree
> to the strength of Stalin's people shall the drought bow.
> Forests, forests of green, when the belts connect,
> against the evil crop failures, they shall stand like castle walls.
> Plum, pears, apple trees, poplars and ears of grain,
> protect and adorn our fields with green borders.
> During infertile droughts, fires shall not threaten,
> reservoirs will supply us, with perhaps an entire sea.
> And when in the skies the rain is out of order,
> we will water our fields, just like green gardens.
> Nightingales shall sing in silent young groves,
> and in the ponds, carps shall be in our land.
> When you head out to the fields, where you are free to breathe,

our golden grains are the pride of the country!
Wheat stands like a wall, the grass smells of honey,
where sand once scraped, rustle fertile orchards!
Apples we shall plant and ponds build,
the fruits we will soon collect, and the harvest celebrate!
In the entire countryside, a new life begins,
we will live more merrily, richly and happily.[65]

New Methods, New Nature

Viliams's Grass-Crop Rotation System

One of the first items on the agenda of the proceedings of the board of the Ministry of Agriculture on 1 December 1949 was a brief report on the significance of the grass-crop system. The report was given by Josef Smrkovský, a top Communist functionary and chairman of the National Land Fund, who in 1949 became the deputy minister of agriculture and the general director of state farms. However, this was not for long, as in 1951, during the biggest Stalinist political trials held in Czechoslovakia, and which focused on removing the general secretary of the KSČ, Rudolf Slánský, he was sentenced to prison. He was released in 1956 (and rehabilitated in 1962). In 1968, he became one of the main figures of the Prague Spring.

It was, however, the same Josef Smrkovský who introduced the grass-crop rotation system to Czechoslovakia. His report entitled *The Significance and Benefits of the Grass-Crop System* contained practical points aimed at extending this system in Czechoslovakia. According to the schedule for introducing the grass-crop system, all contracts with plant breeders were to be closed by January 1950, so that the plan could be started up in 1951 without any problems. Economic issues posed a significant problem to the expansion of the use of the system, and therefore Josef Smrkovský proposed a 70 percent discount on grass seed for collective farms, where as private landowners would receive only a 30 percent discount. These preferable conditions for collective farms were accompanied by a proposal to increase fertilizer supplies to them, which would ensure greater success.[66]

Later in September 1953, the development of the grass-crop system was established as one of the focal points of agricultural research in Czechoslovakia in the greater plan for scientific and research work of the Ministry of Agriculture. At the time, the first results from the introduction of the grass-crop system were available, and basic soil research and microbiology analysis had been conducted. This means that Stalinist methods were not introduced without taking into account local conditions; instead, careful analysis preceded large-scale planning. The work plan for 1954 counted on the expansion of the grass-crop system to cover 6,213 hectares (plus 436 hectares of experimental area); this work was supposed to be conducted by 272 people (of whom 48 were scientists).[67]

The comprehensive plan proposed by the Ministry of Agriculture was broken up as follows: after determining the status of fertile and infertile land in Czechoslovakia, methods for the best way to introduce the grass-crop system were to be specified (1953-1957), and parallel to this in the second phase of the plan, suitable crop compositions were supposed to be specified (1955-1960).[68] Various expert committees followed this work very closely, especially the special Committee for the Grass-Crop System, which was created at the CAAS.[69] At the same time the CAAS followed the procedures used to introduce this agricultural management system in other countries of the Soviet Bloc, such in Romania.[70] The grass-crop system was uniformly applied throughout all of Central and Eastern Europe.

The grass-crop rotation system was, however, just the most visible element of new methods used within the implementation of Stalin's Plan that garnered political support both in the Soviet Union and in Czechoslovakia. Besides the grass-crop system, Czechoslovak agriculture also adopted a new system for cultivating land created by the Soviet agronomist Terenty Simonovich Malstev, known as Malstev tillage. A Czechoslovak delegation learned about this new procedure at the All-Union Agricultural Exhibition held in Moscow in 1954. As opposed to the situation during the Stalinist era of building new landscapes, this delegation did not promote the immediate mass introduction of the Malstev system; instead it proposed to determine experimentally if this method, which was developed for use in the Siberian Kurgan Oblast, would be also useful in a temperate climate. For its time, the proposal (written in 1955) contains a surprisingly open warning that a large-scale adoption of the entire system "could cause great damage" if it was not determined beforehand how and to what degree it could be used on Czechoslovak land.[71]

Czechoslovak Rice and the Socialist Horse
As part of the plan, not only were new methods supposed to be introduced, but so too were new crops. The plan even assumed new breeds

of farm animals would be developed. Growing Czechoslovak rice became one of the main points of introducing new methods. Just as the Siberian plains were to be planted with a special pear tree bred by Michurin, Czechoslovakia was to be planted with rice, which, according to Soviet experts, was excellently suited to local environmental conditions.

In mid December 1952, guidelines for preparing rice paddy sites were written that were supposed to be sent to the agricultural and technical departments of national committees.[72] In September 1953, a plan for rice production was ready.[73]

Work plans developed in late 1953 envisioned within ten years the development of a special variety of "Czechoslovak rice requiring less water and a shorter growing season, featuring higher yields than the imported Dungan-Shaly variety."[74] In comparison to the grass-crop system, this was only a pilot project, which was not supposed to be widespread for the time being; in 1954, this project was to cover only forty-two hectares (with only twelve of them being productive area) and employ eighty-eight people (including nine scientists). This project was declared to be the work of Michurinist people's researchers. These Michurinist clubs produced noticeable results on their rice paddies: Jan Mrška grew on average forty-five quintals of rice on a 45-hectare field on a collective farm in southern Slovakia.

Besides rice, which was supposed to be grown in the cooler regions of Bohemia, in southern areas, such as in southern Slovakia, cotton was supposed to be introduced on a mass scale.[75] Between 1953 and 1954, samples of various varieties of cotton were supposed to be obtained from Hungary and the USSR.

The fascination with rice did not fade in later years, although a certain pragmatism and concern for economic issues was present. The prognosis for rice, however, was not rosy. According to pilot estimations, the area sown to rice would double by 1960, and yields would also increase very slowly.

Table 1.2: The prognosis of cultivating rice in Czechoslovakia

Year	Area (ha)	Yield q/ha
1954	1,600	32.5
1955	1,700	32.5
1956	2,900	33

In 1960, rice was supposed to be grown over an area of 5,000 ha which would allow the Czechoslovak state to provide each of its citizens with approximately 1.5 kg of rice per year. The problem, however, was that rice production was not profitable, as existing prices—and import policies—meant that Czechoslovakia would lose about 175 Kčs per quintal of Czechoslovak rice (in comparison with imported rice).[76] The capital needed to construct paddy fields in the temperate climate of Czechoslovakia, which involved the investment of hundreds of millions of Czechoslovak crowns, was not included in these calculated losses. Early on, growing rice in Czechoslovakia was marked as being "highly loss-making," and it was abandoned in favor of cultivating traditional crops such as beets, sugar beets, and other temperate vegetables.[77] At the beginning of the de-Stalinization process, the economic inefficiency of rice cultivation was mercifully covered up by the Stalinist idealism of the first generation of Communist agriculture.

In Czechoslovak livestock production, the hybridization of traditional Czechoslovak livestock breeds with livestock imported from the USSR was expected. It was also expected that efficiency would be increased while keeping breeding costs low. Therefore, the "creation" of new sheep breeds, which were supposed to increase wool production (by producing wool that was thicker and longer), was planned. In horse breeding, the goal was to breed a "suitable high-performance horse for socialist agriculture and a small mountain horse."[78] The horse "suitable for socialist agriculture" was supposed to be developed through the crossing of warm-blooded mares with Lipizzaner and Kladruber stallions; the small mountain horse was to be bred by crossing hucul pony mares with fjord horse and Kyrgyz horse stallions.[79]

The Michurinist Movement and Its Varieties

The Communist regime placed great emphasis on "people's research," which was also developed based on the Soviet model. The "Michurinist movement" originated in the USSR shortly after the October Revolution, when the first timid attempts were made at establishing Michurinist centers in, for example, Moscow, Novosibirsk, and Semipalatinsk. The people's researcher movement spread very quickly; by the time the USSR entered World War II, there were more than seven hundred Michurinist people's researchers in the Ural region. They founded more than fifteen hundred fruit orchards.[80]

Almost everyone in the USSR became a Michurinist: in reaction to Michurin's statement in which "he claimed with certainty that it is absolutely possible to establish and maintain industrial fruit cultivation in the Urals, but only on the condition that it is grown in situ from the seeds of local varieties," Michurinist cells were developed in five workers' neighborhoods in the industrial city of Magnitogorsk.[81] At the same time, a movement was created in the north Urals and in Siberia, where the people's research of Michurinist A.M. Lukashev, a schoolteacher who developed several large fruit-bearing pear trees intended for eastern Siberia, was spread. As a period publication states: "Out of respect, the people of Siberia have named all of these pear varieties after their developer and call them *Lukashevka*."[82] The story of this Michurinist was repeated in all corners of the USSR: A.I. Olonichenko applied his successful experiments in Krasnoyarsk and N.A. Ivanitstky in Tomsk; V.V. Spirin introduced Michurinist methods just below the Arctic Circle; and F.M. Zorin applied these same methods on thermophilic plants in the area of Sochi.[83]

This model of spreading Michurinist methods was adopted in all Communist countries including Czechoslovakia. To promote these new "progressive" forms of cultivation, breeding, and production, "Michurinist people's researcher" clubs were introduced in all regions, districts, cities, and villages in Czechoslovakia in 1949. They were supposed to work as small centers from which the findings of Soviet agrobiology would be spread throughout the world. These small, ideologically-based work groups allowed amateur breeders and gardeners to conduct their own research, but their main goal was to implement new Soviet methods, such as Viliam's grass-crop system, in Czechoslovakia. Just like in the USSR in the 1920s, the local intelligentsia (i.e., teachers, agronomists, etc.) played an important role and usually led these local groups and clubs.

These "people's research clubs" were massively popular and three years later the Czechoslovak government decided that "people's research" would be organized into a mass movement.[84] This decision provided an enormous stimulus for these clubs to spread to all corners of the Czechoslovak Republic. In practice these clubs were active as part of collective farms, whose employees participated in the activities of these clubs alongside local farmers and gardeners.

The work of these clubs was primarily focused on local problems. "Michurinists" conducted experiments, testing which fodder crops and new plant varieties were successful in certain areas and which were not, as well as on how to best provide nutrients to plants. Experiments in livestock breeding focused mostly on increasing milk production from cows, and on improving feed mixes.

The results were promising: for example, people's researcher Josef Veselý successfully solved the problem of ensuring enough cauliflower seeds in the Czechoslovak Republic; these seeds had been lacking and therefore needed to be imported. As was written in a report from the time, "people's researcher Veselý has achieved results that not even our research institutes should have to be ashamed of."[85] Great successes in introducing exotic crops such as cotton and rice were also celebrated.[86]

In the future, the activities of people's Michurinists were supposed to be expanded and professionalized. In each collective farm an experimental laboratory was supposed to be established that was to demonstrate new methods and to repeat already-conducted experiments. These laboratories were also supposed to offer opportunities for self-education and for farms to conduct their own experiments.

In addition to improving the quality of these clubs, people's research also made its way into schools. In 1950, a countrywide competition was established for Michurinist clubs based at primary schools. This competition was very successful and grew quickly. In 1954, seventy thousand primary and secondary school students were involved in organized clubs.[87] The motivational motto behind the activities for young people was Michurin's motto: "We cannot wait for favors from nature—it is our task to take them from it!"

In 1955, the Czechoslovak Scientific Gardening Society of I.V. Michurin was founded under the sponsorship of the CAAS in Prague. The goal of this society was to join together all members of the gardening community in Czechoslovakia, both experts and amateur gardeners, to collectively achieve better results in gardening and crop production.[88] In the mid-1950s, interest in the Michurinist movement culminated in Czechoslovakia. The Czechoslovak variant of Michurinism was however a purely domestic movement; although it derived its methods from Soviet methods, they were applied in Czechoslovakia to domestic crops and livestock. This is the case of the "Prosenice Movement," which was modeled on the Michurinist movement and was influenced by the Stakhanovite ideals of an early communist society. Last but not least, the spread of this movement symbolized the definitive end of using obsolete, capitalist forms of land management in Czechoslovakia.

The Prosenice Movement was born in the small village of Prosenice in central Moravia; its motto was: "For higher yields of sugar beet per hectare and for higher sugar yields." It was highly influenced by the Soviets as well: following the Soviet model, Czech agricultural workers decided to solve the problem of insufficient sugar supplies in Czechoslovakia by introducing new methods in growing sugar beet. Mariya Lysenkova, a member of the Supreme Soviet of USSR and the chairwoman of the "Victory of Socialism" Kolkhoz in Ukraine, provided a great model for agrotechnical methods for growing sugar beet. She was able to harvest 758 q of sugar beet per hectare in 1947.[89] Olga Ganazhenkova's method for cultivating sugar beets served as another major model. She achieved even higher yields: from two hectares, she harvested 3,030 q of sugar beet (i.e., 1,515 q/ha).[90] However, she did not hold on to the record for long: it was broken in 1950 when, using new agrotechinal methods, Darikha Jantokhova was able to grow 1,892 q of sugar beet on one hectare of kolkhoz land.[91]

In Czechoslovakia, this competitive method for gaining higher yields was adopted in the spring of 1950 and promoted by Bohumil Penek, the director of the sugar refinery in Prosenice, and also the founder of the Prosenice Movement. In early 1950, he challenged Czechoslovak agricultural workers in the newspapers and over the radio to achieve yields of 500 q of sugar beet per hectare. In the eyes of many local farmers this seemed to be an extraordinarily high quota, as the average sugar beet yield between 1918 and 1950 ranged from 121 to 285 q/ha.[92]

Penek's strategy however imitated that of the USSR, copying methods used there before the war. At a congress of kolkhoz Stakhanovites held in 1935, Maria Dyemchenkova, a Ukrainian kolkhoz member, promised to attain a yield of 500 q of sugar beet from one hectare.[93] Penek's thinking was like this: if in the mid-1930s the Soviets could achieve such great success, there were ideal conditions for Czechoslovak agriculture in the late 1940s and early 1950s. On his part, this was a well-thought-out act: he planned everything well in advance. He knew quite well that in his region sugar beet yield was only 193 q/ha. In 1949, he conducted an experiment in collaboration with the Sugar Industry Research Institute in which fields were sown with twenty different varieties of sugar beets from different European countries in order to find out how adaptable each variety was to this new environment. In Prosenice alone, Penek succeeded in achieving a yield of 469.39 q from one hectare, an increase of 235 percent.[94] He was able to attain this yield by using a special method of tilling, fertilizing, and stabilizing the number of beets per hectare; this yield was taken as a starting point for further improvement.

Nearly six hundred collective farms and an additional nine hundred other subjects from all over Czechoslovakia signed up for Penek's challenge. The Kyselovice Collective Farm in the Kroměříž District achieved the best results, having grown 706.5 q/ha.[95] However, not even this result prevented Penek from constant comparison with the extreme results from the USSR, and in 1953 he wrote on the still low results: "Here, man has yet to control nature. We have yet to fully transfer and utilize expertise from the Soviet Union, the expertise of Soviet sugar beet growers, sugar refiners, and scientists."[96]

Therefore, methods for growing sugar beet in Czechoslovakia that were based on Stakhanovite methods were developed soon after.[97] A publication penned by Bohumil Penek on the idea of forming the Prosenice Movement won a state prize in 1952 and the Order of Work in 1953; it was then distributed throughout Czechoslovakia as a practical agricultural manual intended to educate agricultural workers, both young and old. It was a carefully written guidebook that compared Soviet and Czechoslovak methods and contained a large number of practical illustrations, photographs, and technical specifications. The Prosenice Movement became a part of the Czechoslovak sugar industry, and from the mid-1950s it led to reforms in existing methods for growing sugar beets and mechanization. This was an unplanned, yet welcomed, result of people's research following the model of Michurinist clubs.

The Prosenice Movement was an interesting variant on the Michurinist movement in Eastern Europe, but despite becoming such an important part of Czechoslovak agriculture, only a few firsthand witnesses and a handful of historians are familiar with the ideological implications it had in the 1950s.

The Colorado Potato Beetle

One aspect of daily life in Stalinist Czechoslovakia was exposure to propaganda that effectively linked various aspects of the Cold War together. The Colorado potato beetle was—and in the minds of many from the postwar generation still is—inextricably linked to the image of American imperialism. This image was purposely created by Communist propaganda.

Nonetheless, as literary historian Vladimír Macura notes in his study of semantics, the ideological image of the Colorado potato beetle as the "American beetle" bent on destroying socialist agriculture was portrayed similarly to another myth of a harmful insect in European culture. In

Le mythe et l'homme from 1938, Roger Caillois deals with the myth of the praying mantis, and it can be said that without question the myth of the Colorado potato beetle was in many ways based on the same premises of collective memory, but of course with the addition of the ideological rhetoric of the Cold War and high Stalinism. In Czechoslovakia, the negative image of the Colorado potato beetle was in many ways purposefully transformed into the opposite of the great Socialist plan for the transformation of nature: whereas the Stalinist plan was aimed at utilizing nature for the good of the people, the United States with its Colorado potato beetle which it intentionally spread throughout Eastern Europe—at least as it appeared in Communist propaganda—used nature to destroy, and to strengthen imperialism in the world.

Following World War II, the Colorado potato beetle, which was later faithfully accompanied by the American fall webworm, spread as a result of the intensification of agriculture. When the swarms of this striped beetle hit their peak, the destruction they wrought could be equaled to that of locusts in southern areas.

Starting in 1950, the presence of this beetle in Czechoslovak fields was casually linked to American imperialism. The beetle mainly spread throughout western and southwestern Czechoslovakia, areas that directly neighbored the newly established Federal Republic of Germany. A press statement made by the Czechoslovak government on 28 June 1950 even included this claim: "In the western areas of our country the potato beetle has not only appeared on fields, but also on town squares, in streets, in urban inner courtyards, and in many districts they can be found mainly along roads. Boxes and bottles filled with the beetle have also been found. This all proves without a doubt that the current danger caused by this beetle did not, and could not, originate in a natural and normal way, but that this dangerous pest was artificially and intentionally mass-imported via clouds and wind by Western imperialists, as well as by the agents they have sent to our country."[98]

The spread of the potato beetle was thus seen as an unprecedented "biological attack" on Czechoslovak agriculture. It was not long before official reactions were heard. As not only Czechoslovakia but all of Central and Eastern Europe suffered from invasions of the potato beetle, the Soviet Union protested against the intentional spread of the potato beetle in East Germany in a harsh diplomatic note to the United States. The government of Czechoslovakia also sent a similar note criticizing this American attack on Czechoslovak agriculture to the US embassy in Prague.[99]

Why were there such harsh reactions from the Soviet Bloc in 1950? There may be several reasons for this, ranging from the necessity of reacting to a new method of warfare that uses insects instead of a classic army to attack the enemy, to turning people's attention away from the international situation. Just three days before the government of Czechoslovakia made its press statement, North Korea, supported by the USSR, attacked South Korea (on 25 June 1950).[100] Both blocs were at war: while a true war was raging in Korea, a "biological" one was underway in Europe. In the eyes of the countries of the Soviet Bloc, the same country was guilty for starting both wars: the United States of America.

Soon thereafter, the fight against the potato beetle was transformed to the fight against the United States in the media, with Czechoslovakia standing alongside the USSR. The potato beetle was personified as an imperialist similar to a Wall Street banker: it wore a hat or top hat and carried a walking stick. The potato beetle, that wretched bug, whose only aim in life was to eat potato leaves and multiply, became an "ally of the imperialists," an "enemy," a "six-legged Wall Street agent."[101] Collective farms, volunteer work crews from various companies, and above all school-aged children who set out to the fields to collect the enemy bug, joined in on the fight against the potato beetle. Entire generations were mobilized.

The potato beetle appeared regularly in Central and Eastern Europe from the end of World War II until the 1980s. It was, however, politically exploited only in the 1950s; the visual rhetoric of the beetle later weakened. The beetle further spread, deeper into the East, in 1958–1959. The fall web worm followed in its devastating footsteps. Czechoslovakia, Romania, Hungary, and the USSR (especially its eastern republic of Ukraine) were affected. Joint committees were created (e.g., a Czechoslovak–Soviet committee, a Hungarian–Soviet committee, a Romanian–Soviet committee) that were supposed to coordinate the battle against these two pests.[102]

In contrast to the ideological image of the American beetle planted in Eastern Europe by "insidious imperialists," archival materials from the Ministry of Agriculture merely describe the hard data on the fight against these pests, ranging from qualitative observations of infestations, to detailed analyses of materials used to eliminate them, to quantitative analyses and comparisons of their abundance in relation to previous years. Thus, ideology stayed outside the realm of practical activities.

The Colorado potato beetle remained in Central and Eastern Europe for the entire second half of the twentieth century. The consolidation of land and the creation of massive fields that were within "flying range" of each other helped the beetle spread. The opportunities for using chemical sprays were limited, which also played an important role. Although the Colorado potato beetle was not dealt with in the Stalinist Plan, it was one of its consequences—both on the practical level (as agricultural intensification allowed it to spread more easily), and on the ideological (as it represented a targeted attack on the natural resources of the new Socialist bloc).

Building Socialism

New Cities

Just like shelterbelts, which were not such a novel element in the Czechoslovak landscape, unlike in the steppes and deserts of Central Asia, new cities built according to the principles of Socialist architecture as new places for the new Socialist man to live changed the face of the landscape. Dictators loved trees,[103] but it could take up to several decades for shelterbelts to achieve their desired effects. The process of building cities was much faster. This is likely why dictators planned and built new cities with greater enjoyment than they planted trees.

There were essentially two phases in creating a new urban network in Communist Czechoslovakia. During the first phase, cities whose names and legacies were strongly related to the past of the first Czechoslovak Republic (1918–1938) were renamed; during the second phase, new cities were built which indelibly changed the face of the landscape.

Baťa versus Gottwald

The city of Zlín, which is inextricably linked to the businessman Tomáš Baťa, was the first to fall victim to urban renaming. In the 1930s, Baťa built, modernized, and generally made a modern, industrial city out of Zlín, where light industry dominated, with his Baťa shoe company leading the way. World-famous architects, including Le Corbusier, took part in building the physical fabric of the city in the interwar period. After 1945, Baťa's factories were nationalized. Tomáš Baťa died in 1932, but his stepbrother, Jan Antonín Baťa, who took over the business that same year, did not return to Czechoslovakia after the war.

A name change was proposed in 1948, the same year the Communists came to power in Czechoslovakia. The idea originated as a gift of the city to the new Communist president, Klement Gottwald, for his 52nd birthday (born on 23 November 1896). The new official name replacing Zlín was Gottwaldov. In official speeches the people of Zlín were promised that they would live, create, and work according to new principles following the model of Gottwald himself. Ivan Holý, the new general director of the former Baťa shoe factory, emphasized in one radio broadcast that naming the city Gottwaldov was inspired by the renaming of Stalingrad.[104] Such honorific renaming was not unheard of, but thanks to Baťa's influence on Zlín, this new name not only had symbolic meaning, but it also gave some indication of the depth and scope of the total ideological transformation that Czechoslovakia was about to go under. The official name of the city was Gottwaldov from 1 January 1949 until 1 January 1990.

The Steel Heart of the Republic
In the classical sense of the word, however, the creation of new cities, which were supposed to not only embody but also be socialist utopias, were a part of the transformation of nature. The construction of new housing developments near large industrial centers in the late 1940s and early 1950s is a typical example of this. In Czechoslovakia, the northwest Ostrava region bordering with Poland became the largest center of heavy industry; it was nicknamed "the steel heart of the republic."

The portrayal of topos Ostrava, a traditional industrial center, was based on it being the steel heart of the republic in Communist rhetoric, in which the New Steelworks of Klement Gottwald (NHKG) played a key role. "They are right to call New Steelworks a city of steel. It is a city of steel that supplies steel. A city built by people of steel."[105] This portrayal emphasized just how vital the function of Ostrava's industry was for the new Communist regime, and by referring to the republic, the fact that the Czechoslovak state was the new owner of these mines and steel works in place of private owners was stressed.

Thus, in the early 1950s, Ostrava was the heart of the republic. The construction of the two most important industrial complexes in the region—Nová huť Klementa Gottwalda (NHKG) [Klement Gottwald Nová huť Steel Mill] and the Vítkovické železárny Klementa Gottwalda (VŽKG) [Klement Gottwald Vítkovice Ironworks], as both of these complexes were named soon after February—followed in the industrial tradition of the region. The core of the area was primarily the

Jižní závod [Southern Works] in Kunčice (consisting of a machine works and a tube rolling mill). Ammunition was made here during the war. It was supposed to be exempt from being taken as a spoil of war, upon which the Czechoslovak government and the Soviet army soon agreed.[106] The Vítkovická železárna [Vitkovice Ironworks, later VŽKG, Klement Gottwald Vítkovice Ironworks] was another key part of this "steel heart." It was predicted that in 1953 the production of pig iron would reach 3,550,000 tons and that the coke plant would supply 450,000 tons of coke annually.[107]

All top political leaders and members of the Central Committee of the KSČ took part in building Ostrava: "Building up the Ostrava region will be an entire-party affair. Ostrava must become an axis of interest for the whole republic and we must cultivate a sense of pride in our people that they are creating something that exceeds in scope and importance anything that we have yet done over the course of building socialism in Czechoslovakia."[108] Thus, the plan to build up the Ostrava–Karviná mining district "was a comprehensive plan of development and reconstruction of the region" that led to a newly conceived regional plan for Ostrava and to the "erection of new socialist cities in the Ostrava region."[109]

For this purpose, the Vládní komise pro výstavbu Ostravska [Government Commission for Building Up the Ostrava Region] was established with the goal of creating and implementing such plans. These were grandiose plans: in 1950 almost one-fifth of all state investments (17.5 percent) were made in building up this area.[110] The demographic changes that this transformation would bring about were well thought over in advance. For example, it was assumed that the number of agricultural workers in the region would drop as industry strengthened, whereas the number of people working in industry would complementarily rise slowly.

The creation of a new industrial complex also required the nationalization of land on which new factories could be built. In the Ostrava region, 365 hectares of land, 176 buildings, and another 187 residences were nationalized.[111] Affected property owners were recompensed with new houses in the area.

The creation of the Ostrava–Karviná industrial center took place in many stages. First, infrastructure needed to be built, and therefore in the spring of 1949 railway tracks were laid for the first trains, which would arrive in the spring of the following year. These provisionary tracks were replaced in 1951 with permanent railway tracks as work on the industrial complex was being completed. Construction continued at lightning speed: by New Year's Day of 1952, the first blast furnace in the new Kunčice steel works was already in operation.[112] At the end of January, the first coke came off the line from the coke plant of the Nová Huť Klementa Gottwalda.[113] At the same time, as part of the effort to construct industrial buildings in the Ostrava region, 2.5 million cubic meters of earth had been removed, 700,000 m³ of concrete had been used, more than 400 kilometers of train tracks and 3,000 tons of electric cable had been laid.[114]

A large portion of the postwar population of Czechoslovakia contributed to the construction of the Ostrava–Karviná industrial center. A large majority of them participated in various mass programs. In 1949, a preliminary half-year program called Stavba mládeže Ostravska [Construction of the Young of the Ostrava Region] was implemented; 545 members of the Communist youth league took part. This was followed by one of the largest programs of this type to take place in Czechoslovakia: the central Stavba mladých budovatelů socialismu [Construction of Young Builders of Socialism] program, operated from January 1950 until 1953; more than seventy thousand people took part.[115]

Considering the fever pace at which things were being built, mainly young people participated, and "youth camps" were located nearby the construction sites of these new steel mill complexes.[116] For example in 1951 alone, more than thirty thousand young men and women took part in building permanent railway tracks; these included both university and high school students. During the summer of that year, on 6 August 1951, the Milión hodin pro výstavbu Nové hutě Klementa Gottwalda [Million Hours for the Construction of the Nová huť Klementa Gottwalda] program was announced. Tens of thousands of people from the Ostrava region contributed their time, so much that the planned one million man hours of work was exceeded by one hundred thousand hours.[117]

These mass actions were reflected in contemporary socialist realism literature. In these books, the significance of construction projects was exalted and taking part in them was compared to serving on the front in the war:

> When on one winter night in early 1952 we departed for Ostrava, we felt as if we were headed for the front. And right to the front line where they were fighting, where the battle was raging, where

the future of the entire country was going to be decided. Their battleground was the entire northeast wedge of the Czech lands: the Ostrava region – the steel heart of the republic. And what a battle it is! It is about the construction of new factories and new cities. A new, better order! A new man! New magnificent perspectives!"[118]

The construction of an industrial center in the Ostrava region is described in these novels and reports as a battle: mainly a battle against time, a battle against nature, a battle with the capitalist remnants of "black coal Ostrava" that Socialist hands would transform into a working industrial center.

The pace of construction sped up: whereas the first blast furnace took twenty-eight months to build, the fourth one was completed in seven months. The first four coke ovens were built in four years, while the last four were built in two and a half years. In total, between 1949 and 1953, 2,615 billion crowns were invested in constructing the NHKG.[119]

However, the fact of the matter still remains that neither the region, nor Czechoslovakia as a whole, was prepared for such helter-skelter construction. There was a lack of experienced designers, construction engineers, and technicians. Most of the first generation of construction foremen were recruited from the graduates of a six-month course held in Kružberk, where one of the first industrial reservoirs in Czechoslovakia was built from 1949 to 1950.[120]

The first crisis appeared quite quickly and was related to the biggest political trials of the day. In early 1951, construction slowed down. Discussions regarding the delays were led with Minister of Heavy Industry Gustav Kliment by the General Secretary of the CC of the KSČ, Josef Frank. Josef Frank would later be accused, convicted, and executed in the Slanský trials. In the hunt for traitors and spies in the Ostrava region, the secretary of the regional committee of the KSČ, V. Fuchs, was accused, as was Cadre Secretary R. Peschel along with several journalists from the newspaper *Nová svoboda*.[121] Since most of them were involved in the construction of the industrial complex in Ostrava in some significant way, those guilty of causing the construction slowdown were found. Instead of focusing on real problems related to the construction project, top regional political leaders were opportunistically claimed to have sabotaged the project. Most of those arrested and imprisoned were rehabilitated in the 1960s.[122]

As important members of KSČ were arrested, this political trial resulted in priorities being shifted away from the building of new mines, which the arrested secretaries were in charge of. The new communist leadership and the Government Commission for Building up the Ostrava Region thus focused on dealing with planning the entire Ostrava region. Thus plans were developed not just to build new housing, but new cities and reservoirs as well in the surrounding area. Ostrava and the entire Ostrava region were supposed to become a "great and modern city of Socialism."[123]

Havířov

As the Ostrava–Karviná region was being built up, the decision was made that new housing developments would be built in areas not more than a 30–40-minute ride to the factories.[124] In 1947, the first housing developments began being erected in close proximity to existing centers, but it soon became clear that the growing heavy industrial complex would need many more people than originally estimated. Instead of just building housing developments, it was determined that entirely new cities would be constructed. In Upper Silesia, the most important industrial center in Eastern Europe, the new cities that would be built were: Poruba, which would later be annexed to Ostrava, as well as Havířov[125] on the Czech side of the border, and Nowe Tychy on the Polish side.

All of these cities have similar characteristics: they were all built in the style of socialist realism, and the principles of socialist urban planning were applied, although a moderate amount of traditional methods of the day were still preserved. They are all based on a dense plan and are clearly separated into various sectors, which was still not commonplace at the time they were built, and have well-planned infrastructure.

Havířov was built on greenfield land, and according to original plans it was supposed to be the new home of 84,100 people.[126] In 1947, the first buildings began being constructed nearby the settlements of Bludovice and Šumbark; in 1949, the first 96 apartments were ready. Then the city began to grow rapidly: 1,715 apartments were completed in 1953, and after the budding city was joined with surrounding villages, Havířov had a population of over 16,000 at the end of 1955. Fifteen years later, the population had already hit 80,000.

Just like in Zlín/Gottwaldov, controlling the space of the new city symbolically through a striking name was important. In contrast to Gottwaldov, however, there was a public competition to name Havířov. By early 1955, more than 2,350 names had been proposed, and despite the fact that surrounding cities and regional authorities desired to see the city named Bezručov after the Silesian poet Petr Bezruč, political authorities suggested the more striking Havířov—a toponym derived from the Czech word *havíř* (miner).[127] It is certainly not without interest that the historical center of Havířov is today one of few cultural monuments in the style of socialist realism.

Great Works of Socialism

What does the term "great work of socialism" actually mean? What it meant in Czechoslovakia can be answered by examining the definition given in propaganda. In such literature, "great works of socialism" were defined as the following: "They are construction projects that turn economically undeveloped countries into progressive countries and agricultural and semi-agricultural countries into industrial countries . . . Thus, we consider works of socialism to be such works that in their scope and significance change the economic composition of the country towards socialism."[128]

Thus, these works were key construction projects that were initiated as part of the restoration of postwar industry, yet when placed in the new ideological framework of the Communist regime they became reference points for the postwar transformation of society and nature. During the first five-year plan alone, Czechoslovakia invested more than 558 billion Kčs into these large construction projects. In the first stage of investment, building up heavy industry and constructing new steel mills was focused on; the second stage of the plan involved constructing reservoirs and hydroelectric power plants.[129]

This is why the regime focused mainly on the industrial areas of North Bohemia and North Moravia and Silesia in the immediate postwar years. In the early 1950s, an enormous industrial complex, Nová huť Klementa Gottwalda, arose near Ostrava. This truly was a vital construction project that during its first three years consumed 700,000 m^3 of concrete and 35,000 tons of steel for construction purposes, and replaced 2,250,000 m^3 of earth. The steel and iron used in the equipment and machinery located in these new buildings weighed 45,000 tons.[130]

All processes were mechanized, and new, large machines were introduced to the production process. "Socialism is the unyielding enemy of drudgery. It replaces hard physical labor with labor conducted by all manner of complicated and powerful machinery."[131] Young people played a major part in constructing these large projects: in Czechoslovakia, there were programs dedicated to building them, such as the previously mentioned Stavba mladých budovatelů socialism [Construction Work by Young Builders of Socialism], and Přehrady mládeže [Reservoirs of the Young].[132]

Concurrently, infrastructure was also built, including such projects such as Trať družby [the Railway Line of Friendship] and Trať mládeže [the Railway Line of Young People]. The new political orientation of the Czechoslovak state could be best seen in this new infrastructure: in the past, Czech railway lines generally headed southeast towards Vienna as part of the infrastructure system of the Habsburg Monarchy. Later, German wartime infrastructure focused on incorporating the Protectorate of Bohemia and Moravia into the bosom of the German Reich, and thus focused on the north–south axis. However, transportation infrastructure in postwar Czechoslovakia was oriented in one direction: to the East.

The Railway Line of Friendship
The main axis of this transportation infrastructure became the Railway Line of Friendship, which was an extension of the main line that connected the capital city of Prague with Košice, the regional center of East Slovakia. Propaganda focused on how the main ideas of renowned Czech transportation engineer Jan Perner (1815–1845)—i.e., connecting Bohemia with the greater Slavic world—were only realized thanks to the new Communist regime.

In the first half of the nineteenth century, Jan Perner was responsible for building the railways in Vienna; he built a line from Vienna to Brno which was indirectly then connected to Prague via Olomouc. In 1843, Perner wrote: "A direct rail line between Prague and Vienna will be built in any case, but we would not be connected to the Slavic world. Therefore, I propose a line to Olomouc: thus we shall have a connection to the East, to Poland and Russia."[133] His idea of developing transportation infrastructure in Czech lands so that it would be connected to the East was only realized after 1948.

The Communist regime decided to extend the Košice–Bohumín line, which connected Košice, the main city of eastern Slovakia, with industrial Silesia. This single-track railway was originally completed in 1871; after 1945, the decision was made to extend it and to add a second track.[134] This decision was primarily motivated by the growing burden on this single-track line mainly in its easternmost length—that is, on the border in Čierná nad Tisou. In 1950, work began on the second track; in that year alone nineteen thousand people volunteered to work on it, consisting mostly of young people. The grand opening of the line was held on 5 November 1955.

It was the first construction project completed by the nationalized construction industry in Slovakia, and bore the proud name of Trať družby v Košicích [Railway Line of Friendship in Košice].[135] The Slovak head of transportation, Comrade Bacílek, ceremoniously proclaimed it a "work of socialism," for its aim was none other than to change the face of eastern Slovakia. For its time and for the general technical development of the region, it was an important project that required the work of several tens of thousands of people. Temporary camps were built along the tracks for them to live in, and included sanitation, cultural amenities, and shops. Two-thirds of the workers were not construction professionals. They were often recruited as volunteers for programs such as Velká brigáda [The Great Brigade], which was a countywide project announced for the summer of 1951; seventeen thousand members of the people's militia took part, giving up their own vacations to work on the construction projects.[136] The army and schools also got involved—it was truly one of the largest mass programs of the early 1950s.

Thanks to the involvement of such a great number of workers, a biweekly newsletter entitled Trať družby [The Railway Line of Friendship] was published to share information about progress on the line and the use of new construction methods. It also included greetings from top politicians, criticism of problems that had arisen (such as the poorly conceived use of bulletin boards), letters from workers, and competition announcements. It is worth noting that the publication was trilingual: besides articles in Slovak, it also included articles in Hungarian and Romani.[137]

The construction of the Railway Line of Friendship was accompanied by a massive information campaign, which various artists were involved in. The creative art studio in Košice organized collective group visits to the construction site for Czechoslovak artists. Their goal was to capture the transformation of nature and the growth of the new railway line. Dozens of paintings and sketches were created, including *Tuneláři při práci* [The Tunnel Makers at Work] (Mikuláš Jordán, oil, 1953), *Betonárna* [The Cement Factory] (Július Jakoby, oil, 1953), and *Smělý úkol* [A Bold Task] (Jozef Majkut, oil, 1953).[138] Ladislav Mňačko also became involved in the cultural campaign supporting the construction of the Railway Line of Friendship with a play based on the principles of the production novel. His *Mosty na Východ* [Bridges to the East] (1953) tells the story of the dedication of these builders of socialism, and also takes into account questions of gender (one of the bridge builders, Vilma, is a female member of the Communist youth organization who comes up with a new idea of how to build bridges). Nevertheless it in many ways foreshadows the reversal he would make in his later works *Opožděné reportáže* [A Delayed Report] (1963), a set of reportages on the hidden political and social barriers of such great works of socialism, and mainly in his novel *Jak chutná moc* [The Taste of Power] (1967). At that point, this once enthusiastic author of production novels about great works of socialism had already joined the reformers and would later go on to become a dissident (and would live in exile in Austria).

The Railway Line of the Youth
Whereas the Railway Line of Friendship played an important role in connecting Czechoslovakia to the USSR, its younger sister, the Railway Line of the Youth, was especially important for the internal infrastructure of Slovakia. Railway tracks measuring nearly twenty kilometers were supposed to be built from Banská Štiavnica, where noble metals were mined, to Hronská Dúbrava. Such a project was already under construction in the nineteenth century, but just like other projects of the time, the quality of the original construction work was no longer good enough, so only modernization would suffice to make the line usable again in the future. Ground was broken on the line in 1869, and it was completed in 1873; it operated as a narrow-gauge railway until the 1930s. In 1943, the decision was made to rebuild the railway, but due to the Slovak National Uprising work was halted. The rebirth of the railway and its restoration was politically motivated: it became one of the priorities of the fascist first Slovak Republic, and the groundbreaking ceremony was attended, for example, by Bulgarian officials, who were also allied with Nazi Germany at the time.

After World War II, the project naturally drifted back into the center of attention of the new political elite for strategic reasons: on 1 April 1948, the project was officially named The Railway Line of the Youth, and its date of completion was set for 28 October 1949. This project was the first mass action of this type in Czechoslovakia; youth organizations had debated becoming involved in it as early as 1947.[139] Over the next two years, 47,162 young people from Czechoslovakia, Albania, Algeria, Britain, Bulgaria, France, India, Hungary, Poland, Austria, Ukraine, and Norway participated in the construction project. Fifty-five Norwegians arrived nearly immediately after the project was announced in July 1948. In total, 1,602 foreign students took part in construction work.

According to contemporary announcements and later statements about the project, it was clearly a unique project, both technically and organizationally speaking.[140] There were nine permanent camps, and in the summer nine more fully equipped tent camps (including kitchens, showers, toilets and medical facilities). Up to five thousand people took part in each work camp.

The Railway Line of the Young also became a literary topos. All of the major Slovak daily newspapers wrote about it, and it was even on the pages of countrywide newspapers.[141] Poems regaling the project were written, as were reports and production novels such as *Slunce v údolí* [Sun in the Valley] (1955), in which university students expose a spy leader working on the line, allowing them to finish construction on time.[142]

In Slovakia, where unlike in the Czech lands, the KSČ did not win postwar elections, the Railway Line of the Youth had important political implications: it provided an unrivalled opportunity to create a strong and ideologically stable community of Communist-minded youth.[143] At that time, the Communist youth organization was still not the predominant student group, but this project, its promotion, and the participation of non-Communist young people, turned it into one of the largest indoctrination camps of postwar Czechoslovakia. (In the Czech lands, the construction of the Nová huť Klementa Gottwalda in the Ostrava Region played the same role.)

In this sense, great works of socialism were a continuation of postwar construction projects in which young people had participated since 1946, as young people were involved in restoring the Czechoslovak economy and industry. In North Bohemia, for example, the Lidice–Most–Litvínov construction project was organized and students built a new village in place of Lidice, which had been razed by the Nazis, as well as new buildings in the Sudeten cities of Most and Litvínov.[144] In Slovakia, the razed partisan village of Baláže was rebuilt in the same way. However, once the KSČ came to power, such programs grew on a massive scale, as tens of thousands of people took part.

Constructing Reservoirs and Water Management (1948–1962)

Besides transportation infrastructure and the aggressive expansion of industrial centers, the transformation of nature also affected the landscape in untouched areas. Reservoirs and the construction of dams had the biggest impact on these areas. Original plans presented a bold vision of a reservoir on the Orava River in Slovakia becoming the biggest reservoir in all of Central and Eastern Europe. Indeed, it would accomplish this goal, as there were not many large reservoirs in this part of Europe at the time. Another important factor was a plan calling for growth in electricity generation from 1950 to 1955, which, mainly due to reservoirs, was supposed to increase by 500 percent. For example, hydroelectric power plants in Slovakia were supposed to generate as much electricity as had been generated in all of the Czechoslovak Republic before the war.[145]

These were grandiose plans that were rapidly implemented in quick succession. In the Czech lands, reservoirs were primarily supposed to be built on the Vltava, the longest Czech river, as well as on other lesser rivers. In Slovakia, reservoirs were planned primarily on the Váh River, and a large reservoir was also planned on the Danube, later to be known as the Gabčíkovo-Nagymaros hydropower scheme. The size of this reservoir and the construction technology to be used was supposed to equal that of the Soviet Dnieperstroi Dam. However, the Communist regime only began working on the project after decades of planning, and in 1989 work on the Gabčíkovo-Nagymaros Reservoir was interrupted and then greatly reduced. These big plans were based on important economic justifications in their day: in the early 1950s, it was expected that the industrial centers of central and western Slovakia would be connected to and supplied by Black Sea ports.[146]

Once the political contours of the Great Plan for the Transformation of Nature weakened, its wide implementation was no longer possible. The plan was broken up into smaller projects, and the greater part of the idea of controlling nature was shifted to building reservoirs. Paradoxically, this step back did not much weaken the political power or the ideological aspects of the failed great plan, as massive reservoirs are extremely visible elements in the landscape. Besides reservoirs, drainage and irrigation projects, primarily in South Moravia and southern and eastern Slovakia, also affected the appearance of the landscape. Many of these projects, however, were small-scale efforts that had only local significance.

Planned Water Management
Just as for agriculture, two main strategies used by the Communists for transforming water management and using it for building the new regime can be defined. These projects had two goals: to make Slovakia at least partially energy independent and also to shape the ideological contours of the new society. Two important events happened during the

transformation of water management policies in postwar Czechoslovakia in the 1950s: soon after 1948, the Státní vodohospodářský plán (SVP) [State Water Management Plan – SWMP] began being written and was completed in 1953. This plan not only outlined the main priorities of water management, but mainly it compiled an accurate overview of every water resource and its usage in every drainage basin in Czechoslovakia. As this plan was being drafted, amendments were suggested to the Water Management Act of 1955, which were actually made four years later in 1959.[147] Just as was the case for establishing shelterbelts, we can observe a continual transition and the gradual transformation of one system of administering economic policy to another; we are not the witnesses of a sudden break in development or the abandonment of old ways, as many authors have portrayed the situation.

The SWMP was written between 1948 and 1953, and contained a detailed description and analysis of the overall state of water resources in Czechoslovakia. The plan covered all rivers, lakes, ponds, and groundwater sources. This plan served as a basis for further water management planning. In this sense, the Czechoslovak State Water Management Plan was innovative, as it was one of the first holistic concepts of this type. In order to understand its significance, it is important to add that the SWMP would later become a model for developing similar water management plans in other Communist countries.

The most important innovation it contained was the way it comprehensively dealt with supplying drinking water and sewage coverage. The SWMP was also the first document of its type to systematically map and describe problems with water pollution.[148] For Czechoslovakia, which was known as "the roof of Central Europe" due to the fact that all rivers flow out of the country, yet none flow in, this "inventory" of water resources meant great progress and was a strategic step ahead in terms of long-term water management planning. Moreover, Czechoslovakia did not have any large water resources of its own, and therefore when precipitation was relatively low it suffered drought. Considering the great plan for the transformation of the landscape and the economy also saw the construction of new housing developments and the building up of heavy industry, which in themselves resulted in major growth in the consumption of drinking water and industrial water respectively, meant that this plan was an important conceptual element in the transformation of the Czech environment.

As part of the State Water Management Plan, the very important Act on Water Management (Act No. 11/1955 Coll. of 23 March 1955) was also enacted.[149] This law replaced the Austrian Water Act of 1871 in Bohemia, and in Slovakia it replaced the Water Acts of the Hungarian Kingdom of 1885 and 1913. Thus, in 1955 legislation covering water management was unified. This was important because it modernized existing legislation, adopted it to meet the needs of a planned economy, and defined the responsibilities of the state administration in each aspect of water management.[150]

Generally, it can be claimed that the State Water Management Plan, as well as the Water Management Act were indeed driven by economic demands, but they never lost sight of the natural conditions in Czechoslovakia, particularly the status of water resources. Environmental conditions were taken seriously, which resulted in water management projects that were based on actual environmental conditions in Czechoslovakia. The goal of these projects was not to radically transform the desert into fertile land, but more or less to regulate existing rivers and, on the local level, to build irrigations canals. However, this in no way stopped these projects from being used in Communist propaganda, just like in the USSR.

Constructing Reservoirs in Czechoslovakia

Figures on reservoir construction on the territory of Czechoslovakia up until 1945 are quite clear.[151] When Czechoslovakia became independent in 1918, besides the Příbram and Štiavnica reservoirs, there were only twenty-one other reservoirs, with a total volume of 39.71 million m^3. During the First Czechoslovak Republic and World War II, a total of sixteen reservoirs with a total volume of 218.39 million m^3 were built. The largest and most famous of these is the Vranov Reservoir on the Dyje River. It covers an area of 762.5 ha, is 30 km long, and contains 132.69 m^3 of water. Another well-known water management project built in this period is the Střekov Lock near Ústí nad Labem and the German border. The reservoir it forms is 19.5 km long and has a volume of 16 million m^3, helping make this Czech river navigable.

By just briefly looking at the total reservoir volume constructed between 1945 and 1962, and by comparing it to the historical data above, we can clearly see that in this period there was a massive boom in water management projects. In less than twenty years, twenty-three reservoirs with a total volume of 1147.9 million m^3 were built. After 1962, the breakneck speed of this expansion slowed down significantly. The completion of the Orlík Reservoir in 1961 was a milestone, as afterwards only more or less isolated projects followed, such as the

Nechranice Reservoir in North Bohemia, which began being built in 1961 and was finished seven years later in 1968, and the Liptovská Mára Reservoir in Slovakia (1975). On the one hand, the concurrence of this Czechoslovak boom with the highpoint of reservoir construction in the USSR as part of the Stalin Plan for the Transformation of Nature is certainly no coincidence; on the other hand, in the West we can also observe very similar trends for this period.

The Orava Reservoir: The First Great Work of Socialism and its Baroque Roots

The Orava Reservoir was completed in 1953 as the first "work of socialism" in Slovakia. It covers 35 km² and has a volume of 350 million m³. The size of this Slovak reservoir put it ahead of the Czech lands, where reservoir construction was just starting to get underway. The main purpose of the Orava Reservoir was to improve the flow of the Orava River, and above all of the Váh River, and thus to increase the efficiency of the system of smaller dams and reservoirs on the Váh, known as the Váh Cascade, which was built over several stages in the 1950s and 1960s.

Despite the fact that the Orava Reservoir was the first great work of socialism constructed in Slovakia, in reality the idea of building it was not a concept that the Communists came up with. The Communist regime quickly exploited the ideological potential the reservoir provided and took the idea over as its own, with all the ideological trappings.

The first ideas about creating the Orava Reservoir reach far back into history. The Orava Reservoir was built in northern Slovakia in the Západní Tatry Mountains, where local watercourses collected a large amount of precipitation (on average 1,300 mm per year), and thanks to the impermeability of the bedrock, this water was not absorbed into the ground but remained on the surface and flowed into rivers. Therefore, this poor mountainous region was affected by thirty-nine destructive floods between 1557 and 1953, which recurred every ten years almost regularly.[152] The first plans for creating a reservoir were developed in the early eighteenth century; the year was 1730 to be exact. Even though earthworks were begun, the reservoir did not officially begin being built until 1941. However, due to World War II, construction problems, and the poor organization of labor, work on the reservoir did not progress much until 1949. Only once this project was prestigiously promoted to becoming a "great work of socialism," and only then, when obsolete construction methods were replaced with more modern ones, did work on the reservoir continue at a fast pace; it was brought into test operation on 1 May 1953.

Thanks to its history and its significance, this reservoir became an ideal subject for Communist propaganda. The Communist Party used the following line of argument: capitalists were incapable of completing such a large project, and it was necessary for a revolution of the people to occur and a planned economy to be introduced for such a massive project to be completed. Moreover, the reservoir brought jobs to hundreds of people in this poor region on the Orava River; and in addition to the reservoir, new factories would be built and people from the surrounding area would be able to find work easily. They would not have to search for work in other parts of the country or abroad.[153]

This image was then pushed in all forms of media, particular in the Communist Party's newspapers, the *Rudé Právo*[154] and the Slovak *Pravda*.[155] The story of the construction of this reservoir also served as the basis for a production novel. This special literary genre was developed in the 1920s in Soviet Russia; its goal was to use a simplified plot and text to portray the heroism involved in the building of the new regime and its projects. The plot did not play the main role; the collective effort put into building socialism or completing new projects was the most important element of these novels. This genre was closely associated with the construction of great socialist projects in Czechoslovakia.

The construction of the Orava Reservoir was the subject of Dušan Kodaj's production novel *Oravská priehrada* [The Orava Reservoir],[156] as well as of the film *Priehrada* [Reservoir] by Paľo Bielik.[157] The Orava Reservoir as a model work of socialism also became the subject of a triptych created by painter Dezider Milly.[158] The Stalinist regime in Czechoslovakia faithfully imitated how such projects were used in propaganda in the USSR, although the way the Orava Reservoir was presented in propaganda differed in one crucial point: one would search in vain to find the issue of the transformation of nature explicitly addressed in the two mentioned newspapers. Motives used for propaganda purposes almost exclusively included the "taming of the river"[159] and the fight against natural elements (such as frost).[160] Was the image of the transformation of nature purposefully pushed into the background? Or was it due to the fact that this reservoir was built in the mountains, where nature could only be "tamed" and not "transformed"?

The Váh Cascade Project

Besides the Orava Reservoir, the second most important water management project in Slovakia was the construction of the "Váh Casacade," a series of dams and reservoirs along the Váh River, the longest river in Slovakia; it is still there today. Of the twenty-two reservoirs that were built on the Váh in the second half of the twentieth century as part of the comprehensive water management system, half of them were built before 1961. The main reason why the Váh Cascade was built was for the generation of hydroelectric energy; improving the navigability of the Váh and creating irrigation systems for nearby fields were secondary aims. Just like the Orava Reservoir, the idea and plan behind this project were not brand new: hydroelectric plants on the middle Váh had already been built in Ladce, Ilava, and Dubnica between 1936 and 1948; this fact, however, was not mentioned at all in Communist propaganda.

In comparison to the Orava Reservoir, media coverage of the Váh Cascade was less ideological in nature. The main reason for this is that most of the dams and weirs were not built in the early 1950s at the height of Stalinism, but in the late 1950s. Although, we can find many reports about the construction of various projects along the river, the enthusiasm of the early 1950s is missing. The belief that nature can be transformed also disappeared; there is not even a trace of grandiose or megalomaniacal plans, and the main topic focused on was the taming of the ravenous Váh.[161] The reservoir in Nosice, known as the Reservoir of the Youth, stole the media's attention away from the Váh Cascade; this reservoir was, for example, memorialized on canvas by painter Jozef Šturdík.[162]

It can be said that the Váh Cascade was a strategic construction project for postwar Slovakia; its main goal was to increase and stabilize Slovak energy resources. Slovakia was very poor in coal supplies, and after the war it greatly lacked energy resources. Propaganda describing the significance of the series of hydroelectric plants along the Váh emphasized this fact.[163] If we examine in detail the specific output of each hydroelectric plant on the Váh, we will find that their importance was not based on specific output, which was, relative to other reservoirs, very small: in the 1970s, the eleven hydroelectric power plants on the Váh had a total output of 336.35 MW and generated on average 1097.1 GWh per year.[164] In comparison, the output of the hydroelectric plant on the Orlík Reservoir is 364 MW and the output of the Prunéřov I coal-burning power plant, built between 1967 and 1968, is 440 MW. The Váh Cascade was important primarily for its strategic position as part of the existing water management and power infrastructure of postwar Slovakia.

The Vltava Cascade and the Czech Sea at Lipno

The Váh Cascade had a sister on the Vltava in Bohemia. Today, on this river's journey through nearly all of Bohemia—from the Šumava forest to Mělník—we can find a total of nine reservoirs. The motivation behind exploiting the Vltava, this important Czech river, was the irresistible vision of utilizing its power. Contemporary calculations determined that the energy the Vltava produced was the equivalent of 40,000 Škoda tractors or had the same output as 663 1750 hp train locomotives.[165] Moreover, on the section of the river where the Vltava Cascade was built, between Lipno and Vraný nad Vltavou, the river dropped 525 m over 260 km, and theoretically the potential existed for the river to generate around 800 MW of hydroelectric energy. Besides generating electricity, the system also helps to regulate and stabilize the flow of the Vltava, thus making the Vltava, as well as the Labe into which it flows, more navigable; it also protects its sourroundings, especially the capital Prague, from floods.

Between 1948 and 1962, three important reservoirs were built on the Vltava: Lipno I, Orlík, and Slapy. Lipno is the largest reservoir in area on the territory of the former Czechoslovakia (48.7 km^2), whereas Orlík takes the top spot for volume (720,000,000 m^3). The Slapy Reservoir was built between 1949 and 1956 and was the first of these three giants to be built on the Vltava. This new reservoir was officially opened on 5 April 1954: "In June, the first turbo generator utilizing a Kaplan turbine on the highest drop in the world went into operation," as contemporary propaganda put it.[166]

Slapy

The Slapy hydroelectric plant has an output of 144 MW, generating an average of 300 million kWh per year, and is powered by the water contained in the 43 km-long Slapy Reservoir. The dam is 60 m high and 260 m wide. Not only does this reservoir generate electricity, but it has also become a popular recreation area, especially for people from Prague. During the construction of the Slapy Reservoir, stretches of the river dominated by the feared Svatojánské [St. John's] Rapids were flooded; in the past, this section had complicated the work of timber rafters.

The flooding of the Svatojánské Rapids was not only motivated by geography and the potential electricity that the drop in the river could provide. The Communist regime's active fight against religion and faith in God played an important role as well, as in the nineteenth century religion was the only thing people could turn to, to protect them from the will of the river. It was not by coincidence that work on the Slapy Reservoir began in 1949 in a spot where a statue of St. John of Nepomuk had stood for centuries. After having successfully navigated their cargo through the dangerous Svatojánské Rapids, timber rafters prayed to this statue.[167] The calculated selection of this place indicates that the construction of the reservoir was in the eyes of both politicians and common people a harbinger of a new era in which faith in God would be replaced with the ability of humans and science and technology to fight against nature.[168]

Lipno

The Lipno I Reservoir[169] is also known as the "Czech Sea" or the "South Bohemian Sea" due to its large size. This great work of socialism was built between 1952 and 1959. Just like most of the other reservoirs built in Czechoslovakia, the idea behind building Lipno was not thought up by the Communists. The first plans for creating a dam in this area were made in the late nineteenth century, although it took sixty years before such plans were realized.

Frequent floods had occurred regularly here since the seventeenth century, and were the reason people first began thinking about a reservoir. The first plan was conceived in 1892. The first actual design was proposed in 1930, but the implementation of this design was postponed due to the occupation and World War II. Only in the postwar years, when the transfer of the Sudeten Germans affected inter alia the ownership structure of land covered by the planned reservoir in favor of the Czechoslovak state, could the reservoir be built. A background study was conducted in 1948 that incorporated a hydroelectric plant. Construction began in 1951.[170]

This fifty-year delay was immediately utilized in Communist ideology. The Communist regime explained that construction began only in 1951 because the coal barons and paper mill owners who traditionally transported wood along the Vltava using timber rafts had prevented the project from being built, as it was not in their economic interests.[171] In contrast to this attitude, propaganda emphasized that the reservoir would bring benefits to poor villages, which would otherwise suffer nature's whims and frequent local floods.

Behind a 25 m-high and 296 m-wide dam, the Lipno Reservoir covers an area 42 km-long and up to 5 km-wide. Besides generating electricity and protecting against floods, the reservoir, just like the Slapy Reservoir, is also used for recreation. The hydroelectric plant has an output of 120 MW, and in its day it was one of a kind, not only in Czechoslovakia, but in all of Eastern Europe: the turbine hall is located underground and water flows in and flows out through shafts that are several kilometers long. This design was chosen in order to utilize the 130 m drop between Lipno and Vyšší Brod. The Vltava descends this elevation over a very short distance, and therefore a special turbine hall, 60 m long, 34 m wide, and up to 21 m high, was built into a cliff.

Orlík
The third and largest reservoir is the 68 km-long Orlík Reservoir, formed by a 91 m-high, 450 m-wide dam. It is not only the largest reservoir on the Vltava, but also the largest in Czechoslovakia. The hydroelectric plant on this reservoir with its output of 364 MW surpasses that of the Váh Cascade in Slovakia. Orlík was also the last great project to be completed on the Vltava; it was opened after eight years of work in 1961.

The Orlik hydroelectric reservoir was designed to be the largest reservoir in Czechoslovakia at the time of its creation. Its construction was viewed as part of the civilization process: in contemporary materials, it is often described as bringing civilization to the neglected villages surrounding the reservoir (such as Zbenické Zlákovice). More than two hundred wells that were 70–100 m deep were dug for preparatory testing, all without the use of electricity.[172]

Up to 3,500 workers worked on the project, which consumed nearly one million cubic meters of concrete and reinforced concrete. Besides just the size of the project, the technology employed by the power plant was unique for its time: Kaplan turbines were installed in the high head power plant, which required the turbine blades to be specially modified to resist pressure. One of these turbines was even exhibited at the EXPO 58 in Brussels before it was installed. Although the technologies used for the construction were pretty advanced, the transfer from the USSR and the support of Soviet experts was rather small. In archival material from the "Commission for the Help for the Great Works of Socialism" we learn that the experts from the Soviet Union made no remarks on the overall conception of the design, but only proposed some minor changes on the design in order to save some funds.[173]

An interesting bridge was also built nearby Žďákov; it is 542.9 m long and spans 379.6 m. This reservoir was named after Orlík Castle, which was build during Charles IV's reign on a cliff high above the Vltava.

Today, it is more of a water castle, being surrounded on three sides by water.

Industrial Reservoirs
The last group of reservoirs that are included in the early history of Socialist water management in Czechoslovakia were not only the reservoirs in the industrial area of the Ostrava-Karviná mines and in the foothills of the Beskids and Jeseníky mountains, but also the Vír Reservoir on the Svratka River. These reservoirs were created in the 1950s mainly to ensure water supplies for rapidly expanding industrial plants and the population of the Ostrava-Karviná coal basin; the Vír Reservoir was created to supply water for farms in Moravia.

The two main reservoirs in the Ostrava region are the Kružberk Reservoir on the Moravice River and the Žermanice Reservoir on the Lučina River. The Kružberk Reservoir was built between 1948 and 1955. Its dam is 34.5 m high and 280 m wide, and holds a volume of 35.5 million m^3. The Žermanice Reservoir began supplying water to factories in the Ostrava region eight years after construction began in 1958. It is formed by a 314 m-long dam that it is 38 m high and retains 25.3 million m^3 of water. The Vír Reservoir was built between 1949 and 1957 and features a 390 m-wide dam face, which was later surpassed by the dam on the Orlík Reservoir.

The general overall plan for building reservoirs also included small hydroelectric power plants that were intended to help local economic development in regions where it was impossible to directly conduct energy (such as in industrial regions). Such regions included East Bohemia (Křižanovice Reservoir and hydroelectric plant on the Chrudimka River completed in 1954) and poorly accessible areas of the Vysočina Highlands in Moravia (the Vír Reservoir on the Svratka).

The regime did not pass up on the chance to emphasize that the construction of the relatively small Vír Reservoir marked a significant parting from feudal and bourgeois habits. To popularize the reservoir, the Communist regime used an old legend, which folk imagination attributed to Sibyl herself. The land on which the reservoir was to be built once belonged to the Pernštejn family, who built several castles nearby. The most palatial castle of them all, the never-conquered, marble Pernštejn Castle, built in the sixteenth century on a high cliff, was the most prominent building in the area. According to Sibyl's prophecy, however, "once the joyful times come, the local people shall erect another bigger white citadel behind Pernštejn."[174] During Communism, however, this citadel did not become a palatial unconquerable castle, but a dam. It was this dam that fulfilled Sibyl's prophecy of a better tomorrow.

A Comparison of Water Management Projects in Czechoslovakia and the Soviet Union

If we compare reservoirs in postwar Czechoslovakia and in the USSR, it is clear at first glance that besides sharing common ideological underpinnings, these reservoirs do not have much in common. Merely comparing the figures that we have presented demonstrates just how large the quantitative differences between the concepts behind reservoir construction in both countries were. The explanation is simple: sparsely settled regions of the USSR could not compare with densely settled Czechoslovakia. There was not enough space nor were there enough suitable locations to implement such large construction projects. For example, flow rate of the most important Czech river, the Labe, is 300 m³/s, whereas the Volga River's flow rate is 8,000 m³/s.

The differences between these projects are not only found in their size, but also in their goals. Using the example of the Grand Turkmen Canal, it is clear that Stalin and hydrological engineers such as Davydov were truly guided by a grandiose plan to transform and control nature in the spirit of the Michurinist motto: "We cannot wait for favors from nature—it is our task to take them from it!" Through their plans, they attempted to transform infertile deserts into usable land; as we know today, however, this often had catastrophic consequences, as can be seen from the fate of the Aral Sea.

New reservoirs in Czechoslovakia also contributed to a large extent to transforming the landscape of the Vltava River and the Orava region in Slovakia. However, the scale of this transformation was much smaller and occurred without catastrophic impacts on nature. Although biotopes were destroyed locally, soil was not degraded and did not transform into desert. Nonetheless, there were projects in Czechoslovakia that could have fundamentally damaged the environment if they were carried out. Such projects included a planned reservoir in the Křivoklát region[175] and the Danube-Oder-Elbe canal, which is still under consideration today. Building such a canal was already considered in the fourteenth century; today however there are concerns that its construction and operation could result in fundamental changes in the hydrology of Moravia; the consequences of such changes are not predictable.

The fact that these projects were not carried out cannot just be attributed to the small size of Czechoslovakia (in comparison to the total area

Table 1.3: An overview of the largest water management projects in Czechoslovakia[176]

Name	Period	Output/generation	Length	Area
Orlík	1954–1961	364 MW/ 372.2 GWh	68 km	27.3 km^2
Lipno	1952–1959	120 MW/ 143.4 million kWh	48 km	48 km^2
Slapy	1949–1956	144 MW/ 287.7 GWh	42 km	13.92 km^2
Váh Cascade	1954–1961	336.35 MW/ 1091.1 million kWh	–	–
Orava Reservoir	1941–1953	21.75 MW/ 25.6 million kWh	–	35 km^2

of the USSR). For example, in 1949, Romania, which is not much larger than Czechoslovakia, began work on a canal connecting the Danube and the Black Sea. Today, this canal is 64.2 km long and 60 m wide. In 1953, project finances were cut off, and it was completed only in 1984. In order to build this canal, 300 million cubic meters of earth were moved, which is more than twice as much as for the Panama Canal. Therefore, the question remains: What exactly influenced the approaches taken in different countries? Perhaps, following Karl Wittfolgel's[177] concept of oriental despotism, it could be argued that in Czechoslovakia, with its longer tradition of democracy, it was more difficult to implement such megalomaniac plans than in the Soviet Union or Romania.

Therefore, the greatest similarity between Soviet and Czechoslovak projects is found in their significance for the Socialist national economy and in their use for spreading ideology through propaganda. Both countries faced a lack of electricity in the postwar period. Hydroelectric plants therefore became a welcome and relatively cheap way to produce the electricity necessary to industrialize regions that in the past did not have access to sources of energy, such as the Orava and Váh regions. Thus, in this light, these truly were "great" works.

The Transformation of Nature in the Media

The transformation of nature was a big topic for the media, and they portrayed it with full support from the KSČ. The party papers of the KSČ and the Komunistické strany Slovenska (KSS) [Communist Party of Slovakia – CPS]—*Rudé právo* and *Pravda*, respectively—were used for this purpose. Emphasis was placed on "great works of socialism," including the construction of reservoirs, which implicitly referred to the implementation of the Great Plan for the Transformation of Nature in the USSR.

The media's interest was important because thanks to simple propagandistic slogans and methods, the "great plans" for transforming Czechoslovak society were presented to the people. Paradoxically, in doing so, they humanized these monstrous projects. Daily party newspapers featured articles that attempted to clearly relay the political goals of the great plan to readers, and mainly to inform them of the plan's Soviet model—the policies of the Communist Party of the Soviet Union. The media approached the topic very selectively, and emphasized only the aspects of the plan that were relevant to party priorities. This approach mostly affected environmental protection, which in the 1950s was completely pushed aside in favor of transforming nature.

Some Statistics

Statistically speaking, 1952 was the absolute highpoint of media coverage of water management and nature transformation projects. In that year, a total of fifty-four articles were published in *Rudé Právo* on the transformation of nature in either the USSR or in Czechoslovakia. This means that roughly one article covering this topic was published per week. In comparison, just twenty-one such articles were published in 1951, and twenty-nine in 1953.

Another high point was 1956, when the first large projects of the Stalinist Plan were completed. In that year, fifty-two articles appeared in *Rudé Právo*, although their emphasis had significantly changed: in 1952, during the early stages of the Stalinist Plan for the Transformation of Nature, of the fifty-four articles published that focused on it, thirty-eight dealt with the situation in the USSR and only sixteen were about Czechoslovakia; whereas in 1956, forty-five of the fifty-two articles were about Czechoslovakia, and only seven dealt with the Soviet Union.

The same phenomenon can be observed in the analysis of the Slovak newspaper *Pravda*, which was, however, to a large extent dependent upon what was written in the Prague-based newspaper. Just like in the Czech lands, 1952 was the high point of propaganda in Slovakia, when a total of thirty-seven articles were published on this topic. In contrast to the Czech reports, however, the articles contained in *Pravda* were almost evenly divided between Soviet and Slovak coverage. But in 1956, the situation was similar to the one in the Czech lands: of twenty-seven articles, only three dealt with the implementation of the plan in the USSR, whereas the remainder all focused on Czechoslovakia.

This reversal of perspective can be easily explained. At the height of the cult of Stalinism in Czechoslovakia, the Stalin Plan for the Transformation of Nature in the USSR, which Stalin himself was believed to have developed, was an integral part of the Stalinist vision of the world and nature. For this reason, it was also a very interesting topic for the media in Communist-dominated Eastern Europe. The scope of the Stalinist Plan and the possibilities, and above all the perspectives and gains (i.e., planned gains) it could offer demonstrated and confirmed Stalin's wisdom and intelligence. In Czechoslovakia, the Soviet plan for the transformation of nature and its portrayal in the media was one of the main pillars of the creation of a new socialist society. It was a reference point that attracted everyone's attention.

It is also important to realize about our study of media coverage that the first great work of socialism in the USSR, the Volga–Don Canal, was completed in 1952, when the transformation of nature dominated the media. This was the first actual accomplishment of the Stalin Plan, and its success was unprecedented. However, in 1952, Czechoslovakia had yet to fully implement the Stalinist Plan for the Transformation of Nature; the regime only started up preparatory studies and began taking steps leading towards the creation of infrastructure necessary for the plan's centralized administration. The media support that this project received was, nonetheless, enormous. In following years, when the plan was quietly abandoned, the number of articles focusing on the Soviet Union significantly decreased—in *Rudé Právo* there were no more than twelve, and in *Pravda* no more than nine.

The quality of these articles fluctuated. Propaganda concentrated on photography and visuals, as well as on short descriptions of new projects and extensive reports from construction sites. An example of the former is a photograph from the Tsimlyansk Hydroelectric Plant construction site, featuring a caption that reads: "The lights of the Tsimlyansk Hydroelectric Plant reservoir shine at night far into the steppe."[178] Usually, these were just very short texts celebrating the far-reaching importance of new projects by reprinting large-format photographs of great works of socialism.

Visual Propaganda
The media focused primarily on visual aspects of propaganda when presenting great works of socialism. Most articles of this type were based on photographs showing large water management projects in the USSR and lacked any deeper textual information and analysis. It seems that the importance of these projects lies in their existence, which moreover strengthens the overall context of the great transformation of nature.

Another significant and typical element of media coverage of the construction of reservoirs was the great speed at which they were built. An article on the Slapy Reservoir emphasized this, and thus symbolically also its importance: "The construction of the massive reservoir in Slapy is progressing along at a great pace. The builders of this reservoir have obliged themselves to push forward deadlines in honor of the Nineteenth Congress of the Communist Party of the Soviet Union. Socialist competition, which has successfully been developed on the construction site, helps workers to exceed the plan."[179]

The same assumption, that speed equals importance, was of course also valid in Slovakia, which is particularly obvious in media coverage of the USSR. The article "Rozsah prací na Hlavním Turkmenském průplavu vzroste letos na čtyřnásobek" [The Scope of Work on the Main Turkmen Canal Has Grown by Four Times This Year] describes the rapid pace of work on this new project and on creating social infrastructure for the workers.[180]

However, by the mid-1950s, speed began to be taken over by quantity and innovation. In 1956, an article entitled "Na Dněpru rostou vodní stavby"[181] [New Water Projects Grow on the Dnieper] outlined new hydroelectric plant projects on the Dnieper and included calculations of how much electricity would be generated. Similarly, in 1959, another article bearing the title "Sedmé vodní dílo na Volze" [A Seventh Reservoir on the Volga] described the construction of a new hydroelectric power plant on the Volga nearby Saratov that stood out from other projects mainly due to the fact that it was to be built from 50 percent pre-fabricated concrete elements.[182]

The Czechoslovak media faithfully copied the Soviet model, and therefore, quantity and innovation began being written about in Czechoslovakia. For example, in 1960, an article appeared in *Pravda* entitled "Dvanáct na Váhu"[183] [Twelve on the Váh] about the construction of the twelve existing water management projects on the Váh, and describes new methods in water management project construction in Slovakia.

Socialist Reportage

Just as the production novel portrayed the heroic efforts of the builders of reservoirs, the construction of great works of socialism gave birth to a new form of "socialist reportage" on the pages of daily newspapers. For example in "Vyprávění mladých budovatelů Volžsko-donského průplavu" [What the Young Builders of the Volga–Don Canal Have to Say], people could read a detailed report about the lives of young workers on this large project. In articles such as "Světlé dni mého života" [The Bright Days of My Life] and "Komsomolec se musí starat o všechno" [Komsomol Members Must Take Care of Everything] engineers Lyubov Zajninova and Mikhail Komarov wrote about the time they spent on construction sites.[184] It is not very surprising that these articles celebrate the size and progressiveness of these projects. After 1956, however, these types of articles would be printed only exceptionally, and after 1959 they disappeared from the pages of *Rudé Právo* and *Pravda* entirely.

Reportage on these great projects was very widespread in Czechoslovakia. For example, in 1952, an article entitled "Za čestné plnění úkolu na stavbě přehrady ve Slapech"[185] [For the Honorable Completion of Tasks in the Construction of the Reservoir in Slapy] described the construction of the Slapy Reservoir and how it was based on the Soviet model. The report, however, also contains descriptions of problems on the construction site such as an inadequate supply of equipment and an absence of "party work" (which was quite bold for the time). Even after 1956, one can find articles such as "Orlík kontra Bratsk" [Orlík vs. Bratsk] (1959),[186] which, in the spirit of socialist reportage, deals with the construction of the Orlík Reservoir. Besides traditional topics such as socialist competitions, we can also find an instructive tale of the power of Czechoslovak–Soviet friendship, describing how Czechoslovak workers wrote to the builders of the Bratsk Hydroelectric Power Station and asked them for advice about their reservoir on the Vltava.

These Czechoslovak articles were also laced with explicit references to the Stalinist Plan for the Transformation of Nature, which at that time practically no longer existed. References to Soviet models, however exemplarily, show that these articles and reports were written in the right context.

Transformation or Subjugation?

Without a doubt, the biggest difference between Czechoslovak propaganda about the transformation of nature in Czechoslovakia and in the Soviet Union is found in semantics: whereas texts about the USSR were based on and actively used the phrases "transformation of nature" or "reconstruction of nature," those on Czechoslovakia weakened this radical approach by the use of the term "subjugation." Thus, we find differences between both countries, not only in the scope of the actual projects but also in how they were presented in media.

For example, in an article about the Soviet Union entitled "Donské stepi mění svou tvář" [The Don Steppes Change Their Appearance] (1952), we can read about how "districts that will be irrigated in the first stage of work will quickly change in appearance. Where there was once bare steppe, pretty workers' settlements will grow... The land is changing right before our eyes. Man, inspired by the ideas of communism, is transforming it. The fertile areas of the Don are covered in a thick network of canals; new young forests, orchards, and vineyards shall grow."[187] The article continues with further description of the Stalin Plan for the Transformation of Nature, which includes the construction of irrigation canals, the establishment of shelterbelts, and the transformation of the steppe into a "garden of Eden" with vineyards. These and other literary topoi, such as "being/becoming the masters of nature," "with the help of Stalin," or "creating new seas," were characteristic of articles about the Soviet Union.

In contrast, these topoi were rarely used to describe the situation in Czechoslovakia. One example can be found in an edition of *Pravda* from 1952, the year the propaganda campaign supporting the Stalinist Plan for the Transformation of Nature reached its highpoint. The article is entitled "Premena prírody v ZSSR a u nás" [The Transformation of Nature in the USSR and in Our Country] and states:

> We are standing before the problems of increasing the productivity of agriculture, of interfering with nature and transforming it, of managing water following the model of the USSR in the form of retaining water in reservoirs and in the form of building a hydrological hub in Bratislava with the goal of regulating the Danube and irrigating the Danubian Lowlands... On the path to socialism, we shall no longer exploit our natural wealth and live under the yoke of the forces of nature, but, with the help of the USSR and alongside it, we too shall influence nature.[188]

An article in *Rudé Právo* entitled "Krocení řeky" [Taming the River] was, of the few expressive articles ever written, the most expressive of all: "However, the Váh, just like every wanderer, is not very stable. When snow melts in the mountains, waves of muddy water maliciously roll through its channels. In the summer though, the Váh is thirsty. How can this wild, evil river be tamed?"[189] Nonetheless, this article was rather an exception and told the reader more about subjugating nature than about transforming it.

Thus, auxiliary constructions related to subjugating nature, such as "taming rivers," prevailed in accounts of the transformation of nature in Czechoslovakia. In the article "V Považí před 1. Májem" [Along the Váh before 1 May], this literary topos is employed: "Kilometers, many long kilometers ahead rage the untamed rapids of the thundering Váh ... In coming years, the river will give us benefits. New factories will be built in the Váh region as there will be a lot of electricity. Drudgery will disappear from the fields, and at night every village will be lit up with electric lights."[190]

Subjugating nature in the Czechoslovak understanding, however, is not violent, rough subjugation, but instead it is useful subjugation, and moreover, as contemporary journalists wrote, it is a kind of subjugation that could successfully lead to the creation of a new, socialist esthetic. For example, in the article "Lipno – nové vodní dílo na Šumavě" [Lipno – A New Reservoir in Šumava] readers can discover that "the Vltava, a river that poets have sung its praises, Smetana's Vltava, will become a river that will power our factories. Nothing will be taken away of its beauty; high dams and big lakes will create completely new conditions in the Šumava Mountains, and tourists and recreation seekers will make their way to the 'Šumava Sea,' because this formerly barren region will be settled."[191]

In 1959, reports on the transformation of nature slowly disappeared, and after Stalin's death, de-Stalinization, and the beginning thaw, descriptions of such projects were no longer the bombastic portrayals of the past. Journalists still wrote about taming rivers but employed phrases that were commonly used in Western Europe as well: "No one could manage to utilize the wealth of wood the taiga provides until the new representative of the pioneering era came along—the Soviet initiative of transforming and industrializing Siberia ... The river is tamed. The gigantic task of utilizing its power for industry, however, continues."[192]

The Decline of "Building Socialism"

In 1960–1962, the heroic age of transforming nature and building great works of socialism was on the decline. Propaganda was mainly disseminated by publishing various large-format photographs of construction sites with short descriptions of their progress. The decline in the amount of text was compensated by a growth in the number of photographs. Although the completion of the largest reservoir in Czechoslovakia, the Orlík Reservoir, would have in earlier years undoubtedly led to bombastic descriptions of achievements in transforming nature, between 1959 and 1962 this motive does not appear in a single article. Could it be that the Communists and their propagandists shrunk from taming and martyring Smetana's celebrated and praised river?

The Orlík Reservoir, the largest project to affect the landscape in the 1950s (with the exception of the creation of industrial areas), was still, however, the pride of the Communist government, as is demonstrated by the delegations that came to view it: in 1960, Cambodian Prince Norodom Sihanouk admired the reservoir;[193] in 1961, Leonid Brezhnev[194] viewed it; and in 1962, even Ernesto "Che" Guevara admired it.[195]

"In the Name of the Great Work..."

The Stalin Plan for the Transformation of Nature, which was introduced in hard-line Stalinist Czechoslovakia during the first phase of Sovietization in the country, never had the chance to fully develop. Nonetheless, the plan had important impacts on society. Some of these impacts have been examined, others have remained covered behind the merciful curtain of forgetting, and some have become such a fixed part of people's collective memory that many people do not even realize that they were in fact outcomes of this plan.

Today, the implementation of the plan can be divided into several phases. The first preparatory phase occurred mainly in the late 1940s after the Communists took over. This was followed by a period when the main executive bodies involved in the plan were institutionalized, culminating in 1953 with the establishment of the CAAS. The next phase involved the earliest attempts at mechanically adopting Soviet models, which was however interrupted by the death of Stalin and then of Gottwald in 1953. The Twentieth Congress of the Communist Party of the Soviet Union in Moscow definitively brought an end to this period. The last phase, which occurred roughly between 1956 and 1962, involved the abandonment of directive Soviet orders and the implementation of original Czech plans that in many ways were founded on Czechoslovak interwar traditions in agriculture and nature conservation. In 1960, the fanatical approach to building socialism largely disappeared from the great plan.

Changing the Macroclimate

But still, in 1955, the implementation of the Stalin Plan would see its last great rise in Central and Eastern Europe. At a conference of the CAAS entitled "On the Coordination of Scientific Research Work and on Cultural Educational Activities in Agriculture" (*O koordinaci vědeckovýzkumné práce a o kulturně osvětové činnosti v zemědělství*) held on 6–10 December 1954 in Prague and attended by delegations from all of the people's democracies, Czechoslovak delegates proposed the central coordination of all projects involving the "transformation of nature" in people's democracies.

The main instigator of the plan to coordinate activities, Karel Růžička, secretary of the Czechoslovak Government Commission for Improving the Management of Agriculture, Forestry, and Water, made the argument that in carrying out the transformation of nature in Communist countries it was already clear that "activities leading to the transformation of nature are undertaken in accordance with local conditions".[196] The goal of these activities was to have a positive effect on nature, by being in accordance with natural laws, such as aiming to prevent erosion and to promote the planting of suitable species.

For the first, we find in Karel Růžička's speech a positive definition of just what the "transformation of nature" means:

> The transformation of nature is a great and complex activity, consisting of a wide range of measures, each with the goal of improving: *vegetation*, including species, quality, the speed of ripening, etc.; *soil*, including structure, the permeability of water and air, the amount of nutrients and humus, fertility, etc.; *surface water*, including regulating flow throughout the year, flow conditions including drop and speed, temperature, quality, evaporation, etc.; *groundwater*, including increasing and decreasing levels, creating conditions for accumulating groundwater, potability and curative qualities, etc.; *atmospheric moisture and climate*, including water vapor in the air, the creation of dew, evaporation rates, the influence of drought, etc.; and finally *improving the use of natural resources and forces for the service of*

man, for acquiring energy, for transportation, for utilizing subterranean mineral resources, peat, oil, water, coal, for medical services, etc. The above-described activities can be referred to as a "transformation of nature" only when they are carried out on a large scale and comprehensively. Individual measures, locally limited, might mean improvement and the positive development of using each component [of nature], but they cannot influence nature.[197]

The Czechoslovak delegation then made the justifiable complaint that due to the small size of all of the socialist countries—with the exception of the USSR and China—it would be impossible to accomplish such feats as in the USSR. This meant, for example, that within the Plan for the Transformation of Nature in Central and Eastern Europe it would not be possible to change macroclimatic conditions.[198]

In 1955, the plan was still supposed to copy how the plan had been implemented in the USSR, where initial work took place between 1928 and 1939, and where it had been fully implemented since 1946. The USSR had managed to meet several smaller goals and make the first steps in implementing this great plan in the mid-1950s. Considering that decisions were made to implement similar measures in Hungary, China, Bulgaria, Romania, East Germany, and Poland, the Czechoslovak delegation proposed that the activities of all of these countries be coordinated and that there be an exchange of information between them. This was supposed to take place through both official channels (i.e., through ministries and other authorized institutes), and informally (i.e., through the exchange of publications, printed materials, and academic literature). It was exactly the intention of state dirigisme in Communist countries for a central international scientific institute for implementing the transformation of nature in each country to be created. This institute was supposed to keep a record of research projects, methods, and results, and, at the same time, coordinate research activities. It was also supposed to be responsible for dealing with problems related to international projects, such as the issues of implementing "the Danube–Oder Canal, the complexity of issues involved in the joint plan for the Miskolc–Nyíregyháza–Mukachevo–Košice region, water management issues in the Ostrava–Racibórz–Rybnik region, the Görlitz–Zittau–Liberec region, along the entire Labe/Elbe in between Kolín and Hamburg, etc."[199]

Soviets in Their Own Trap

We can truly only speculate as to why exactly the overall implementation of the Great Plan for the Transformation of Nature was abandoned. Stalin's death put a definitive end to the promotion of the plan. One of the possible explanations of why this plan came to an end, which is nothing more than a mere piece in the colorful patchwork of various influences, may be the simple fact that, during the implementation of the plan in Czechoslovakia, the Soviets fell into their own trap. Besides Klement Gottwald, the Stalinist president, who was mainly just the public face of the propaganda surrounding the implementation of the plan, essentially two other figures professionally instigated and supported the Great Plan for the Transformation of Nature: Rudolf Slánský and Josef Smrkovský. Rudolf Slánský, the first secretary of the Central Committee of the KSČ, was executed in the political trial in December 1952. Josef Smrkovský, the man who instigated the initial work on implementing the great plan at the Ministry of Agriculture, was arrested and imprisoned until 1955 as a result of this trial. Although the establishment of the Committee for the Transformation of Nature was announced with great fanfare in January 1953, by March of 1953, Klement Gottwald was dead, Rudolf Slánský was also dead, hanged as a traitor months before, and Josef Smrkovský was in prison as his accomplice. The Great Plan for the Transformation of Nature therefore lost political support before it had even actually begun. After all, who would want to return to the thoughts and ideas of political traitors?

And thus the plan was started up in Czechoslovakia; however, it was never implemented in its original political scope. It seems that it was implemented in its original scientific scope—but only in its preparatory phase: a detailed soil map of Czechoslovakia was created; areas to be afforested and where reservoirs could be built were determined; and, after much experimenting with Soviet crop varieties, the decision was made to return to growing traditional temperate crops.[200]

The original remnants of the Stalinist Plan for the Transformation of Nature were quickly absorbed by the system and became an integral part of it. At the CAS, the Committee for Aiding Great Works of Socialism, which had been established in 1952, became the Committee of the Board of the CAS for Water Management,[201] from which after the events of 1968 the independent Committee of the Board of the CAS for Environmental Issues (1972–1977) was created. This continuity illustrates the social transformation of ideas about the relationship between man and nature, and demonstrates that Stalinist patterns mechanically adopted from the USSR no longer worked in the 1960s, and therefore a fundamental transformation occurred in how nature was approached.[202]

This change was confirmed, for example, by the creation of the Institute for Landscape Planning and Protection of the CAS,[203] which was established in 1962, and disappeared in 1971 when it was merged with the Cabinet for the Theory of Architecture and Environmental Planning of the CAS. This merger resulted in the creation of the Institute of Landscape Ecology of the CAS (which existed until 1993). Complementarity in institutional development—when in the 1950s active landscape transformation was emphasized, whereas in the 1960s landscape "planning" and protection were discussed, and in the 1970s landscape ecology was at the center of attention—clearly demonstrates continuity in how people thought about the landscape, although the transition from two extremes in ideas (from the transformation of nature to its protection) leads more towards an interpretation emphasizing discontinuity in postwar development.

Continuity and Discontinuity of Nature Conservation

The final question could be—or rather should be—also a question of continuity and discontinuity of prewar and postwar developments, and also developments between 1948 and 1989. In the 1930s and 1940s, two main approaches to nature conservation can be identified. The first approach was strongly technocratic and anthropocentric. The approach emphasized the potential of man to change (and essentially control) nature to his own benefit. The second approach can be characterized as being proto-environmental, and was based on contemporary biology and ecology. It tried to find a balance between the interests of society, the economy, and the laws of ecology. Within the Stalinist Plan, the technocratic approach to nature protection flourished. Continuity in Czechoslovakia can be observed in the person of Antonín Klečka, the founder of the Committee for the Transformation of Nature of the CAAS, who in the 1930s and 1940s was one of the most active contributors to the conservationist magazine *Krása našeho domova*.

The second approach to conservation, which today can be considered to be proto-environmentalist, valued nature in its own right to a certain extent, and attempted to join the interests of nature with the economic interests of society. This approach also survived in the 1950s and 1960s in state nature conservation offices (which fell under the authority of the Ministry of Education until 1958, and afterwards under the authority of the State Institute of Historical Preservation and Nature Conservation). It was also present to a lesser extent in the CAS and CAAS. After 1956, when de-Stalinization was commenced (and the pseudoscientific theories implemented after the Soviet model were refuted), and especially after 1962, this approach began to gain new ground. The establishment of the Institute for Landscape Planning and Protection of the CAS in 1962 was indeed an expression of significant changes in thinking about nature conservation, and on the symbolic level it represented a synthesis of both approaches to nature conservation: the technocratic approach (in the 1950s represented by A. Klečka) and the proto-environmental approach, which was widespread in the 1930s and 1940s. In the 1950s and 1960s, even this group of conservationists used the terminology of the Stalin Plan (see the semantic field of "planning nature," etc.), which legitimized their standing within the system.

Therefore, it is important not to lose sight of nature conservation within the analysis of the Great Plan for the Transformation of Nature. As the Communist regime left behind a legacy of massive environmental degradation, it is not often thought of in connection with nature conservation and landscape protection, yet despite this, state nature conservation in Czechoslovakia grew rapidly between 1946 and 1960. In 1948, a law was enacted establishing the Tatra National Park in Slovakia as the first national park in Czechoslovakia.[204] In addition, dozens of small-scale nature reserves were established, as were the first large-scale landscape protection areas, and work on founding the first Czech national park in the Krkonoše Mountains was underway (it would be established in 1963). In 1956, at the height of the original hard-line approach to applying the Stalinist Plan, the first law on nature conservation was approved (Act No. 40/1956 Coll., on state nature conservation). The Sovietization of landscape management, which was inspired by the Stalin Plan, was seemingly in conflict with the aims of nature conservation, but in the end, conservationists utilized the failure of this plan in their favor.

Kafkaesque Paradigms

The way the Stalinist Plan for the Transformation of Nature was reflected in art gives the most accurate picture of developments in Czech society, including all of the Kafkaesque paradigms of the history of Eastern Europe. We can thank this great plan for the creation and mass spread of new literary genres in the region. Here we find socialist journalism and the production novel; what else could production novels be written about if not great works of socialism? The authors of such literature also underwent their own development full of disillusion. After 1968, Jan Trefulka, a writer and poet, who in the 1950s wrote *chatushkas* about afforestation, wrote the novel *Veliká stavba* [The Great Work] (1973). It tells the fantastic tale of a former people's judge, who works with a group of former convicts whom he had sentenced, on a "great work" in the middle of a virgin forest. He refuses to leave the construction site, even though work on the project had long ago been stopped. The construction project of the novel and the motives it contains are a reference to Kafka's *The Castle* and *The Trial*. The staged "people's trials" that the judge instigates against workers on the project who refuse to meet the deadlines of the "great work" receive a death penalty that the judge proclaims is "In the name of the great work..." Thus, Czech literature found a literary parallel to the Stalinist Plan for the Transformation of Nature: great works of socialism like Kafka's *The Castle* and *The Trial*. Czech society is still looking for a way to relate to this plan today.

Doubravka Olšáková is a senior researcher at the Institute for Contemporary History of the Academy of Sciences of the Czech Republic in Prague, where she leads the working group on contemporary environmental history. She has published a book titled *Science Goes to the People!* (2014, in Czech) about the dissemination of science in Communist Czechoslovakia and the indoctrination of the masses.

Arnošt Štanzel is a PhD student at the Graduate School for East and South East European Studies of Ludwig Maximillian University in Munich, Germany, where he is working on his PhD thesis entitled *An Environmental History of Water Management in Romania and Czechoslovakia during State Socialism (1945–1989): A Comparison.*

Notes

1 "Osnova hlavních směrů výzkumu do roku 1960 v oboru vodního hospodářství (důvěrné)," 14 June 1954, National Archives, ČSAZV, 45, sign. A3A.

2 L. Kalinová, *Východiska, očekávání a realita poválečné doby. K dějinám české společnosti v letech 1945–1948*, Prague, 2004, 43.
3 Ibid., 41–51.
4 See Kalinová, *Východiska, očekávání a realita poválečné doby*; K. Kaplan, *Proměny české společnosti 1948–1960*, Prague, 2007.
5 See *Německé menšiny v právních normách 1938–1948: Československo ve srovnání s vybranými evropskými zeměmi*, Prague, 2006, 171–235.
6 The US Army stopped in front of the demarcation line running between Karlovy Vary, Plzeň and České Budějovice.
7 Slovak results differed: Demokratická strana [Democratic Party] 61%, Komunistická strana Slovenska [Communist Party of Slovakia] 30%, Strana svobody [Liberty Party] 4% and Strana práce [Labor Party] 3%. J. Cuhra et al., *České země v evropských dějinách. Díl čtvrtý od roku 1918*, Prague 2006, 172–73.
8 Cuhra et al., *České země v evropských dějinách*, 183–92.
9 M. Blaive, *Une déstalinisation manquée: Tchécoslovaquie 1956*, Brussels, 2005; and (in Czech) M. Blaive, *Promarněná příležitost: Československo a rok 1956*, Prague, 2001.
10 A. Kubačák and M. Beranová, *Dějiny zemědělství v Čechách a na Moravě*, Prague, 2010, 374.
11 A. Kubačák, *Zemědělství v letech 1918–1948: Osudy úřadu a jeho ministrů*, Prague, 2005, 103–6.
12 Ibid., 381.
13 Ibid., 399.
14 Ibid., 393.
15 J. Hofman and J. Jindra, *Za sovětskými zkušenostmi v lesnictví (K týdnu lesů a stromů 1951)*, Prague, 1951, 39.
16 "Zápis ze 20. schůze kolegia," 2 March 1949, National Archives, fond Ministerstvo zemědělství II – schůze kolegia ministrů zemědělství, Praha, 1, sign. 19.
17 R. Slánský, Projev, in NS RSČ 1948–1954, 51 Meeting, část 7/9, 21 December 1950 – http://www.psp.cz/eknih/1948ns/stenprot/051schuz/s051007.htm (2 May 2013).
18 Ibid. (http://www.psp.cz/eknih/1948ns/stenprot/051schuz/s051007.htm (2 May 2013).
19 See J. Connelly, *Captive University: The Sovietization of East German, Czech, and Polish Higher Education, 1945–1956*, Chapel Hill, 2000.
20 *Marxisticko-leninskou ideovostí a stranickostí proti kosmopolitismu a objektivismu ve vědě. Sborník dokumentů I. ideologické konference vysokoškolských vědeckých pracovníků v Brně 27. února – 1. března 1952*, Prague, 1952. See also M. Štorek and Paleček, "Jak to začalo v Československu," in V. Soyfer, *Rudá biologie – pseudověda v SSSR*, Brno, 2005, 311–27.
21 See M. Franc, *Úderná skupina? Výprava českých lékařů a přírodovědců do SSSR v roce 1950 ve světle dopisů Ivana Málka*, Prague, 2009.

22 "Rozmluva čsl. vědecké delegace s Lysenkem," 14 November 1952, National Archives, ČSAZV, 369, (no sign.).
23 See T. Ježdík and V. Maděra, eds, *Proti kosmopolitismu a objektivismu ve stavitelství: Sborník projevů z ideologické konf. fakulty inž. stav. ČVUT v Praze*, Prague, 1953.
24 "Zpráva o činnosti Československé akademie zemědělských věd a jejích členů," 11 July 1955, National Archives, ÚV KSČ, Politbyro 1954–1962, vol. 50, arch. unit 68, point 7, f. 16.
25 Ibid.
26 F.J. Vohryzka-Konopa, *Venkov v Temnu. Násilná socializace zemědělství*, Munich, 1989, 126.
27 See "Hlásenie k bodu 2. Zprávy ČSAZV č. 4/54," 8 May 1954, National Archives, ČSAZV, 30 (no sign.); "Zápisnica napísaná dňa 6.8.1954 na porade skupiny šlachtitelov kukurice," 6 August 1954, National Archives, ČSAZV, 34, (no sign.). Cf. J. Scholz, "Rozšiřování pěstování kukuřice na zrno, zelenou hmotu, siláž a zvyšování jejích výnosů," *Věstník ČSAZV* 2, 1955/1, 32–38. General reports on the corn production in Hungary, Romania, Bulgaria, and Poland, cf. *Věstník ČSAZV* 2, no. 2–4 (1955), 130–34.
28 "Věc: Sovětská kukuřice, v Lednici," 8 July 1957, National Archives, ČSAZV, 522 – Komise pro kukuřici (Agenda 1953–1958, 1. část).
29 "Korespondence s Pavlem Valouškem," 17 August 1958, National Archives, ČSAZV, 34, (no sign.).
30 "Předběžná zpráva ze studijní cesty do sovětského svazu," 14 September – 14 October 1956, National Archives, Ministerstvo zemědělství – zahraniční vztahy, 55, sign. 146.
31 "Zápis ze závěrečné besedy u soudruha ministra se sovětskou zemědělskou delegací," 11 May 1957, National Archives, Ministerstvo zemědělství – MZ-ZV, 55, sign. 146.
32 Ibid.
33 See "Informace pro soudruha ministra k dnešnímu přijetí sovětského velvyslance," 27 May 1960, National Archives, Ministerstvo zemědělství – MZ-ZV, 55, sign. 146.
34 D. Olšáková, "Zur Geschichte der Tschechoslowakischen Akademie der Landwirtschaftswissenschaften 1952 bis 1962," in *Agrarwissenschaften in Vergangenheit und Gegenwart*, ed. S. Kuntsche, Diekhof, 2012, 65–82.
35 In January 1953, twelve Soviet advisers were employed at the Ministry of Agriculture. They co-authored the proposal of the Czechoslovak Academy of Agricultural Sciences. See K. Kaplan, *Sovětští poradci v Československu, 1949–1953*, Prague, 1993, 113 (Příloha č. 6: Doplněk seznamu poradců v letech 1952–1954).
36 See A. Klečka, *Rozpravy o mičurinské agrobiologii*, Prague, 1951.
37 *Československá akademie zemědělská a její nástupnické organizace 1924–2004*, Prague, 2004, 38.
38 Ibid.

39 "Návrh usnesení vlády o nové organisaci zemědělské propagace, vědy a výzkumu," 21 August 1952, National Archives, ÚV KSČ, Politický sekretariát 1951–1954, vol. 37, arch. unit 111, point 12, f. 6.
40 Ibid., 9.
41 Ibid., 13.
42 Ibid.
43 "Návrh zákona a statutu Československé akademie zemědělských věd a návrh politicko-organisačního zajištění slavnostního zahájení Československé akademie zemědělských věd," 27 October 1952, National Archives, ÚV KSČ, Politický sekretariát 1951–1954, vol. 43, arch. unit 126, point 6, f. 1.
44 In German: die Deutsche Akademie der Landwirtschaftswissenschaften zu Berlin.
45 "Návrh zákona a statutu Československé akademie zemědělských věd a návrh politicko-organisačního zajištění slavnostního zahájení Československé akademie zemědělských věd," 27 October 1952, National Archives, ÚV KSČ, Politický sekretariát 1951–1954, vol. 43, arch. unit 126, point 6, f. 3.
46 Ibid., 12.
47 Ibid., 13.
48 "Zvolení nových zahraničních členů ČSAZV," 7 November 1958, National Archives, ÚV KSČ, Politbyro 1954–1962, vol. 195, arch. unit 267, point 9, f. 3.
49 Ibid.
50 "Zvolení nových čestných členů Československé akademie zemědělských věd," 16 October 1959, National Archives, ÚV KSČ, Politbyro 1954–1962, vol. 240, arch. unit 321, point 18, f. 4.
51 "Začlenění Československé zemědělské akademie věd do Československé akademie věd," 20 January 1962, National Archives, ÚV KSČ, Politbyro 1954–1962, vol. 338, arch. unit 428, point 14, f. 6.
52 Ivan Brauner, Augustin Kalandra, Jiří Kořátko, Emil Kunz, František Mareček, Jan Plesník, Václav Stehlík, Jaroslav Svoboda, Jaroslav Šimon, and Emil Špaldon. From the ten members, eight were members of KSČ, and two had no political affiliation. See "Začlenění Československé zemědělské akademie věd do Československé akademie věd," 20 January 1962, National Archives, ÚV KSČ, Politbyro 1954–1962, vol. 338, arch. unit 428, point 14, f. 15.
53 See *Zachování alejí jako typického prvku české krajiny: sborník referátů z odborného semináře konaného dne 29. dubna 2010 v Praze*, ed. J. Esterka, Prague, 2010; J. Hendrych, "Aleje a stromořadí v kontextu klasické krajiny," *Acta Pruhoniciana* 95, 2010, 123–38.
54 "Zápis ze 20.schůze kolegia," 2 March 1949, National Archives, Ministerstvo zemědělství II – schůze kolegia ministrů zemědělství, Praha, 1, sign. 19.
55 Hofman and Jindra, *Za sovětskými zkušenostmi v lesnictví*.
56 J. Čížek, *Sovětská věda o úloze lesa*, Prague, 1952, 11.
57 Ibid.

58 Ibid., 9.
59 Kaplan, *Sovětští poradci v Československu, 1949–1953*, 113 (Příloha č. 6: Doplněk seznamu poradců v letech 1952–1954); S. Brain, "Stalin's Environmentalism," *Russian Review* 69, no. 1 (2010), 93–118.
60 "Zápis ze 20. schůze kolegia," 2 March 1949, National Archives, fond Ministerstvo zemědělství II – schůze kolegia ministrů zemědělství, Praha, 1, sign. 19.
61 *Příručka pro lidové výzkumníky v pěstění lesů*, Prague, 1955.
62 J. Hofman and J. Mottl, *Náměty pro lidové výzkumníky*, Prague, 1951.
63 *Příručka pro lidové výzkumníky v pěstění lesů*, 58.
64 See *Lesní ochranné pásy*, Prague, 1953.
65 Russian lyrics: M. Čujev, music: V. Ruděnko, Czech translation: J. Trefulka. J. Hofman – J. Jindra, *Za sovětskými zkušenostmi v lesnictví*, (supplement).
66 "Zápis o 66. schůzi kolegia," 1 December 1949, National Archives, Ministerstvo zemědělství II – schůze kolegia ministrů zemědělství, Praha, 5, sign. 66.
67 "Návrh plánu hlavních úkolů vědecko-výzkumných prací na rok 1954," 10 September 1954, National Archives, Ministerstvo zemědělství II – schůze kolegia ministrů zemědělství, Praha, 1, sign. 19, fol. 34.
68 Ibid., 46.
69 "Zápis z porady komise pro trávopolní soustavu," 19 October 1954, National Archives, ČSAZV, 503, (no sign.).
70 See "Písemné sdělení rumunské delegace – Travopolní osevní postupy v Rumunské lidové republice," *Věstník ČSAZV* 2, no. 2–4 (1955), 116–22.
71 See "Dopis Jaroslava Šimona z února 1955," February 1955, National Archives, ČSAZV, 503, sign. 461 (Komise půdoznalecká, porady 1953–1958, 1.část).
72 "Zápis z 300. schůze kolegia," 19 December 1952, National Archives, Ministerstvo zemědělství II – schůze kolegia ministrů zemědělství, Praha, 37, sign. 316.
73 "Zápis ze schůze kolegia," 17 September 1953, National Archives, Ministerstvo zemědělství II – schůze kolegia ministrů zemědělství, Praha, 40, sign. 359.
74 "Návrh plánu hlavních úkolů vědecko-výzkumných prací na rok 1954," 10 September 1954, National Archives, Ministerstvo zemědělství II – schůze kolegia ministrů zemědělství, Praha, 1, sign. 19, fol. 37.
75 Ibid.
76 "Otázka rentability pěstování rýže v ČSR," 1953, National Archives, ČSAZV, 45, sign. A3A.
77 Ibid.
78 "Návrh plánu hlavních úkolů vědecko-výzkumných prací na rok 1954," 10 September 1954, National Archives, Ministerstvo zemědělství II – schůze kolegia ministrů zemědělství, Praha, 1, sign. 19, fol. 42.
79 Ibid.

80 I.I. Prezent, *Ruku v Ruce s přírodou*, Děčín, 1950, 131 (cf. chapter *Mičurinské hnutí v SSSR*, 131–47).
81 Ibid., 132.
82 Ibid., 133.
83 Ibid., 135–36.
84 "Zpráva o mičurinském hnutí v ČSR," 15 March 1955, National Archives, Ministerstvo zemědělství – MZ-ZV, 55, sign. 146, 1.
85 Ibid., 2.
86 Ibid., 3.
87 Ibid.
88 B. Kavka, "Význam a úkoly Čs. vědecké společnosti zahradnické I.V. Mičurina," *Věstník ČSAZV* 2, no. 6–7 (1955), 451–60.
89 B. Penka, *Prosenické hnutí*, Prague, 1952, 22.
90 Ibid., 25.
91 Ibid., 28.
92 Ibid., 30.
93 Ibid., 32.
94 Ibid., 38.
95 Ibid., 68.
96 B. Penka, *Stachanovské pracovní methody v řepařství a cukrovarství*, Prague, 1953, 10.
97 Ibid., 9–11, 144–50.
98 V. Macura, *Šťastný věk (a jiné studie o socialistické kultuře)*, Prague, 2008, 62.
99 Ibid.
100 Ibid., 63–64.
101 Ibid., 66.
102 See "Zpráva z druhého zasedání smíšené československo-sovětské komise pro boj proti mandelince bramborové a přástevníčku americkému," 1–18 August 1959, National Archives, Ministerstvo zemědělství – MZ-ZV, 55, sign. 146.
103 S. Brain, "The Great Stalin Plan for the Transformation of Nature," *Environmental History* 15, October 2010, 670.
104 J. David, *Smrdov, Brežněves a Rychlonožkova ulice – Kapitoly z moderní české toponymie*, Prague, 2011, 132.
105 J. Brabenec, *Velká proměna: Reportáž o stavbě čs. přehrad, hutí, dolů, elektráren, cementáren, naftových polí, železnic a sídlišť*, Prague, 1954, 107.
106 *Historie výstavby a vzniku NHKG*, Prague, 1966, 30.
107 Ibid., 54–55.
108 Ibid., 72.
109 Ibid., 73.
110 Ibid., 76.
111 Ibid., 66.
112 *Veliké dílo – Fotografická reportáž o první československé stavbě socialismu Nové huti Klementa Gottwalda v Ostravě Kunčicích*, Prague, 1952.

113 Ibid.
114 *Historie výstavby a vzniku NHKG*, 126.
115 Ibid., 95.
116 Brabenec, *Velká proměna*, 102.
117 *Veliké dílo*.
118 M. Svácha, *Země na úsvitu – Reportáž o budování Ostravska*, Prague, 1953, 9.
119 *Dějiny NHKG*, Prague, 1981, 23.
120 Ibid., 22.
121 *Historie výstavby a vzniku NHKG*, 112.
122 Ibid.
123 Ibid., 119.
124 Ibid., 68.
125 See "Postavíme hornické město Šumbark-Bludovice," Švácha, *Země na úsvitu*, 149–54 and also 64–75.
126 *Historie výstavby a vzniku NHKG*, 116. (See O. Káňa, "Zrod koncepce strukturálních přeměn Ostravska a počátky sídlištní a průmyslové přestavby kraje na začátku první pětiletky," *Slezský sborník* 63, no. 2 (1965), 192–213.)
127 David, *Smrdov, Brežněves a Rychlonožkova ulice*, 134–55.
128 J. Dubský, *Stavby socialismu v lidově demokratických zemích*, Prague, 4.
129 Ibid., 7. See J. Dubský, *Naše stavby socialismu*, Prague, 1952.
130 Dubský, *Stavby socialismu v lidově demokratických zemích*, 9.
131 Dubský, *Naše stavby socialismu*.
132 Ibid.
133 Brabenec, *Velká proměna*, 113.
134 See R. Převrátil, *Trať družby – Z histórie Košicko-bohumínskej železnice*, Martin, 1953.
135 Ibid., 6.
136 Ibid., 10.
137 See *Trať družby – Časopis pracujúcich v ZTD-ČSSZ*, n.p., 1, 1951. The first issue was out on 15 April 1951.
138 *Trať družby vo výtvarnom diele*, Košice, 1953.
139 See E. Sýkora, "Trať mládeže po 30 rokoch," *Trať mládeže a súčastnosť*, Nitra, 1980, 53; also B. Toman, "Dvadsať mesiacov na Trati mládeže," *Trať mládeže a súčastnosť*, 75.
140 Ibid., 76.
141 D. Hernychová, "Stavby mládeže v knižnej produkcii Smeny a Mladej fronty," *Trať mládeže a súčastnosť*, 271–76.
142 A. Pludek, *Slunce v údolí*, Prague, 1955.
143 See M. Matula, "Trať mládeže a vysokoškoláci – očami dneška," *Trať mládeže a súčastnosť*, 134–51; M. Ričalka, "Rozvíjanie širokej sociálnej aktivity mládeže po roku 1945," *Trať mládeže a súčastnosť*, 279–97.
144 M. Malík, "Rozvoj pracovných hnutí mládeže v ČSM," *Trať mládeže a súčastnosť*, 178.
145 Dubský, *Stavby socialismu v lidově demokratických zemích*, 10.

146 Ibid., 11.
147 K. Růžička, *Vodní Hospodářství*, Prague, 1962, 29–30; J. Rosík, "Státní vodohospodářský plán a zásady pro další plánovité řízení vodního hospodářství," *Vodní hospodářství* 4, no. 4 (1954), 99–103.
148 J. Říha and V. Plecháč, *Význam vody v životním prostředí*, Prague, 1973, 113.
149 Z. Mařík, "Nový zákon o vodách," *Vodní hospodářství* 24, no. 1 (1974), 1–4.
150 A. Novosád and J. Jiroušek, *Československé vodní hospodářství: Základní organizační, právní a ekonomické vztahy*, Prague, 1960.
151 Pavel, a kol., *Stavebníctvo na Slovensku. Vodohospodárske stavby*, Bratislava, 1991; S. Kratochvíl, *Vodní nádrže a přehrady*, Prague, 1961.
152 Huba, "Oravská priehrada – dielo socializmu," *Krásy Slovenska* 60, no. 5 (1983), 10.
153 People from Orava preferred to leave the region and emigrate, very often to the United States; see M. Rusnák, *Orava na starých pohľadniciach*, Bratislava, 2010, 13.
154 "Nikdy nebylo lepší na Oravě," *Rudé Právo* 30, no. 1, 1 January 1950, 4; "Roste Nové Slovensko," *Rudé Právo* 51, no. 1, 1 January 1952, 6; "Nad Oravou vyšlo slunko," *Rudé Právo* 33, no. 25, 25 January 1953, 1.
155 "Dni na Oravskej priehrade," *Pravda* 33, no. 33, 8 February 1952, 5; "Předvečer Dudkov na Oravě," *Pravda* 35, no. 86, 28 March 1954, 3.
156 D. Kodaj, *Oravská priehrada*, Bratislava, 1953.
157 *Priehrada*, režie Paľo Bielik, 1950.
158 D. Milly, *Triptych Oravská priehrada* (1949).
159 "Nová země – noví lidé," *Rudé Právo* 32, no. 347, 25 December 1952, 3.
160 "Fotka Oravská Priehrada," *Pravda* 33, no. 25, 25 January 1953, 1.
161 "Krocení řeky," *Rudé Právo* 36, no. 121, 1 May 1956, 4.
162 J. Šturdík, *Priehrada mládeže* (1951).
163 "Nad mapou budoucnosti Slovenska," *Rudé Právo* 36, no. 152, 1 June 1956, 1.
164 L. Tekeľ, "30 rokov hydroenergetických vodných diel na Slovensku," *Vodní Hospodářství* 25, no. 10 (1975), 263–66.
165 Brabenec, *Velká proměna*, 15.
166 Ibid., 48.
167 Ibid., 45.
168 http://praha.idnes.cz/potapeci-vytahli-ze-dna-slapske-prehrady-zbytky-sochy-jana-nepomuckeho-1n7-/praha-zpravy.aspx?c=A130512_194944_praha-zpravy_aha (accessed 1.4.2016).
169 There is also a smaller auxiliary dam, Lipno II, at Vyšší Brod.
170 I. Beneš, *Lipenská přehrada*, Prague, 1965.
171 Brabenec, *Velká proměna*, 34.
172 Ibid., 29.
173 "Pracovní konference o vědeckých problémech při výstavbě vodních děl na Dunaji a Orlíku," Archives of CAS, Komise presidia pro vodní hospodářství, vol. 25, 25.

174 Ibid., 55.
175 http://kostelik.cz/wp-content/uploads/2012/03/Krivoklatska_prehrada.pdf (accessed 9 April 2013).
176 Tekeľ, "30 rokov rozvoja hydroenergetických vodných diel na Slovensku," 263–66; S. Kratochvíl, "Vývoj využití vodní energie v ČSSR za poslední čtvrtstoletí a jeho perspektivy," *Vodní Hospodářství* 25, no. 5 (1975), 127–32.
177 K.A. Wittfogel, *Die orientalische Despotie: eine vergleichende Untersuchung totaler Macht*, Cologne and Berlin, 1962.
178 "Fotografie Cimljanská přehrada," *Rudé Právo* 32, no. 170, 2 July 1952, 1.
179 *Rudé Právo* 32, no. 275, 15 October 1952, 2.
180 *Rudé Právo* 32, no. 51, 29 February 1952, 1.
181 *Rudé Právo* 36, no. 84, 24 March 1956, 3.
182 *Rudé Právo* 36, no. 84, 24 March 1956, 3.
183 *Pravda* 41, no. 295, 23 October 1960, 2.
184 *Rudé Právo* 32, no. 146, 8 June 1952, 4.
185 *Rudé Právo* 32, no. 12, 12 January 1952, 3.
186 *Rudé Právo* 39, no. 184, 5 July 1959, 1.
187 *Rudé Právo* 32, no. 13, 16 January 1952, 1.
188 *Pravda* 33, no. 179, 30 July 1952, 6.
189 *Rudé Právo* 36, no. 121, 1 May 1956, 4.
190 *Rudé Právo* 32, no. 97, 20 April 1952, 3.
191 *Rudé Právo* 32, no. 167, 29 June 1952, 3.
192 *Rudé Právo* 32, no. 176, 27 June 1959, 3.
193 "Princ Norodom Sahanuk na Orlíku," *Rudé Právo* 40, no. 329, 27 November 1960, 1.
194 "Srdečné a radostné uvítaní delegace KSSS a soudruha A. Novotného v jižních Čechách," *Rudé Právo* 41, no. 138, 19 May 1961, 1.
195 "Kubánští hosté na Slapech a Orlíku," *Rudé Právo* 42, no. 245, 5 September 1962, 1.
196 K. Růžička, "Koordinace akcí „přeměny přírody' ve státech lidově demokratických," *Věstník ČSAZV* 2, no. 2–4 (1955), 154.
197 Ibid.
198 Ibid., 155.
199 Ibid.
200 "Stenografický zápis z diskuse k XVII. valnému shromáždění ČSAZV," 21 December 1960, National Archives, ČSAZV, 73, (no sign.), 70–74; "Zpráva o dosavadním průběhu komplexního průzkumu půd v ČSR," (after 19 July 1961), National Archives, ČSAZV, 507, (no sign.).
201 The Committee of the Board of the CAS for Water Management was dissolved in 1979, and a special Committee of the Board of the CAS for Orlík Dam was created in 1954.
202 We would like to thank Nicholas Orsillo for his contribution to this part of our text.

203 In Czech: Ústav pro tvorbu a ochranu krajiny ČSAV [Institute for Landscape Planning and Protection of the CAS].
204 The Tatras National Park (TANAP) was created by the Act of the Slovak National Council no. 11/1948 Coll., 18 December 1948, taking effect from 1 January 1949.

Bibliography

Beneš, I., *Lipenská přehrada*, Prague, 1965.
Blaive, M., *Promarněná příležitost: Československo a rok 1956*, Prague, 2001.
———, *Une déstalinisation manquée: Tchécoslovaquie 1956*, Brussels, 2005.
Brabenec, J., *Velká proměna: Reportáž o stavbě čs. přehrad, hutí, dolů, elektráren, cementáren, naftových polí, železnic a sídlišť*, Prague, 1954.
Brain, S., "Stalin's Environmentalism," *Russian Review* 69, no. 1 (2010), 93–118.
———, "The Great Stalin Plan for the Transformation of Nature," *Environmental History* 15, October 2010.
Československá akademie zemědělská a její nástupnické organizace 1924–2004, Prague, 2004.
Čížek, J., *Sovětská věda o úloze lesa*, Prague, 1952.
Connelly, J., *Captive University: The Sovietization of East German, Czech, and Polish Higher Education, 1945–1956*, Chapel Hill, 2000.
Cuhra, J., J. Ellinger, A. Gjuričová, and V. Smetana, *České země v evropských dějinách. Díl čtvrtý od roku 1918*, Prague, 2006.
David, J., *Smrdov, Brežněves a Rychlonožkova ulice – Kapitoly z moderní české toponymie*, Prague, 2011.
Dějiny NHKG, Prague, 1981.
Dubský, J., *Naše stavby socialismu*, Prague, 1952.
———, *Stavby socialismu v lidově demokratických zemích*, Prague, 1952.
Esterka, J., ed., *Zachování alejí jako typického prvku české krajiny: sborník referátů z odborného semináře konaného dne 29. dubna 2010 v Praze*, Prague, 2010.
Förster, H., J. Herzberg and M. Zückert, eds, *Umweltgeschichte(n): Ostmitteleuropa von der Industrialisierung bis zum Postsozialismus: Vorträge der Tagung des Collegium Carolinum in Bad Wiessee vom 4. bis 7. November 2010*, Göttingen, 2013.
Franc, M., *Úderná skupina? Výprava českých lékařů a přírodovědců do SSSR v roce 1950 ve světle dopisů Ivana Málka*, Prague, 2009.
Gestwa, K. *Die Stalinschen Großbauten des Kommunismus: sowjetische Technik- und Umweltgeschichte, 1948–1967*, Munich, 2010.
Gottwald, K., *Pro štěstí lidu: přetváření přírody ve stalinské epoše*, Prague, 1951.
Hendrych, J., "Aleje a stromořadí v kontextu klasické krajiny," *Acta Pruhoniciana* 95, 2010, 123–38.
Hermann, T., and D. Olšáková, eds, *Plánování socialistické vědy: dokumenty z roku 1960 ke stavu a rozvoji přírodních a technických věd v Československu*, Červený Kostelec, 2013.

Historie výstavby a vzniku NHKG, Prague, 1966.
Hofman, J., and J. Jindra, *Za sovětskými zkušenostmi v lesnictví (K týdnu lesů a stromů 1951)*, Prague, 1951.
Hofman, J., and J. Mottl, *Náměty pro lidové výzkumníky*, Prague, 1951.
Ježdík, T., and V. Maděra, eds, *Proti kosmopolitismu a objektivismu ve stavitelství: Sborník projevů z ideologické konf. fakulty inž. stav. ČVUT v Praze*, Prague, 1953.
Kalinová, L. *Východiska, očekávání a realita poválečné doby. K dějinám české společnosti v letech 1945–1948*, Prague, 2004.
Kaplan, K. *Proměny české společnosti 1948–1960*, Prague, 2007.
———, *Sovětští poradci v Československu, 1949–1953*, Prague, 1993.
Klečka, A., *Rozpravy o mičurinské agrobiologii*, Prague, 1951.
Kodaj, D., *Oravská priehrada*, Bratislava, 1953.
Kovda, V.A., *Velké stavby komunismu a jejich národohospodářský význam*, Prague, 1952.
Krásy Slovenska 60, 1983.
Kratochvíl, S., *Vodní nádrže a přehrady*, Prague, 1961.
Kubačák, A., *Zemědělství v letech 1918–1948: Osudy úřadu a jeho ministrů*, Prague, 2005.
Kubačák, A., and M. Beranová, *Dějiny zemědělství v Čechách a na Moravě*, Prague, 2010.
Kuntsche, S., *Agrarwissenschaften in Vergangenheit und Gegenwart*, Diekhof, 2012.
Lesní ochranné pásy, Prague, 1953.
Lysenko, T.D., and V.N. Sukačev, *J.V. Stalin a mičurinská agrobiologie*, Prague, 1951.
Macura, V., *Šťastný věk (a jiné studie o socialistické kultuře)*, Prague, 2008.
Marxisticko–leninskou ideovostí a stranickostí proti kosmopolitismu a objektivismu ve vědě. Sborník dokumentů I. ideologické konference vysokoškolských vědeckých pracovníků v Brně 27. února – 1. března 1952, Prague, 1952.
National Archives, ČSAZV.
National Archives, Ministerstvo zemědělství – MZ-ZV.
National Archives, Ministerstvo zemědělství II – schůze kolegia ministrů zemědělství.
National Archives, ÚV KSČ, Politbyro 1954–1962.
National Archives, ÚV KSČ, Politický sekretariát 1951–1954.
Němec, M. *Jak les pomáhá přetvářet přírodu*, Prague, 1953.
Německé menšiny v právních normách 1938–1948: Československo ve srovnání s vybranými evropskými zeměmi, Prague, 2006.
Novosád, A., and J. Jiroušek, *Československé vodní hospodářství: Základní organizační, právní a ekonomické vztahy*, Prague, 1960.
Olšáková, D., *Věda jde k lidu!: Československá společnost pro šíření politických a vědeckých znalostí a popularizace věd v Československu ve 20. století*, Prague, 2014.

Pavel, P. et al., *Stavebníctvo na Slovensku. Vodohospodárske stavby*, Bratislava, 1991.
Penka, B., *Prosenické hnutí*, Prague, 1952.
———, *Stachanovské pracovní methody v řepařství a cukrovarství*, Prague, 1953.
Pludek, A., *Slunce v údolí*, Prague, 1955.
Pravda 33, 1953; *Pravda* 35, 1954; *Pravda* 41, 1960.
Prezent, I.I., *Ruku v ruce s přírodou*, Děčín, 1950.
Převrátil, R., *Trať družby – Z histórie Košicko-bohumínskej železnice*, Martin, 1953.
Příručka pro lidové výzkumníky v pěstění lesů, Prague, 1955.
Říha, J., and V. Plecháč, *Význam vody v životním prostředí*, Prague, 1973.
Rozkvět lesů a přeměna přírody v Sovětském svazu, Prague, 1953.
Rudé Právo 30, 1950; *Rudé Právo* 32, 1952; *Rudé Právo* 33, 1953; *Rudé Právo* 36, 1956; *Rudé Právo* 39, 1959; *Rudé Právo* 40, 1960, *Rudé Právo* 41, 1961, *Rudé Právo* 42, 1962.
Rusnák, M., *Orava na starých pohľadniciach*, Bratislava, 2010.
Růžička, K., *Vodní Hospodářství*, Prague, 1962.
Slezský sborník 63, 1965.
Soyfer, V., *Rudá biologie – pseudověda v SSSR*, Brno, 2005.
Strakoš, M., *Nová Ostrava a její satelity. Kapitoly z dějin architektury 30.–50. let 20. století*, Ostrava, 2010.
Švácha, M., *Země na úsvitu – Reportáž o budování Ostravska*, Prague, 1953.
Trať družby – Časopis pracujúcich v ZTD-ČSSZ, n.p., 1, 1951.
Trať družby vo výtvarnom diele, Košice, 1953.
Trať mládeže a súčastnosť, Nitra, 1980.
Veliké dílo – Fotografická reportáž o první československé stavbě socialismu Nové huti Klementa Gottwalda v Ostravě Kunčicích, Prague, 1952.
Veliké přetvoření přírody SSSR, Prague, 1952.
Věstník ČSAZV 2, 1955.
Vitha, O., *O velikých stavbách komunismu v Sovětském svazu*, Prague, 1950.
Vodní hospodářství 24, 1974; 25, 1975.
Vohryzka-Konopa, F.J., *Venkov v Temnu. Násilná socializace zemědělství*, Munich, 1989.
Wittfogel, K.A., *Die orientalische Despotie: eine vergleichende Untersuchung totaler Macht*, Cologne and Berlin, 1962.

 CHAPTER 2

Untamed Seedlings
Hungary and Stalin's Plan for the Transformation of Nature

Zsuzsanna Borvendég and Mária Palasik

Characteristics of the Dictatorship

By 1947, Eastern and Western Europe had become politically separated with severe consequences for the East European sphere. In Hungary, the democratic coalition government that had been more or less functional since December 1944 was abolished, and when the Independent Small Holders' Party [Független Kisgazdapárt, FKGP] won the November 1945 National Assembly elections, the most influential politicians of the party—such as party secretary Béla Kovács and Prime Minister Ferenc Nagy—were removed from power with the covert and sometimes direct aid of the Soviet Union due to pressure from the upper leadership of the Hungarian Communist Party [Magyar Kommunista Párt, MKP].[1] By fall 1947, it had also become clear that the ratified peace treaty consolidated the Soviet occupation of Hungary, and by the second half of 1948, the Hungarian Communist Party had completely absorbed the Social Democratic Party [Szociáldemokrata Párt, SZDP] to unite as the Hungarian Workers' Party [Magyar Dolgozók Pártja, MDP] and initiate radical political change. These changes were implemented slowly between 1947 and 1949, and included the abolishment of the multiparty system, the nationalization of private property, the establishment of a totalitarian system and the centralization of the economy. Power was assumed by a small group of the Hungarian Workers' Party lead by Mátyás Rákosi,[2] who had his own cult of personality.

On 18 August 1949, the parliament formed by new, rigged parliamentary elections had approved the constitution of the People's Republic of Hungary.[3] As the name implies, the republic had been changed to a people's republic where, following the Soviet model, the president of the republic was replaced by the institution of the Presidential Council headed by a chairman. The constitution had formally kept the separation of legislative, executive, and judicial powers and thus the institutions of parliament, government, and the court system, but in practice, all power was concentrated in the hands of the Hungarian Workers' Party, leading to a dual system in which the party became the initiating, governing, and controlling body, and legislation, execution, and administration were merely mediating channels of party policies. Parliament assembled for two or three short sessions per year and its only task was to approve the party's initiatives without any actual debate; systematic legislative work was thus replaced by statutory orders issued by the Presidential Council and approved by parliament without debate. Aside from the formal separation of powers, the constitution also introduced the Soviet system of administration, and the politicization of the private and public spheres began. State party decision making and control thus extended to all areas of daily life, leading to the rise of the totalitarian state.[4]

Legal security and official authority bodies became unreliable for the average citizen as the once separate powers were merged; checks and balances between legislation, execution, and jurisdiction were erased, and laws became secondary to statutory regulations. In the constitution, "freedom of speech, freedom of press, and freedom of assembly" were only ensured "in accordance with the interests of the working class," and the multiparty system ended without any legislative closure, rendering the right to vote meaningless in the absence of alternative choices. Nevertheless, parliamentary and municipal elections were held repeatedly and participation—as well as casting one's vote for the nominated or supported candidates of the Hungarian Workers' Party—was obligatory.[5]

The dictatorship of Mátyás Rákosi was sustained and eventually destroyed by terrorism and the principle of "constant vigilance." By this time, Hungary was "building socialism" and faithfully adopting the same mistakes Stalin had made in the Soviet Union in the 1930s; accordingly, once Rákosi's leadership had disposed of every enemy among the bourgeoisie, the petite bourgeoisie, right-wing social democrats and the clergy, they began purging their own ranks within the party as well. This process began with the Rajk trial[6] and silencing pre-1945 Western Communist emigration, and was then followed by trials against social democrats, during which Rákosi's government charged and sentenced people who had previously been paving the Communists' path to power.[7]

The concept of equality had played a crucial role in the way Rákosi and Gerő's leaderships envisioned the future; in fact, from the great French Revolution onwards, all left-wing political movements were characterized by treating social equality as a positive value and sought, first and foremost, to reduce unjust inequality. From World War II until the mid-1970s, Western societies were also preoccupied with demanding equality, but within a plural society, other political forces tend to shift the focus to other values such as freedom and fraternity, leading to a certain balance without any major system breakdowns. In the case of Hungary, the totalitarian leadership of the Hungarian Workers' Party had completely disrupted this balance of values, and thus began a forced downward "leveling" in the name of equality.[8]

The large-scale socialization of the economy (first coal mines and major banks, then factories employing more than a hundred, and eventually more than ten, workers, of wholesale and retail, then finally of the service sector and private property ownership)[9] eventually led to the process of categorizing social groups into two "classes" and one "stratum"—workers, peasants, and intellectuals. However, the natural process of developing a middle-class peasantry due to post-1945 land reforms was brought to a halt by mass collectivization and high produce-surrender demands beginning at the end of 1948. Peasants began to abandon their lands at the beginning of the 1950s, and these were promptly swallowed by gargantuan industrial investment projects using mostly Soviet resources (at cities like Sztálinváros, Kazincbarcika, and Komló—more on these later). Sped-up and forced industrialization led to high levels of downward and upward mobility, shattering the formerly closed, traditional structure of Hungarian society.[10]

The former ruling classes had all but vanished within a few years. The aristocracy, the middle-class landholders, state officials and high-ranking members of the armed forces had left the country in large numbers,[11] while those who stayed and lost their former positions and property sought a living as workers or employees – assuming they were not incarcerated or interned and deported from their homes.[12] Due to the Rákosi government's measures liquidating the private small-scale industry and retail, these classes—approximately 350,000–400,000 citizens—were forced into a complete social shift.

The new political system attempted to completely eliminate the former intellectual stratum on account of their being politically untrustworthy, and replace them with a wave of new intellectuals coming from peasant and worker backgrounds, declared faithful to the system. They thus began producing cadres en masse to fill leading positions in the party, in government, in the armed forces, and in the economy and the intellectual–cultural sphere. With family members included, the numbers of these former workers almost reached one million, and appointing them to leading positions gave the illusion of the working class genuinely becoming the ruling class. However, as a myriad of individual and family tragedies have shown, many of them were unable to fulfill the professional and political requirements of their newfound positions.

While the Hungarian Workers' Party was seeking power, a part of society—those who were successfully convinced that they really were building a state of peasants and workers—willingly assisted the rise of their dictatorship. However, the grand promises of "socialism" and the reality of its actual existence slowly became questionable, even among the majority of those who had first seen hope in the Rákosi system. The system, of course, had abused people's trust by giving them false hopes about a better world, but public opinion can only be led astray for so long, and by 1951–1952 the dictatorship had become an obvious burden on workers and peasants alike.

The omnipotence of Mátyás Rákosi (having been prime minister since 1952 and number one in the party) came to an end for a short time. Following the death of Stalin, the Hungarian party leadership was summoned to Moscow by the Soviet party leadership in June 1953, where the guests received a slew of mortifying criticism aimed at Rákosi. The Hungarian party leadership was blamed not only for excessive industrialization, the army development plan and the forced organization of producers' cooperatives, but also for the show trials, the low standards of living and the personal cult of Rákosi. The Soviet hosts thus suggested a new Hungarian government with Imre Nagy as prime minister, and determined the key aspects of the new government program, the most important aim of which was to correct these mistakes. Although all important changes were thus decided in Moscow, the positive changes between 4 July 1953 and spring of 1955 that the new prime minister whole-heartedly supported against Rákosi came to be known as the politics of the "new block" led by Imre Nagy. However, the leadership power struggle between Ny. Khrushchev, L. Beria, and G. Malenkov to continue Stalin's legacy worked in Rákosi's favor, who immediately opposed Imre Nagy upon receiving word of the arrest of L. Beria, his strongest attacker at the Moscow meeting. Then, as soon as Khrushchev managed to remove Prime Minister Malenkov from the Soviet government, Rákosi led a direct assault against Imre Nagy, and not only dismissed him from his position as prime minister but also ostracized him from the party. Rákosi's resumed dictatorship may not have reached its pre-1953 power, but still lasted until 18 July 1956 when, with Soviet approval, the Central Committee of the Hungarian Workers' Party relieved him of his position as party secretary and member of the political committee on account of his state of health; however, during that time, his dictatorship had had catastrophic consequences for the whole country.

(Un)Planned Economy—New Socialist Cities
Transforming the Economy According to Soviet Models

Dividing Europe into East and West also had economical consequences. Central and East European countries experienced similar political processes to what transpired in Hungary, regardless of whether they had emerged on the winner's or loser's side, since their economic policies had been similar from the start. Following Stalin's plans for industrialization in the 1930s, the development of the industry, and above all the heavy industry, by huge investments became the primary goal of the state. At the time, industrialization became synonymous with the development of the heavy industry and especially of mining and mechanical engineering[13]—except Hungary did not have the raw material resources of the Soviet Union. In an increasingly bipolar world, their primary goal from 1949 onwards was to prepare for the next world war, and to do so, the Soviet Union demanded that its allies prepare their heavy industry for war. However, the national economic policies of the region were all characterized by limited resources, and in practice, they neglected production costs, profit-oriented economics, the modernization of technologies, and the protection of the environment, and attempted to compensate by excessive labor force demands.[14]

Following Stalinist directives, the Hungarian Workers' Party led by party secretary Mátyás Rákosi began to build "the country of iron and steel," despite the fact that there were no iron ore deposits in Hungary, nor was there enough coking coal for their purposes, which questions the entire raison d'être of these energy-consuming and highly expensive investments.

The three-year plan between 1947 and 1949 successfully restored all war damage in Hungary. Then came the first five-year plan, which was to start on 1 January 1950 pressured by the Budapest session of the Communist Information Bureau on 16 November 1949, which declared Yugoslavia, Hungary's southern neighbor, an aggressive power preparing for another war. Party leaders were thus justified in their demands of developing the army and devising plans to develop mining and metallurgy, and constructing a corresponding heavy industrial combine. They also envisioned the simultaneous fast-paced development of industry and agriculture as well as comprehensive modernization of the whole economy.

On 10 December 1949, the plan was approved by parliament. However, this plan was practically never followed as the designated (51 billion HUF)[15] investment budget was increased as early as the first quarter of 1950 by 14 percent of that period's designated budget, and in December 1950 the investment budget of 1951 was raised by 35 percent,[16] showing how much the plan was influenced by a sense of voluntarism, where leadership believed that meeting these unrealistic figures was only a matter of will.

Hungary's political and economic–political ambitions were fueled further with the turn of events between the two superpowers in 1950; once the Korean War had begun in July, Cold War threats gave way to actual war and action. Consequently, Stalin developed a new strategy and invited the number one leaders and ministers of war of all European Socialist countries to Moscow on 8 January 1951 to discuss the state of all Socialist armed forces. This date was *the* turning point in the economic policies of the region: after the Moscow meeting, the already strenuous plans had become completely irrational and unfeasible. Most well-known economic–political accounts hint at this turn of events but offer no explanations beyond the Korean War and do not mention this meeting.[17] So far, no archive documents have been found pertaining to the Moscow meeting. All we have at our disposal of the meeting and what transpired are the notes, interviews, and memories of four participants, and chronologically, the first two are secondary sources. One of them is Czechoslovakian Minister of Defense Alexej Čepička relating the events of the secret meeting to Karel Kaplan, party historian of the Czechoslovakian Communist Party, who included these memoirs in his book on the era after emigrating to the West;[18] and the other is an interview with Polish party leader Edward Ochab.[19] The third source—deemed primary—is the systematic minutes of the events by Romanian Minister of Defense Emil Bodnăraş;[20] and the fourth source is Mátyás Rákosi's *Memoirs*.[21] All four sources confirm and supplement one another, which allows us to safely state that the official meetings took place between 9–12 January, during which the Soviet Union was represented by Stalin and members of the Politburo such as Molotov, Malenkov, and Beria, as well as Minister of Defense Marshall Vasilevsky and Military Chief and General Shtemenko, both high-ranking military officials. On behalf of the people's democracies, the following number one leaders and party secretaries of the Communist parties and ministers of war were present: Edward Ochab, Konstantin Rokossovsky (Poland); Rudolf Slánský, Alexej Čepička (Czechoslovakia); Mátyás Rákosi, Mihály Farkas (Hungary); Gheorghe Gheorghiu-Dej, Emil Bodnăraş (Romania); and Valko Chervenkov and Petar Panchevski (Bulgaria). The meetings were also joined by Soviet military advisors of the four occupied countries: Gusev (Czechoslovakia), Boyko (Hungary), Kolganov (Romania), and Jemeljanov (Bulgaria).

Stalin opened the Moscow meetings by stating that although they had initially thought that the United States of America was an invincible power preparing for World War III, it had become clear that they were not only unprepared to initiate another world war, but unable to tackle even a relatively small war like the Korean War. In his view, it was obvious that the Korean War would occupy the United States for at least two or three years, creating an excellent opportunity for the global revolutionary movement. According to Stalin, the main task of the people's democracies was to use this time to build strong, modern, ready-for-combat armed forces, and Soviet military leaders were invited to the meetings to present fully devised plans on how many troops the armed forces should contain in times of peace and times of war:

> Poland: 350,000 troops in times of peace, 900,000 in times of war.
> Czechoslovakia: 250,000 troops in times of peace, 700,000 in times of war.
> Hungary: 150,000 troops in times of peace, 350,000 in times of war.
> Romania: 250,000 troops in times of peace, 700,000 in times of war.
> Bulgaria: 140,000 troops in times of peace, 350,000 in times of war.[22]

The above-mentioned figures meant that Hungary was charged with the task of building a 150,000 troop army by the end of 1953, an incredibly high number compared to the existing army of 40,000, but according to Rákosi, the plan was accepted without any opposition. According to the minutes made by Bodnăraş, the participants considered the 350,000 troop requirements of Hungary and Bulgaria to be insufficient in the event of military mobilization and ended up modifying both to 400,000. It was also made clear that these armed forces were to be supplied with rations, equipment and fuel by their respective countries. When Polish Minister of War Marshall Rokossovsky opposed the idea that Poland would have to create the army planned for the end of 1956 within three years, Stalin dismissed him saying that "if Rokossovsky can guarantee that there wouldn't be a war until 1956, then they may proceed with their original development plan, but if he cannot, then he had better accept" the Soviet suggestion. Bulgarian party leader Chervenkov then suggested that Bulgaria did not have a heavy industry and therefore the suggested military development could only be envisioned with necessary supplies from the Soviet Union, which Stalin declined stating that Bulgaria had to develop its own military industry.[23]

According to Mátyás Rákosi, upon their return from Moscow, he as chairman of the Defense Committee[24] had immediately begun to devise a military development plan aided by Lieutenant General Vasily Romanovich Boyko, leader of the Hungarian Soviet military advisors, who had also attended the Moscow meeting. Rákosi related the issue they were facing as follows:

> It was immediately clear that the sum we had designated for the next few years of the five-year plan had to be raised substantially. Simultaneously, we had to raise figures in industrial production as well, especially in the heavy industry . . . slowly we realized that we had to establish a military industry that would not only supply our own armed forces but help to supply that of the other Socialist countries as well within our future cooperation, and to do that, we would have to build the barracks and airports of these divisions within three years. Furthermore, we would have to gather those strategic supplies and materials that would be able, for at least a minimal amount of time, to supply the necessities for such an army and military industry during war. When I read the first approximated figures of how much food, textile, non-ferrous metal and fuel, etc. we would need, it had immediately become clear to me that this would not fit within the scope of the five-year plan that we had approved a year before. These plans would have to be updated accordingly.[25]

This excerpt highlights the context of the resulting economic political crisis—Rákosi and his select few had begun working on the new challenge, as a result of which they legally modified the five-year plan in May

1951 to raise the designated investment budgets by 60 percent. This also meant that almost 90 percent of steadily rising industrial investments had gone to the heavy industry. The following figures show the absurdity of the entire process: the originally designated figures of the first five-year plan were raised so that, compared to 1949, production by 1954 would go up by 200 percent instead of the original 86 percent, heavy industry would improve by 280 percent instead of 104 percent, and the light industry would go up by 145 percent instead of 73 percent.[26] Leading politicians addressed this issue by saying that this would be the prerequisite of developing the whole economy, and both their rhetoric and the plan stated the necessity of developing and collectivizing agriculture. However, all development goals had become completely unrealistic and in practice served only as a tool for propaganda. Instead of investing in agriculture, the agricultural sector was reduced to nothing and most of its resources were reassigned to the heavy industry.

The Establishment of the New Socialist City

The modified five-year plan had prioritized mining and material production, and the settlements that housed these investment projects received special support and were promoted to become "Socialist cities," prominent political settlements and the main beneficiaries of redistribution. This was, of course, based on the Soviet model as the number of cities in the Soviet Union nearly doubled within thirty-five years, from 709 in 1926 to 1,696 in 1959.[27] The best-known ones include Vorkuta and Pechora in the North, Magnitogorsk in the Urals, Bratsk in Siberia, Zhigulyovsk by the Volga, and Dushanbe in Middle-Asia, which was also called Stalinabad between 1929 and 1961.[28]

If we accept Pál Germuska's categorization, we could say that, aside from their prominent roles, another common characteristic of Socialist cities was that they were built to supply a single enormous production factory or a branch of heavy industry (leading to the tragic end of these settlements decades later when that leading industrial branch suffered an economic crisis). The majority of Socialist city-dwellers were employed by the industry while the service sector employed fewer people due to its low standards. Another characteristic of Socialist cities was that both the city and its industrial compounds were based on unified building plans,[29] and yet the majority of these still do not have a city center to this day. In fact, most of them had become cities before they became functionally able to fill the role of a regional center. The difference between

Socialist cities and West European or North American "new" cities that were similarly built around industrial investments was that the development of Hungarian Socialist cities went hand in hand with acquiring certain privileges to the detriment of other settlements. Socialist cities were established to serve centralized goals, while Western new cities were built in order to decentralize, and the service sector was not neglected the way it was in Hungary.[30]

The Socialist cities established in Hungary at the beginning of the 1950s were Sztálinváros (today called Dunaújváros), Kazincbarcika, Komló, and Tatabánya. Kazincbarcika was established from three small villages between 1947 and 1954, which themselves had been industrial settlements in the Eastern Borsod coal basin of northeastern Hungary: Sajókazinc and Berente were mining villages, and Barcika had been housing a power plant since the 1920s.[31] Kazincbarcika became a city in 1954 after the Borsod Chemical Combine,[32] Hungary's first major chemical industrial compound, had been built here between 1949 and 1954 (its successor, BorsodChem Rt., is still operating to this day). The small villages and their surrounding hills and valleys soon changed drastically as industrial compounds sprung up and spread, their silhouettes "adorned" by tall houses and factory chimneys.

Similarly to Kazincbarcika, Komló was also a small mining settlement in the Mecsek Hills of southern Hungary where black coal mining dated back as far as the eighteenth century. It was treated as a major investment from 1949, became a city in 1951, and absorbed four surrounding mining villages in 1954 (Kisbattyán, Mecsekfalu, Mecsekjánosi, and Mánfa, the latter of which was separated in 1992 and became its own municipality), thus becoming the industrial center of the region.

Tatabánya was established in 1947 by combining four settlements (Alsógalla, Bánhida, Felsőgalla, and Tatabánya), and breaking historical and cultural traditions, it was made the county center of Komárom-Esztergom County in 1950. This decision was economically, socially, and politically loaded as the thousand-year-old former royal residence of Esztergom, home of King Saint Stephen, was stripped of its administrative status, and its powers were transferred and entrusted to a working city with no bourgeoisie.[33] Due to forced coal mining, there were periods when 20 percent of Hungary's coal—or two-thirds according to other sources—came from Tatabánya. However, Hungarian leaders were well aware that Hungarian coal deposits were diminishing and thus introduced other industrial branches and developments to the region as well. In 1950, the production of mining machines began, and the light industry was represented by sewing facilities, wool workshops, and a telecommunications factory.

Sztálinváros had no industrial roots like the above-mentioned cities and was built completely from scratch on an agricultural site near the border of the village of Dunapentele. Its history is unique even within the history of Hungarian settlements, and thus deserves its own discussion in the next section.

The Exemplary Socialist City: Sztálinváros

Hungary had built its iron and steel factories in Ózd, Salgótarján, Diósgyőr, and Csepel during the second industrial revolution in the period of the Austro-Hungarian Monarchy, at a time when the country was two-thirds larger than it is today and was able to supply the necessary raw materials for the blooming heavy industry. However, forty to fifty years later these factories became insufficient, and plans for the construction of a new iron industrial combine were made in 1937 and then during World War II at the insistence of the German ally. Eventually, a government decree was issued in 1943, initiating a new heavy industrial investment that was to be realized in Mohács by the Danube River. Preparation began at the beginning of 1944, but then all activities ceased on account of the country becoming a warzone, and it was only at the beginning of 1949 that the Political Committee of the Hungarian Workers' Party—with the help of Soviet advisors—decided that the iron factory was indeed to be established in Mohács, some eleven kilometers from the Yugoslavian border. Work then resumed on 12 February 1949 in accordance with Government Decree 1400/1949. Mohács was a good choice due to the proximity of the Danube and the availability of black coal in the vicinity from the mines of Pécs and Komló; however, by the end of 1949, Hungarian–Yugoslavian relations had deteriorated to the point that the investment had to be terminated since Hungarian leaders were afraid that President Tito's Yugoslavia would attack Hungary. This called for a change of plans, and Ernő Gerő, chairman of the National Economic Council, had made a proposition on behalf of the State Economic Department of the Hungarian Workers' Party that the iron industrial combine should be transferred from Mohács to Dunapentele, another settlement located by the Danube River.[34] When the case was discussed by the Council of Ministers, there were political as well as other reasons for the change, including the fact that the recently discovered soil conditions north of Mohács would generate too much soil movement in case of an earthquake, and constructing an earthquake-proof base for the iron factory would come at huge additional costs.[35] However, it is important to note that the newly designated area did not receive a thorough geological examination during the preparation of this decision, and it is still unknown why the decision did not take soil conditions into account. Regardless, on 27 December 1949, the Political Committee of the Central Committee of the Hungarian Workers' Party had approved the proposition for transferring the iron factory.[36]

The outskirts of Dunapentele, a settlement established in the side valley by the raised beach of the Pentele loess-plateau on the right side of the Danube, were chosen as the new site of the iron factory. Through millions of years, water and wind had combined to form the surface of the Danube wash-lands, where the wind created stretches of quicksand by shifting the sand deposits of accumulated alluvia. Dust-like alluvium travelling through air had thus settled thickly over the intersecting surface areas, and through different processes of wearing away and soil formation, this substance became loess, the dominant loose and porous rock of the area.[37] Regarding the case of Mohács, we had seen that one of the cons of building, even if it was a late observation, was the threat of earthquakes; however, when Dunapentele was chosen, the decision disregarded the fact that the loess plateau was fundamentally unsuitable for an industrial investment of this magnitude, and even the locals and geologists were well aware of recent landslides. During the winter of 1943/44, the bank sank after the winter floods had passed, and seven family houses cracked and collapsed. In 1947 a major landslide occurred on a 250-meter-long section of the bank, followed by another one on a different section of the Danube bank in 1948.

Furthermore, decision makers completely disregarded the fact that the only available resource in Hungary was coking coal and only 70 percent of the necessary amount was available, forcing the country to supplement its resources from the Soviet Union, Poland and Czechoslovakia. Iron ore itself had to be brought from a location some 2,000 kilometers away from Krivoj Rog, and transported over water to the Danube dockyard at Dunapentele.[38]

In January 1950, a land survey took place on the loess plateau south of the border of Dunapentele, a village of 4,200 inhabitants. The surveyors designated 690 hectares of agricultural estate where the iron factory and the city were to be built, each on 345 hectares of land.[39] On 2 May 1950, the actual land survey began without telling farmers of the monopolization of their lands, so the surveyors were forced—with a heavy heart—to level already seeded soil. Farmers received minimal financial compensation for their appropriated property, and the state never compensated them for the incurred horticultural damage.

The Socialist city and the iron works were both treated as priority investments in Hungary's economic planning until the end of 1953. The new Socialist settlement was built simultaneously with the iron works, then joined with the previously existing settlement and declared a city in April 1951. By 1 May 1951, the first street of the city was built, and on 7 November 1951 the new city was named Sztálinváros [City of Stalin] and the iron factory was named dunai Sztálin Vasmű [Danube Stalin Iron Factory], thus adding to the number of "Stalin" settlements in the region (Volgograd to Stalingrad between 1925 and 1961, Varna to Stalin between 1949 and 1956, Braşov to Orasul-Stalin between 1950 and 1960, Eisenhüttenstadt to Stalinstadt between 1953 and 1961, and Katowice to Stalinogród between 1953 and 1956).

It is no exaggeration that the new city was constructed on the grain and corn fields by the already existing village of Dunapentele. In István Örkény's vision of Sztálinváros, he wrote:

> What was in the making there was truly a monumental enterprise. Nothing of the sort has ever happened in Hungary before: to build part of a city, move in the factory builders, begin constructing the factory and at the same time develop the city, too, while the factory keeps spreading—all of this had a sense of great romance to it. At the same time, the fact that the whole thing was a fundamental mistake—a national economic one—well, we did not know that, it was discovered later that we shouldn't have built on a loess plateau where every house required a special foundation. We didn't know it then, that we could have built it better and at a better place in a different way; and the mass of people who banded together there was awfully exciting. Some were the cream of the crop, while some were the scum of the country—for example, they bought whores from Pest by the trainload [after the monopolization of the brothels – the authors] to work there, and sometimes excavators would unearth infant corpses from the ground.[40]

Soon after erecting a barracks for the builders on the outskirts of Dunapentele, city houses were built to provide homes for four thousand families, then a workers' hostel capable of accommodating five hundred people, as well as schools, kindergartens, nurseries, a hospital, a council headquarters, a cultural institute, a cinema, a mall, a bakery, and a slaughterhouse. This future city also needed public utilities, roads, and parks—the only thing left out of the Socialist city plans was a church. The iron industrial complex was built on the grain fields, and the first finished structure was the blast furnace in 1954.

The newfound city of Sztálinváros soon came to life as people from all over the country had come to live there. By the first half of the 1950s, it had 28,000 inhabitants plus 4,200 workers who were still building the city and the iron factory living in the barracks and were thus considered temporary residents.[41] Following Stalin's death, new policies had abandoned forced industrialization, a decision that affected Sztálinváros as well, but the investments were not terminated as they had been in other parts of the country; they simply continued at a slower pace. In 1956, Stalin Iron Factory became the Danube Iron Factory, and in 1961, Sztálinváros became Dunaújváros [New Danube City].

The political decisions made at the end of 1949—without any environmental impact assessment—had severe and long-lasting consequences.[42] In the past decades, there have been countless accounts of building damage, slides, and public utility damage from the loess surface, as well as design and implementation errors that have fundamentally affected the city to this day.[43]

The Sciences and the Academy Under Pressure

New Research Methodologies

As has been discussed above, the single-party system and Stalinist dictatorship that prevailed by 1949 had a hitherto unseen impact on the natural environment. At the beginning of the 1950s, they began, based on Soviet practice, to implement nature transformation policies the likes of which had never been seen in Hungary before—and nor since. Their decisions not only affected water-supply management, soil cultivation, and forestry, but also led to the introduction of new crops completely foreign to Hungary's climate—in other words, politics began to formulate expectations for the natural sciences as well. The causes and arguments behind introducing new crops go back as far as the 1920s and 1930s when the production results of Soviet agriculture failed to reach even pre-1917 levels, and scientists, especially geneticists, came under enormous social pressure. The majority of them accepted (or were forced to accept) that the new Socialist biology had to serve national agriculture in direct and immediately profitable ways. In the words of Tamás Bereczkei, quoting one of the Soviet geneticists, "the party and the state were facing the grandiose task of improving production"[44]—and a key figure of this plan was Trofim Denisovich Lysenko.[45]

How Lysenko's faux scientific doctrines had become the norm is summarized by Bereczki as follows:

> The pressing issue was that basic genetic research had fallen far behind the megalomaniac ambitions of plant improvement. Michurin, who was treated as a veritable culture hero, simply crossed different fruits with one another with a lot of diligence but no scientific background to go with it—it would have been up to genetics to supply effective theoretical basics and methodological principles for creating new forms of plants. What is more, they constantly demanded that he put his theories into practice to boost stagnating agricultural production when it was obviously not possible. Although there were brilliant professionals among Soviet geneticists, they could perform no miracles. Only Lysenko believed himself capable of performing miracles and thus created "Michurinist Biology" in order to connect the practice of plant improvement with his own genetic theory. As a result, he made promises regarding agriculture that those "wanting to believe" received with standing ovation, promising that by directly changing the gene pool he would produce plant species with greater crop yields within two or three years—something that would take classic genetics 15–20 years based on Mendelian inheritance and the laws of Darwinian selection. Classic geneticists knew that these laws could not be bypassed and one cannot achieve sudden changes in species due to the conservatism of genetic material and the relative slowness of natural selection. However, the future of Soviet genetics became more and more dependent on what results it could achieve in agricultural practice, and obviously, it was Lysenko who had received all official funding on account of his grand promises, regardless of the fact that he could not deliver any results. People simply believed him because they wanted to believe him, and all Lysenko did was to fuel an already dominant small-minded and practice-oriented value system.[46]

Nature transformation had been an ongoing process in the Soviet Union since the 1920s and was present in everyday propaganda as well, but it received a boost with the great Stalinist Plan for the Transformation of Nature in 1948.[47] Following Lysenko's presentation at the 1948 August session of the Lenin Academy of Agricultural Sciences, Lysenkoism became an official school of biology. Alternative research was completely silenced in the Soviet Union, journals were discontinued, and the few opposing researchers who remained were—at best—discharged.[48] Lysenko's practice was based on the idea that altering the environment would result in new species, and with the right procedures the conscious manipulation of environmental factors would result in changes in the genetic pool.[49]

Although the Hungarian sciences had been experimenting with new plants that were not native to temperate climates, these experiments were always determined by the economy and economical considerations rather than ideological agendas.[50] However, this situation changed in the fall of 1948 when the August–September 1948 issue of *Társadalmi Szemle*, the scientific journal of the Hungarian Workers' Party, published the Soviet Communist Party's Central Committee-approved version of Lysenko's August 1948 presentation at the Lenin Academy of Agricultural Sciences. The journal editor's introduction clearly marked the path to be followed by Hungarian scientists, stating that Lysenko speaks of the "victory of the progressive sciences" against the "reactionary schools of science" and the "apolitical bourgeois sciences," and Hungarian agricultural experts must heed the "call of the times." Consequently, the first paper to be published in the new scientific journal *Agrártudomány* in February 1949 was titled "A szovjet agrobiológiáról" [On Soviet Agrobiology]. According to Irén Klára Szabó, author of this paper, "Michurin and Lysenko had proven, in the face of the inheritance theories of Mendel and Morgan, that transmission is accomplished not only through chromosomes but also through every part of the body, and thus circumstances can change hereditary features. It was by recognition and goal-oriented application of these two theses that Soviet agriculture achieved the results that all of the world looked at in well-deserved awe."[51] The next step was a volume of selected papers by Lysenko published that same year (1949) by the party's publishing house,[52] followed by a published volume of selected papers by Michurin in 1950, and the complete works of Lysenko in the same year.[53] Between 1949 and 1951, a host of Soviet scientific literature appeared, all of it following Michurin's and Lysenko's theories, and the propagation of Lysenko's doctrines became a requirement in scientific information dissemination.[54]

Lysenkoism was soon introduced in education and became course material at colleges and universities as early as 1949. In the beginning, it was taught by lecturers appointed by the party, who were not only charged with propagating Lysenkoism but also with offering "borderline vilifying criticism of genetics."[55] Lysenkoism was introduced to elementary and middle-school curricula by 1950[56] and teachers had no choice but to teach Lysenkoism as the curriculum demanded. Schools even turned part of their school yards into "Michurin gardens"[57] where selected pupils could work under supervision—today these would be called "practice gardens."

The new scientific policy was forcefully introduced to scientific research by plans created and approved by central politics,[58] whereby following Lysenkoism and questioning the fundamental principles of genetics became a central directive. Soviet scientists were invited to Hungary in order to advise Hungarian agricultural and biological research institutes, starting with the "Month of Soviet Culture" held in February 1949, which was attended by a Soviet delegation headed by Deputy Minister of Health Alexander Schabanov (physician) along with agrobiologist Ivan Gluschenko, member of the Soviet Academy of Sciences and close colleague of Lysenko himself, who held many lectures for Hungarian experts on Lysenko's basic principles.[59] Hungarian experts, too, paid more frequent visits to Soviet scientific institutes and acquired scientific degrees at Lysenkoist institutions. Delegations of two hundred peasants were sent to the Soviet Union for experience exchanges every year,[60] but agricultural scientists were also sent there for similar purposes in 1950 and 1951.[61] In fact, during the late summer visit of 1950, Lysenko himself greeted the Hungarian scientists at the hall of the Lenin Academy of Agricultural Sciences. "Ever since we had found out that we would be meeting Lysenko, we had lived in happy excitement, discussing and arguing about what sort of issues we should raise and seek his advice and guidance in. Our excitement was tinted by gratitude for the great son of the Soviet people who would take the time to directly aid us in accomplishing our tasks," wrote Andor Bálint, director of the Zsámbék Institute of Agrobiology in his report of the meeting.[62]

Dr. Sándor Igali (1926–2003), a contemporary researcher, began his career at the beginning of the 1950s at the same Institute of Genetics that opposed Lysenkoism.[63] In 2002, he, as a trustworthy witness, gave the following account of how the obligatory propagation of Lysenkoism affected scholars:

The forced propagation of Lysenkosim was done by the most loyal champions of the Communist Party, banding together with opportunistic "non-party" cohorts. After the turning point [1948], they occupied all leading positions, from ministers to heads of department, and only left a few indispensable and internationally renowned experts in place. The directors of research institutes, the head professors of university departments, the department secretaries of the Hungarian Academy of Sciences and all ministry and authority decision makers were Lysenkoists. They had also occupied leading positions in Hungarian scientific associations and represented Hungary in international institutions, committees, and associations so that Lysenkoism could also be disseminated through administration. There were, of course, exceptions, thanks to whom genetics—though forced into illegality—had survived even the darkest years of the dictatorship. The infectious seeds of Lysenkoism had spread to varying degrees, and the severity of the damage and the susceptibility of experts involved were different as well. It advanced most rapidly and most thoroughly in agriculture, thus placing the greatest pressure on the agricultural sciences. This disease was worst among plant improvers but did not affect stock breeders too badly. Although vets were closely involved in agriculture, they were less susceptible to this "illness" and recuperated relatively quickly. Biologists were also mostly unaffected. Doctors, aside from a few isolated cases, remained immune to it. . . . Lysenkoism, aside from the lecturers of the Marxist and Leninist faculties of universities, was only mentioned in publications by a few scholars, and even then was given as brief and unimportant a part as possible.[64]

The Institutional System of Nature Transformation

Much like other institutions, the Hungarian Academy of Sciences [Magyar Tudományos Akadémia, MTA] was also Sovietized in 1949. The Hungarian Workers' Party was suspicious of the "reactionary" association of scholars, and decided that the organization led by conservative thinker Zoltán Kodály[65] was unsuitable for the representation and implementation of Soviet scientific policies—and their fears were not unjustified as the Hungarian Academy of Sciences had a unique role in national thinking on account of its history. It was established in 1825 as a publicly funded national initiative at a time when Hungarians did not have national autonomy, and since the academy did not come from a dominant authority, the whole population viewed it as a unique representation of national identity. It enjoyed autonomy in the scientific sphere and retained its independence even once the state had assumed all financial responsibility for its maintenance from the 1920s. This is when they began to build a network of research institutes inspired by the German model, where these institutes were established independently of universities and gained complete autonomy in pursuing scientific activities on state funding.[66]

In 1948, the Stalinization of sciences began almost immediately after abolishing the multiparty system. The Hungarian Workers' Party established in June 1948 had decided within the first few weeks to destroy the Hungarian Academy of Sciences, and while it could not be abolished due to the above-mentioned reasons, they had attempted to curb its importance and role in scientific life by relegating its scope of activities to a newly established body. On 1 July 1948, the Secretariat of the Hungarian Workers' Party approved a legislative proposal concerning the establishment of the Hungarian Science Council [Magyar Tudományos Tanács, MTT],[67] led by President Ernő Gerő, which would systematically govern scientific life as a subordinate body of the prime minister. The scope of the Hungarian Science Council included determining the general course of scientific research, devising research plans, and supervising their activities. University professors and the leaders of scientific institutions could only be appointed with the approval of the council.[68] The Hungarian Science Council[69] was under direct party control—it was, in reality, governed not by the presidency but by the party's own academic body led by the same Ernő Gerő.

Regarding the topic under discussion, it is striking that the Hungarian implementation of Stalin's Great Plan for the Transformation of Nature had received such a prominent role in the research policies determined by the Hungarian Science Council. In fact, agronomic sciences were the second most important issue on the 1949 agenda of the council, right after the development of mining and the heavy industry.[70] Part of the future five-year plan was a five-year scientific research plan devised by the council, where implementing Michurin and Lysenko's new school of agrobiology in Hungarian agriculture received a prominent role. However, new research and experiments could not begin without first establishing a few new institutions. At the time, the number of remaining pre-1945 agricultural research institutes was 121, all of them governed by the Ministry of Agriculture [Földművelésügyi Minisztérium, FM][71] Since the Communist Party was against the autonomy of scientific institutions, coordinating the activities of so many institutions proved to be a difficult challenge, which is why the Agricultural Science Center [Mezőgazdasági Tudományos Központ, MTK] was established by the Ministry of Agriculture on 19 February 1949 to govern all agricultural, farming and forestry-related scientific institutions.[72] However, the Agricultural Science Center was only an administrative body since scientific policy was determined by the Hungarian Science Council.[73]

The Hungarian Science Council was short lived, but still managed to initiate the process of nature transformation that was eventually finished by the Sovietized Hungarian Academy of Sciences. It announced the detailed schedule of nature transformation, established the institutions necessary for its implementation, and created not only the Agricultural Science Center, but also the Institute of Agrobiology, the Institute of Forestry and the Irrigation Research Committee.[74]

By summer of 1949, the party's stance regarding the Hungarian Academy of Sciences had changed; although they had initially planned to destroy the academy from the inside in order to ensure the implementation of Soviet scientific policies, they later realized that this organizational framework does not meet Soviet requirements. When, in the spring of 1949, the above-mentioned Soviet politician Schabanov and the agrobiologist Gluschenko had visited Hungary, they—inter alia—studied the work of the Hungarian Academy of Sciences, and their observations induced Ernő Gerő to travel to Moscow in order to hold a personal conference with the leaders of these scientific policies,[75] where he was promptly informed that the exemplary Communist state insisted that the organizational system of regulating scientific activities be established in accordance with the Soviet model. Ernő Gerő's explanation of the necessity of re-establishing the credibility of the Hungarian Academy of Sciences to the Hungarian leaders went as follows:

> Every people's democracy has academies but none of them has an institution like our own Science Council. And lastly, the Soviet Academy of Sciences (and Gluschenko made this very clear during his visit here) is of the opinion that it cannot maintain contact with official state bodies such as the Hungarian Science Council, it can only do so with the Hungarian Academy of Sciences; and since the current composition of the Hungarian Academy of Sciences is unsuitable for the Soviet Academy's expectations, it effectively cannot maintain contact with either institution.[76]

In the fall of 1949, the reorganization of the Hungarian Academy of Sciences began, and it is important to note that significant political changes had occurred within one year of establishing the Hungarian Science Council. After the parliamentary elections of 15 May 1949, Hungary had officially become a single-party state, and Hungarian society had been terrorized and subjugated to the point where there was no danger of serious resistance. At the party secretariat session of 14 September 1949, a resolution was passed to unite the Hungarian Academy of Sciences and the Hungarian Science Council, summarizing the experiences of the past year as follows:

> Regarding the Hungarian Academy of Sciences, the party has long been of the opinion that . . . we should withdraw and only take care to stifle any open anti-democratic opposition. These tactics have proven correct since, one, it would have been unwise to incite the public opinion by dismantling the academy right after the secularization of all church schools, and two, during this year, the abandoned academy has proven itself to be incapable and incompetent in the eye of the scientific public.[77]

Before the primacy of the Hungarian Academy of Sciences could be re-established in governing scientific life, its political trustworthiness had to be ensured by a drastic cutback in membership.[78] However, come the fall of 1949, academy members had already been rendered unable to seriously oppose party policies since over fifty members of the academy had "fallen out" of the voting loop due to illness, staying abroad, or having been sentenced to prison, and the majority of the remaining voters would have adhered to Communist requirements.[79] However, the party was still unsatisfied and intended to further reduce membership by introducing a new membership category besides honorary, regular, and correspondent membership: advisory membership, which allowed one to use the title of 'academician' but not to participate in the activities of the institution.[80] Of the 258 members, 103 were reelected, 122 members of academy had been demoted to advisory membership, 17 scholars had been rejected from membership due to staying abroad, 13 scholars had lost their membership due to the abolishment of the Sub-Department of Arts, and three well-learned experts had been sentenced to prison.[81] Twenty-seven new members, all of them faithful Communists, had joined the academicians, whose political role had been more prominent than their scientific activities.

In the Sovietized Hungarian Academy of Sciences, mechanical engineering and the natural sciences were prioritized, with four departments out of six belonging to these fields. Department IV of the academy was called Biological and Agricultural Sciences, and had the greatest relevance to the current topic; therefore it is important to discuss its activities. Its role was to determine scientific research policies and prepare scientific drafts. It only influenced research and experimentation indirectly—it was instead charged with the task of "coordinating and leading the ideological struggle."[82] Detailed scientific plans were devised by the Agricultural Science Center and, after its reorganization on 20 May 1950,[83] by the Agricultural Experiment Center [Mezőgazdasági Kísérletügyi Központ, MKK], an institution charged with coordinating all research related to agriculture, and summarizing the results. During inspections, Department IV of the MTA could make observations regarding the operation of the center, but could not issue orders or instructions. The actual superior body was the Ministry of Agriculture, and the Agricultural Experiment Center was classified as one of its main departments.

Regarding the operation of nature transformation institutions, the academy had a smaller role than the National Planning Office, which was established in 1947, the year of announcing the three-year plan. The National Planning Office held audits at experimental sites, various national companies and other institutions to ensure their required operation and measure their progress in complying with their respective schedules.

The Agricultural and Cooperative Committee of the Hungarian Workers' Party had a key role in determining agricultural research policies, and it decided to further centralize agricultural research as of 1 January 1950. It commissioned the Agricultural Science Center to reorganize the farms of all ten Hungarian agricultural schools into experimental farms,[84] placing these new institutions under the center's control, and turning them into sites of soil improvement, crop improvement, and production technology experiments.[85] In reality, however, the institutional system established to conduct these experiments was under constant reorganization, to the point where the already tangled organization system suffered from a complete lack of transparency, even to the scholars and experts working on the experiments. Based on a "snapshot" taken at the beginning of 1950, the Agricultural Science Center operated the following institutions:[86] Institute of Agrobiology, Institute of Agrochemistry, Crop Production and Crop Improvement Research

Institute, Horticultural and Wine-Growing Research Institute, Animal Health Research Institute, Institute of Forestry Sciences, Institute of Agricultural Machine Testing, Institute of Agricultural Organization, Agricultural Documentation Center, Agricultural Museum, and Institute of Meteorology and Geomagnetism.

As stated above, the Agricultural Science Center became the Agricultural Experiment Center on 20 May 1950; however, this transformation was not conclusive and major reorganizations occurred. All research was audited during the summer, and feedback was negative in almost every area; the causes behind these poor results were deemed to have arisen from organizational problems and individual failures. The scapegoat was, of course, sought at the highest level possible, leading to the conclusion that Department IV of the Hungarian Academy of Sciences was inadequate for the task. The department was then split in two, and biology and agriculture were managed separately.[87]

On 23 April 1949, a government decree announced the establishment of the Agricultural Organization Institute[88] in order to "examine the theory and practice of agricultural planning and factory farming, as well as the organization of production work and its statistical results."[89] In other words, it was established to ensure the practical application of the research results of science centers, compare them with the planned requirements of central policy regarding agriculture as well as assess the results of conducted experiments. The above-mentioned document hints at severe deficiencies in the work of the institute because its personnel were unsuitable for the designated tasks, not only due to its size but also because the professional experience of the employees did not meet the set requirements, a fact that had become evident during experiments with grass-crop rotation.

The Crop Production Center was established in the first half of 1950 and was in charge of organizing work involving contractually produced industrial crops (such as cotton).[90] The center was directly under the supervision of the minister of agriculture, and operated a regional body in each county. The center, itself a national company, basically contracted individual farmers, producers' co-operatives, and state farms to grow plants as instructed by national policy.

Another institution worthy of mention is the Science Council established within the Ministry of Agriculture—in accordance with the Soviet model—in charge of the operative management of research. Its establishment was deemed necessary in order to coordinate the extensive and unmanageable organization system, thereby connecting the

Hungarian Academy of Sciences, the Agricultural Experiment Center, and the Ministry of Agriculture.[91]

By fall of 1950, the scientific and experimental institution system intended to implement agricultural innovations in accordance with the Soviet model had been established and become operational. However, the institutions thus far mentioned are not nearly a comprehensive list of all research and experimental sites operating in Hungary, since the all-pervasive bureaucracy had been branching out in this area as well and disputes regarding the scope of these various organizations had still not been settled by the end of the examined period. However, we may safely say that the Hungarian Academy of Sciences, once considered the citadel of scientific research, had relatively little influence on the practical implementation of experiments aimed at transforming nature. It had more of an ideological role, since work conducted on the experimental farms was mostly supervised by leading functionaries of the party through various institutes.

Hungarian Experiments to Transform Nature

The Political Decision

The Stalinist transformation of nature soon affected the rhetoric of Hungarian Workers' Party leaders, a rhetoric that reflected the resolution to adhere to the Soviet model. The topic was first touched upon—very carefully—by Mátyás Rákosi in his pre-parliamentary election campaign speech in Celldömölk on 8 May 1949, saying that "in most producing groups, Socialist peasants are all pondering how they could improve production, quantity and quality; therefore, they are experimenting with new crops and new methods of production." Ernő Gerő put this in more concrete terms at the discussion of the first five-year plan in December 1949 by saying: "We must introduce new crops such as cotton, which we intend to plant on 57,500 hectares in the final year of the planned period; furthermore, we must experiment with the production of rivet wheat, kenaf, tea, citruses and the rubber-yielding kok-saghyz [sic]." However, this was not merely about the introduction of new crops; the approved five-year plan emphasized that "by 1954, agricultural production must reach 142.2 percent of production rates in 1949, crop production must yield 135 percent, and stock breeding must reach a growth rate of 151.1 percent."[92] In other words, a drastic improvement in quantity was also required in accordance with the

Soviet model; however, the act only mentions new crops in a single sentence, merely saying:

> Experiments are to be conducted on producing new crops hitherto unnative to Hungary within the planned period. Cotton production is the highest priority; therefore, efforts must be made to produce Hungarian cotton in increasingly greater quantities to meet the raw material demands of the Hungarian textile industry. In order to do so, 57,500 hectares of cotton are to be planted by the final year of the planned period—for 1954.[93]

The May 1951 modification of the act reinforced these policies by saying:

> Experiments are to be conducted on introducing and producing new crops hitherto unnative to Hungary within the planned period.... In order to meet the raw material demands of the textile industry in increasingly greater quantities, land use for cotton production must be increased from 345 hectares in 1949 and originally planned 57,500 hectares by the end of 1954 to a total of 115,000 hectares by the end of 1954.... Furthermore, the production of kok-saghyz, kenaf as raw material for the textile industry, and peanuts as raw material for the oil and sweets industries shall be implemented.[94]

However, the Secretariat of the Hungarian Workers' Party had passed a regulation on the principles of agricultural research long before the modified act issued on 23 August 1950. Its appendix included a list of each new crop to be introduced to Hungary,[95] and agricultural experts and scholars were required by law to adhere to its regulations. The introduction of the regulation labeled the majority of agricultural researchers as reactionaries[96] who were "disciples of the Mendel and Morgan school of biology, who formally accept the materialist approach but in reality are following an idealist approach."[97] It also stated that the majority of researchers have unfavorable connections to the Soviet Union, oppose Soviet science and look down on its results, and also that experimental institutes are unwilling "to learn about or implement the scientific results of the Soviet Union." Scholars were also accused of detaching agricultural research from practice, that their research departed from the aim of solving the "most crucial" problems of the national economy or worse, and that they concealed their results in order to prevent their implementations in the economy. Failure was blamed on "damage caused by the enemy," faulty management and the lack of politically trustworthy "research cadres." It then listed seven items that were to be implemented in order to solve these deficiencies. First, ensuring the supremacy of materialist biology and instructing the Ministry of Agriculture "to organize the widespread public dissemination of research results produced by the application of Michurinist biological theory and practice." The second item emphasized the need for systematic and production-oriented research. The third item regarded the popularization of Soviet methods, while the fourth item highlighted the need to educate researchers on the agricultural results of the Soviet Union. The rest of the items concerned reducing the administrative burdens of researchers and ensuring the education of new cadres. However, the "Information Appendix" attached to the regulation proposal at the party secretariat is even more telling than the proposal itself. It provided guidelines for agricultural scientific research by field for "the upcoming years," including crop production, crop improvement, crop health, animal health, agricultural machine testing, forestry, horticulture, and factory organization. For the current topic, the most important research provisions regulated the production of rice, cotton, and the *Taraxacum kok-saghyz*, the introduction of citruses,[98] vernalization, and the implementation of irrigation systems.

Crazy about Cotton

According to the 1949 communication of the Hungarian News Agency [Magyar Távirati Iroda, MTI]: "With the exception of saline soil and low, cold, and moist soils, cotton can be grown all year round and anywhere in Hungary, and harvested three times per year."[99] As we shall see later, this propaganda statement was practically true due to the fact that cotton required seven to eight months of warm, even weather, and therefore grew unevenly. Harvests lasted for months, causing severe problems to farms contracted to produce cotton.

Attempts to introduce cotton to Hungary were not completely unprecedented. Hungary had experimented with cotton production as early as the eighteenth and nineteenth centuries, but did so in the southern regions—such as the Temes-Bánát area—that had been separated from Hungary after World War I.[100] Between the two world wars, cotton was one of the more valuable imported products in Hungary; therefore, experimentation began to produce a species with a reduced growing season.[101] At the time, the attempts were conducted reasonably since researchers were well aware that "cotton is the child of sunshine and Hungarian weather is only favorable for its growth three years out of ten."[102] After 1945, voluntarism completely disregarded the natural environment and began large-scale production and experiments simultaneously, without waiting for experiment results.

In 1948, cotton production began on a sparse 22 hectares in the warmest counties of Hungary, and primarily in the counties of Békés and Szolnok.[103] The scientific community did not share the blind optimism of politics and lobbied for moderation with decision makers; even veterans like Ernő Obermayer, spearhead of rice improvement experiments, had warned about the foreseeable failure, saying that "Hungary is further to the north than any other cotton-producing lands in the world; therefore, it would require great efforts to realize large-scale successful public production of Hungarian-grown cotton."[104] He also warned in his paper of how uneconomical the experiments would be; however, the Communist leadership would not bow to either the "wild forces of nature" or to common sense. In fact, the short-lived career of cotton had become the most memorable and most expensive mistake of the completely irrational agricultural craze of the Rákosi era.

At the end of 1948, the National Cotton Production Company was established in order to implement large-scale cotton production.[105] Its tasks included the introduction of cotton as well as other special crops to Hungary, and it was required to plant cotton over approximately 362 hectares of land during 1949, while the five-year plan had envisioned cotton production over 57,500 hectares of land by the end of the planned period. The headquarters of the company was located in Budapest, with eight offices, two factory farms, two plant improvement sites, and four experimental sites under its supervision.[106] Consequently, the directors of the National Cotton Production Company looked upon crop cultivation as their own special privilege and refused to share "the glory" of it on account of its political importance. The director of the company was agriculturalist Ferenc Schüller,[107] who had done everything in his power to bypass the officials of the Ministry of Agriculture as well as the agricultural department of the party, preferring to maintain contact with the National Planning Office on agricultural matters, and the Agricultural Science Center and its successor on scientific matters.[108] He even established a crop improvement department within the company, despite the fact that the cotton production company was not in charge of scientific research.[109] The party, of course, did not approve of the company's independence and decided that its operation would have to be re-evaluated after "the campaign," and that political influence would be re-established through worker and peasant cadres.[110]

The importance of cotton production increased with the "escalation of the international situation" and the desire to become completely independent of the capitalist world; therefore, the Soviet Union required not just Hungary but every other member state of the COMECON as well to produce cotton.

> The office of the COMECON has devised a draft that requires members of the COMECON to become mostly independent in cotton production and completely independent of the capitalist countries in hemp and linen production. To this end, it prescribes the large-scale increase of production of the above mentioned plants. The COMECON wishes to improve production by increasing the area of land used for production, by various ways of agricultural technological regulation (such as irrigation, increased fertilization, and seed improvement), by increasing maximum production rate with these technologies and, finally, by providing producers with various benefits.[111]

Hungary's assigned tasks and requirements were tailored to fit the implemented five-year plan, and stipulated that Hungary should produce cotton on 60,000 hectares of land by 1953, and implement an irrigation system over 10,000 hectares.[112] Although these numbers did follow the envisioned Hungarian figures, they also surpassed previous expectations; therefore, it was necessary to increase land used for cotton production as early as 1950.

The prioritization of cotton was the reason behind the arrival of Soviet advisor A. Skoblykov to Hungary during 1950 to inspect scientific and production results and share his experiences to assist cotton production. He deemed the party resolution of 14 June 1950 unacceptable due to its provision that cotton experiments were to be conducted in Szeged's future Southern Hungarian Research Institute.[113] On the one hand, the Szeged institute had no cotton fields, and so transferring experiments to Szeged from already operational research sites would have been unreasonable; on the other hand, he was of the opinion that the current organizational structure did not provide cotton research with enough independence considering its importance. Based on these observations, he suggested the establishment of a separate research institute in Székkutas, home to one of the largest experimental cotton fields of the country.[114] With the establishment of the Cotton Production Research Institute operating under the Agricultural Research Center, the monopoly of the National Cotton Production Company ended. The new institute not only incorporated the crop improvement department of the company, but the funds required to establish the institute had been taken directly from the national company budget.[115] Miklós Derera, former director of the experimental site in Szentes, became director of the Székkutas institute, and the scope of the National Cotton Production Company was reduced to actual production, the organization of field work and contracting producers.[116]

During 1948–1949, Hungary only experimented with small-scale cotton production. The actual "cotton skirmish" began in 1950 when Hungary began planting the unnative crop despite every alarming figure, and no longer did so primarily on the fields of individual volunteering farmers, but in state farms and producers' cooperatives on a larger scale. In the spring of 1950, 6,200 hectares of land had been planted with cotton despite the fact that 1949 cotton yields had proven that Hungary could only produce cotton with great expenses, hard work, and a very small average crop yield. While the price of 1 centner of ginned cotton was 545.5 HUF in 1950, production costs of one yoke[117] of cotton were as high as 1,400 HUF.[118] In 1949, the almost mature crops looked like they would yield 1.5 centners of cotton per yoke; however, frost in October and rain in November reduced the national average crop yield to a mere 25 percent.[119] In other words, there was a crop yield of only 37.5 kilograms per yoke, which means that the cost of 1 centner of cotton was 3,730 HUF.

Although cotton production figures were clearly discouraging, building socialism knew no bounds. The designated fields were all planted with cotton despite the fact that production conditions were completely lacking,[120] including soil preparation equipment, warehouses to store the harvested crops, and drying equipment; in addition, there were not enough sunflower planting seeds for crop protection, and they had financial difficulties purchasing pesticides and machinery.[121] Consequently, some state farms and producers' cooperatives attempted to avoid these requirements; for example, Szolnok County received orders to produce cotton on 230 hectares of land, but cotton was only planted on a bare 132 hectares, prompting the National Cotton Production Company to recruit individual farmers. Their offer was attractive enough to soon allow them to increase the amount of land designated for cotton production.[122] The farmers were not, of course, informed of the requirements of the foreign crop; they practically had no idea of what they signed up for other than what they had heard from educators visiting the houses of eligible farmers in the target areas, teaching them about the importance of producing cotton and the invincibility of socialism. People thus signed contracts, partly because they had very little choice but also to ensure that they would be compensated in case of an unsuccessful attempt to produce cotton—something that naturally did not happen. A good example is the case of the farmers from Szelevény who complained that the planted cotton did not grow due to cold weather, but on account of administrative difficulties they only received permission to plough out the cotton at the end of May, by which time it was already too late to plant and produce corn.[123] The government, too, forgot to compensate them.

Cotton proved to be a difficult crop that required manual cultivation, but the local labor force was too small in most farms, forcing them to recruit seasonal workers.[124] However, providing these workers with accommodation and supplies, and establishing their working conditions, had been poorly done as shown by the following brief excerpt on the conditions on state farms:

> The plantations were full of weeds, and the head of the local office of the National Company observed during his inspection that some eighty workers were pulling out weeds with their bare hands, their nails torn off or damaged so severely from the thorns of the Tribulus terrestris plant that they were unable to continue their work. Furthermore, the doing of work [sic] was so minor that the cost of unfinished required labor was multiple times greater than the price of one standard scraping machine. Farm working conditions are criminal. There are 430 seasonal workers but only 100 beds were commissioned and no beds have arrived yet; instead, the 430 workers have received 60 blankets. Workers sleep on the floor and have become completely infected with lice. The farm does not have enough transportation power to deliver lunch to the workers, which leads to the workers abandoning work at 3 p.m., even during inspection, due to lack of food.[125]

Cotton production faced a lot of difficulties, including administrative problems. Since cotton required warmth and sunshine, the most eligible areas were the southern regions of Hungary that were in the immediate vicinity of Yugoslavia, one of the greatest enemies of Hungary at the time.[126] It has been discussed in a previous chapter that the heavy industrial investments planned for the southern region had to be transferred further north on account of the unfavorable Tito government; nevertheless, cotton refused to adapt to these political considerations and the southern regions remained the most favorable ones for cotton production. Since it was a top priority crop, it required near-constant "supervision"; however, these borderline fields were not open to just anybody—civilians were forbidden to trespass and border guards were ordered to shoot trespassers on sight.[127] From 1 July 1950, one required a special license—a kind of internal passport—to come within fifteen kilometers of the border, and the area within two kilometers of the border was inaccessible without a permit from the Border Guard under the State Security Authority.[128] These permits were "one use only" and had to be requested for each visit, making the work of cotton producers and workers especially difficult; therefore, the National Cotton Production Company sent a request the minister of agriculture asking for unlimited access to these cotton fields for the "agents" of the company.[129]

Similarly, in order to gain access to weather forecasts, another ministry permit was required; therefore, the cotton production company wrote a letter to the minister of defense requesting that they receive all relevant climate data from the Meteorology and Weather Research Institute.[130] We have no information of the ministry's reply, but the company soon issued another letter explaining that the data was required by Comrade Skoblykov,[131] that knowing the weather forecast would obviously be very helpful to agricultural workers, especially in the case of such "high priority" industrial crops as cotton, and mentioning the name of the Soviet advisor and their hope of success in the struggle for cotton. They eventually convinced the minister, who agreed to the company receiving the classified weather forecast biweekly.

It is probable that harvest reports went through embellishing "touch-ups" since 19 percent of all cotton fields had completely fallen out of production due to bad weather, and yet the national estimate for the cotton harvest was 3.2 centners per yoke[132] and subsequent reports talked about "improved prospects for production."[133] Also, were we to accept the validity of official reports, we would have difficulties understanding the resistance of state farms and producers' cooperatives when faced with the 1951 plans for cotton plantation. Contracts were pushed by agitators across the country, but there was hardly any increase in the amount of contracted land. Farmers talked of a lack of labor force and the unsuitability of their lands, and the chairman of the Iváncsa council outright took the unopened document containing these planned figures, crumpled it up and shoved it in his desk, swearing as he left the company's agent behind.[134] However, figures had to be met, at least on paper, and the number of cotton plantations continued to grow compared to the previous year. However, before cotton was to be planted, several farms attempted to make unilateral modifications to their forced contracts; for example, in Gyula, the Equality and People's Republic producers' cooperative significantly reduced the area of land designated for cotton in their spring plantation plans, causing the Ministry of Agriculture to send instructions to the executive committee of the Békés County Council to investigate why the Gyula City Council had allowed the reduction.[135] The members of the county council were likely agricultural people who had assessed the situation in a rational manner because they completely supported their urban colleagues, saying "it would have been wrong to force a contract that the cooperative would have been unable to fulfill. We believe that the Agricultural Department had made the right decision."[136]

Their case—the Békés County Council supporting farmers—had set a precedent, and in the following few weeks several counties must have shown resistance, since a ministry report in April made the following statement: "Due to deficiencies, the agricultural departments of several county councils have reduced the area of land for cotton production that was contractually agreed upon."[137] The report adds that county councils have no authority to do so, and should only be able to do so with permission from the cotton department of the Ministry of Agriculture. Csongrád faced a similar situation in which the ministry was forced to reduce the area of land designated for cotton and attempted to defend its decision to the party by saying that when the executive committee of the county council were interrogated about the overdue plantation of cotton, it gave instructions for ploughing fields that had already been planted with corn. This was not allowed by the ministry; therefore, fields that had been already planted with something else were excluded from obligatory cotton production.

Regardless of resistance, the central development policy continued to call for increasing the size of cotton plantations. In the spring of 1951, a regulation was issued that surpassed every expectation imaginable, especially considering the situation discussed above, by including cotton production in the agricultural plans of four northern counties—Komárom, Nógrád, Borsod-Abaúj, and Szabolcs-Szatmár.[138] For example, producers in Nógrád were instructed to produce cotton on 6 hectares of land,[139] and since the mountain climate of these northern territories was unfavorable for the tropical crop, there was nothing to harvest in the end.[140] Nevertheless, they were still forced to produce cotton in 1952, and though they attempted to gain the ministry's support, the reply was not very promising: "We are processing your request and will . . . notify you."[141]

The year 1951 did not bring any substantial changes to the way cotton was cultivated; the labor force was still insufficient, cotton fields were weed-infested, and farms prioritized the harvest of grains over de-weeding cotton plantations.[142] They did recruit seasonal workers again, but they also refused to work in poor working conditions. Work hours were raised to ten hours per day, and schoolchildren were taken on field trips to de-weed cotton, but results did not improve. In the fall, however, harvesting cotton became even more problematic; the Hungarian climate caused cotton to mature slowly, and plants had to be checked every two or three days to pick the freshly matured cotton bolls, which meant that harvesting had to be done manually as using machinery was out of the question. Farms attempted to increase the labor force by making Sunday an obligatory working day; however, in villages where church law was still respected, this measure was met with fierce resistance. Participation in the cotton harvest was also made obligatory from the upper grades of elementary school to higher education,[143] and considering that elementary school children had to bend down constantly while picking cotton, we can only imagine how difficult it must have been for an adult to engage in such activity for ten hours every day. Not to mention that the picked bolls had to be examined and sorted into the different pockets of a cotton-picking apron according to its quality. There was still no means of transportation that would have supplied workers with regular meals and hot beverages, despite the fact that harvests lasted until frosty November, and there were still no warehouses to store cotton, let alone dry it.

The 31 October 1951 report of the Association of Working Youth[144] [Dolgozó Ifjúsági Szövetség, DISZ] on the lack of provisions for children sent to the cotton plantations by the Labor Force Reserve Office[145] [Munkaerőtartalékok Hivatala, MTH] is truly appalling.[146] The Association of Working Youth was instructed to mobilize 5,000 students to harvest cotton; however, the association refused and decided to send 4,400 children from the Labor Force Reserve Office instead in order to harvest cotton for four days. The children were transported to the designated farms by special trains, but their transportation was already on a par with deportation conditions. Provisions arrived at the station late so dinner had to be thrown into the trains through the windows, and when the children arrived, their hostel was not heated. Those more fortunate had straw mattresses to sleep on, but most children had to sleep on a layer of hay on the floor. Luckily, they had been required to bring blankets, which helped to keep them from the late fall cold. Catering was on a par with internment camps: only a small number of children received hot meals, and most were only given bread to eat. Lunch was delivered to the fields, and amounted to no more than a ladleful per child. They had no cleaning facilities or even fresh water to drink, which one farm attempted to solve by offering one hundred liters of wine, leading to the intoxication of the majority of children. By the third day, most problems were more or less taken care of and provisions improved, but these working children received very little money, with outrageously high amounts deducted for their alimentation. However, the most shocking fact was that they were working a mere fifty meters away from a minefield by the southern border—and none of them were warned about the danger.

In 1952, social resistance to cotton production was growing, but so was the size of cotton plantations, prompting Comrade Skoblykov to write a letter to Ottó Sándor, voicing his discontent that farms planted other crops instead of cotton, despite their signed contracts.[147] Hungarian peasantry was quite resourceful, as illustrated by the story of Sándor Bozsó, a farmer in Kőtelek who had decided to establish a producers' cooperative in his village in order to save his father-in-law from the disadvantages of being labeled a kulak. The first task of the cooperative was to plant 17 hectares of cotton, but few of them considered it a reasonable task: "Are they out of their minds? This isn't the homeland of cotton and who would tend to it? Do I have to hire workers? But it was no good going to the district so I just traded the cotton with the cooperative at Tiszaföldvár for 9 hectares worth of lettuce seeds. The women said this was better because it won't sprout, but at least the ducks are able to eat the lettuce."[148] However, resourcefulness did not end there; the lettuce was insured and then planted right next to the flooding Tisza River, causing the flood to ruin the crops and the insurance company to compensate them. In the words of the head of the Kőtelek producers' cooperative, "You have to get on by sashaying around all these stupid regulations."[149]

The case described above was an isolated one; most of the nation did adhere to the designated figures of the five-year plan. Cotton had already claimed most of the fertile land from much more valuable crops, causing severe shortages in agricultural production; however, in 1952, irrigated cotton production was introduced as well, meaning that the majority of irrigated fields (and there were not many in Hungary to begin with) had to be planted with cotton.[150] In some parts of the country, cotton had become the most widespread crop in 1952, occupying 5.2 percent of all farmland in the Mohács district and 7 percent of the Villány wine lands.[151] It is important to add that the amount of cultivated land continued to decrease during the Rákosi era while the amount of land used for industrial crops had quintupled by 1953. As a result, the amount of land used for wheat, which was indispensable in sustaining the population, had decreased by 21 percent compared to the pre-war period.[152]

The year 1953 became a turning point in the history of cotton; the seeds had been planted, but due to the cold spring weather, 70 percent of these fields had to be ploughed out[153] and the rationalization of agricultural production eventually shifted the focus away from cotton production. The new government did not force harvests on the remaining cotton fields, and although farms involved in production requested more workers, the minister of education did not allow them to transfer students to the cotton fields.[154]

The Taraxacum kok-saghyz or Russian Dandelion in Hungary

Yellow dandelions—commonly called "dog milk" in Hungary—are native to Hungary and distant cousins of the *Taraxacum kok-saghyz*, which is referred to as the "rubber dandelion" on account of its similar looks,[155] while the rest of the world calls it Russian dandelion or rubber root. The fact that the root of this plant native to Uzbekistan and Kazakhstan yields good quality rubber was discovered in 1932 in Kazakhstan, which at that time belonged to the Soviet Union. The Russian dandelion, then, was rediscovered as a possible raw material in the search for national resources for rubber production in the Soviet Union.[156] Following the discovery, the large-scale production of Russian dandelions began in Uzbekistan and Kazakhstan, and continued until the mid-1950s, serving as the primary material for the production of rubber tyres.[157] However, the *Taraxacum kok-saghyz* was also popular elsewhere; when Japan cut off the USA's rubber supplies during World War II, the United States also began to experiment with the plant they called "Russian dandelions" until synthetic rubber completely replaced it, and its production ceased there several decades ago.[158]

The *Taraxacum kok-saghyz* had not been unknown in Hungary; in fact, it had been cultivated on the estate of Albrech, archduke of the Habsburgs, near Mosonmagyaróvár during World War II when it was produced on some 95 hectares of land. The archduke even established a raw rubber processing factory on his estate in Öntéspuszta (under the management of Barna Sólyom, director of the Hungária Rubber Factory), and the processed raw rubber was transported weekly to the Hungária Rubber Factory in Budapest to produce rubber tires for Archduke Albrecht's vehicles.

According to legend, the *Taraxacum kok-saghyz* was introduced to the archduke's farm by a Russian agronomist and plant improver called Strasovsky, serving in the battalion of General Andrey Andreyevich Vlasov who joined the Germans during the war. Strasovsky deserted and came to the estate in 1942,[159] where he was advised by a Ukrainian professor called Tatochenko on his occasional visits to the estate.[160]

Strasovsky's work with the *Taraxacum kok-saghyz* was assisted by László Kovátsits,[161] a researcher at the National Plant Improvement Insitute in Mosonmagyaróvár. The contemporary government had gleaned the opportunities inherent in producing *Taraxacum kok-saghyz* and appointed Kovátsits the national ministry commissary for rubber plant production under the supervision of the Ministry of Agriculture. Kovátsits became so involved in producing Russian dandelions that he even published a book in 1944 titled *A hamvas gumipitypang termesztése és nemesítése* [The Production and Improvement of the *Taraxacum Kok-saghyz*] (Baross Nyomda, Mosonmagyaróvár, 1944, 79 p.). Processing home-produced rubber had become an industrial branch of strategic importance, leading to the appointment of engineering university professor Barna Sólyom as government commissary to spearhead production in 1944.[162] When the production of *Taraxacum kok-saghyz* surfaced in mainstream politics in 1949, they obviously contacted the professionals who had participated in its production at the former estate of the archduke, which had been nationalized in 1945.[163] Consequently, one of the centers became the experimental farm of the Agricultural Research Institute in Mosonmagyaróvár, another center became the Rubber Plant Research Site of the Mosonmagyaróvár division of the National Cotton Production Company in Püski, a settlement on the banks of the Danube, and the third center became the Experimental Farm of Iregszemcse in Tolna County. Experiments with the *Taraxacum kok-saghyz* were conducted in Mosonmagyaróvár and Püski by deputy head of department László Kovátsits and in Iregszemcse by Dr. Ernő Kurnik, the director of the site.[164] At the beginning, these sites were managed by the National Cotton Production Company until the National Rubber Plant Production Company was established in 1951 in order to manage production.

During 1949–1950 the main goal was crop improvement, and taking pre-1945 experiences into account, producers primarily sought to improve propagation methods and establish suitable conditions for large-scale production in Hungary's climate. Planting seeds, of course, had to be procured from the Soviet Union.[165]

The *Taraxacum kok-saghyz* first emerged in the news of the Hungarian News Agency on 20 April 1949 and was described with contemporary names and terminology. As can be seen below, the role of the Soviet Union, the newest Soviet agrobiological methods, and the interests of the national economy were also mentioned side by side:

> The Soviet Union has been producing kok-saghyz on several thousands of yokes,[166] which supplies a significant amount of rubber to meet the Soviet Union's demands. We have received a large quantity of kok-saghyz planting seeds from the Soviet Union . . . If it is enough, then we shall soon have Hungarian-grown kok-saghyz seeds at our disposal while our scientists are hard at work using Russion agrobiological methods to transform the kok-saghyz and allow it to grow in Hungarian soil in our climate.
>
> If the experiments succeed, kok-saghyz shall be grown by producers' cooperatives and state farms on large masses of land. Kok-saghyz production is of great national economic interest as Hungary has to rely on a significant amount of imported rubber at the moment.[167]

On 12 December 1949, the Hungarian News Agency announced that, at the start of 1949, "Hungary had begun to produce the rubber dandelion based on Soviet experience." It informed the nation that experimentation and improvement had begun simultaneously:

> The best-looking specimen with the highest concentrations of rubber had been selected and propagated in greenhouses. The success of these experiments was ensured by applying Soviet knowledge and experience. During the implementation of the five-year plan, the kok-saghyz, the gift of the Soviet Union, had become a very valuable plant to our developing Socialist industry and agriculture.[168]

The roaring optimism of the Hungarian News Agency was not the reality. Experts at the experimental sites were constantly reporting their difficulties to the Agricultural and Cooperative Department of the Hungarian Workers' Party, to the Agricultural Experiment Center of the Ministry of Agriculture, and to the Cotton and Special Crop Production Department of the Ministry of Agriculture. A confidential report informs that in 1950 the experimental site in Püski produced a rather large amount of *kok-saghyz*, so the Ministry of Agriculture instructed them to pick all of it and transport it to the Organic Chemical Research Institute in Budapest for processing. However, the institute did not have the capacity to process all of it, so the National Cotton Production Company, following the orders of the Ministry of Agriculture, instructed the institute to section the leftover roots for propagation[169]—in the middle of December. While this propagation method was familiar to the experimental site and had been successful before when using premium specimen cut into four pieces for propagation, they now had to use this method on wilted roots that had been stored for too long—inevitably resulting in failure.[170]

The decision to organize the large-scale production of the *Taraxacum kok-saghyz* was made in the summer of 1950. According to the proposal:

> One of the key tasks of Hungarian agriculture in the course of the five-year plan is to supply our growing industry with raw materials; therefore we must not only improve the produced quantity and quality of already introduced industrial crops, but endeavor to produce all currently imported raw materials in Hungary. As in every other field and area, the assistance of Soviet agriculture is indispensable in achieving this goal, and one of the most significant issues was to establish rubber production by applying Soviet knowledge and experience.[171]

Of course, if we consider how the various official bodies regarded experiments to "solve the issue of Hungarian rubber production," we strongly doubt that they had taken it seriously when we already know what had happened with the decent amounts of *kok-saghyz* produced at the Püski experimental site in the fall of 1950. It seems as though these official bodies only followed instructions out of necessity but had no faith in any of it, and had been caught completely off guard when these institutes actually managed to produce results.

Our doubts about the seriousness of *kok-saghyz* production are further justified by the fact that a decision was made in the summer of 1950 to implement the large-scale production of the *Taraxacum kok-saghyz* and 287.5 hectares of land were to be planted in 1950/51, yet there was still no planting seed available by the end of October, which meant that early-winter planting was cancelled. When the deputy head of the Cotton and Special Crop Production Department of the Ministry of the Agriculture submitted a complaint, the referents of the Ministry of Foreign Trade replied that "the case is being evaluated by the highest Soviet authorities."[172]

One year passed and the situation remained unchanged. By that time, the production of the *Taraxacum kok-saghyz* was managed by the National Rubber Crop Production Company established in 1951, their work assisted by a Soviet advisor by the name of Aledikin at the Ministry of Agriculture. The director of the company, Károly Sarlai, described the operation of the company as follows:

> We had received planting seeds from the Soviet Union and had to plant them one centimeter deep. Its outer kitin shell would only pop from moisture and had to be stored at 0°C for ten to fifteen days before planting. During the winter, we dug pits to store snow and ice, but the cargo did not arrive. When the ice melted, we decided to put the seeds in cold storage, but by the time it arrived on 25 May, we had no time to do that. The supply carriage was attached to the express at Záhony, and from there we took it to the farms by taxi.[173]

Sallai had no bad experiences with the Soviet advisor—according to him, all he ever said in a cautious tone was, "This is how we do it, but you should do it the way it's good for you."[174]

The *Taraxacum kok-saghyz* was labor force intensive and required a lot of preparation. It had to be hoed by hand *and* machinery at least three or four times, and seeds, for the most part, had to be collected manually; therefore, once the seeds matured, workers were forced to inspect the fields three times a day. Also, seeds could only be separated from their shells by rubbing them manually on sifters and then using industrial fans multiple times; extracting their long roots was also to be done manually since these roots clung to the soil much like the dandelions infesting Hungarian fields.

In his confidential report of 30 July 1951, Ferenc Balázs, director of the Mosonmagyaróvár research institute, complained:

The fact that the majority of the contracted workers left during harvest season and the rest left because they had been summoned by the military has made the work of the Püski experimental site very difficult. Although the Püski experimental site and the institute did everything in their power to bring the labor force back to the site, it did not receive enough support from local authorities. As a result, it could not meet propagation plans 100 percent, nor was it able to finish weeding in time and had to solve this problem by focusing its remaining work force on managing the most pressing agricultural technological experiments, letting the scarcely growing propagation experiment become weed infested since weeds posed no obstacle in harvesting the seeds.[175]

When the large-scale production of the *Taraxacum kok-saghyz* began in 1951, the National Rubber Plant Production Company had contracted farmers to produce *kok-saghyz* on a total of 615 hectares of farmland in 31 state farms and 152 producers' cooperatives. However, the problems at the experimental sites also surfaced here, and by the time seeds were to be planted, farmers had already been lagging behind, reasoning that certain producers' cooperatives and groups were not politically mature enough, and that they were failing to view the production of the *kok-saghyz* "as their primary objective." The other perceived deficiency was failure to use the proper agricultural technologies and follow instructions. Experts of the National Rubber Plant Production Company inspected producing units every four to five days, and while they were on site, work was going well, but the moment they left, "everything was neglected and 4–5 days of negligence was enough to cause a portion of young and superficially sprouting plants to die."[176] Inefficient production was (also) attributed to the fact that the chosen farmlands were unsuitable for large-scale production. The plant preferred marsh soil; however, there are no marshlands available at state farms and producers' cooperatives since they had either got rid of these marshlands and wetlands through farm reallocation or had traded them with private farmers for better farmland.[177]

Inspectors of the National Rubber Plant Production Company were also limited in terms of mobility because the company had not been appropriately supplied with vehicles upon its establishment. It only received transportation equipment in September 1951 in the form of sixteen motorcycles and the promise of four cars; however, they were not allocated the necessary fuel to operate these vehicles, and when the managers complained, they were told to request a taxi budget.[178]

We have no data regarding the amount of *Taraxacum kok-saghyz* produced. During experimental production in 1949, the plant yielded 22.5–25.5 centners of root per yoke, which was later to be tripled. The rubber content of roots varied between 0.2 and 1.82 percent. Subsequent production reports, however, did not contain annual totals, which leads us to the assumption that the large-scale production of the *kok-saghyz* had never been regarded as a Hungarian agricultural success story.[179] This is confirmed by the fact that in the spring of 1953, the *kok-saghyz* had completely disappeared from agricultural planning, and the Political Committee of the Hungarian Workers' Party, at its 19 March session in 1953, did not even mention the *kok-saghyz* in its plant production development proposal. The National Rubber Plant Production Company was soon abolished in 1953.[180]

Experiments with New Industrial Fiber Crops

By decision of the Hungarian Workers' Party, cotton and the *Taraxacum kok-saghyz* were joined by plants that Hungarian farmers had never even heard of, including the production of *kenáf* (kenaf) and *rámi* (ramie).

Just like cotton, kenaf (*Hybiscus cannabinus*), or "fibrous malvacea" to Hungarians, is a fiber crop that was produced to replace jute as a raw material for the textile industry. Belonging to the malvaceae family, this tropical and subtropical plant grows uncultivated in Africa but had been produced in India since antiquity. The plant first surfaced in Europe—in England—in the form of ropes and sacks made from kenaf, and during World War II when there was not enough jute available, kenaf suddenly became more valuable. Following the war, production experiments began across the world from the Soviet Union to the United States.

In 1949, Hungary commissioned the Iregszemcse Experimental Farm and the Szeged Agricultural Research Institute to research the conditions for introducing kenaf to Hungarian soil and to begin crop improvement as well. Iregszemcse promptly started work, and produced kenaf on 11.5 hectares of farmland in 1950 and 50 hectares in 1951.[181] In its native land, kenaf grows up to 5.0–5.5 meters, while in Hungary, it only grew to 2.0–2.5 meters and produced 15–20 centners of dry stems—the source of fiber—per yoke, and yielded 150–200 kilograms of seeds for propagation the following year. When the fiber content of the Hungarian-grown dry stems was examined in the Textile Industrial Central Research Laboratory, fiber content was found to be 10–14 percent. They also confirmed that it replaces jute perfectly and approaches the fiber quality of hemp, and that its rough fiber is suitable for producing string, ropes, sack cloth, tapestries, and oakum. It could also be dyed.[182] Experiments with kenaf at the Iregszemcse Experimental Farm were supervised by Dr. Ernő Kurnik, who ascertained that both dry stem and seed production were well below "data in international literature" and yielded far less fiber than species grown in the Soviet Union—therefore he deemed it necessary to acquire all species of kenaf from the Soviet Union.[183] They also increased the amount of farm land designated for experiments in 1952, and planted kenaf on 287.5 hectares;[184] however, the large-scale production of kenaf could not begin due to the fact that the weather was not warm enough after the harvest for the plant to dry up properly and thus yield fiber. By 1953, kenaf disappeared from production plans altogether.

Ramie (*Bochmoria nivea*) or "Chinese snow thistle" to the Hungarians, is native to China. In Hungary, the center of experiments with this plant was the Iregszemcse Experimental Farm, although it was also produced by the Cotton Research Institute at its own farm in Tengelic, both sites supervised by Dr. Ernő Kurnik. Ramie was considered the strongest and finest fiber source out of all fiber crops and was produced on 2 hectares of farmland in 1951 and on 11.5 hectares of farmland in 1952. According to the production experiment reports of Dr. Ernő Kurnik, the planting seeds for the experiment came from Bulgaria and Italy.[185]

However, the production of ramie posed a huge issue to experts. According to the notes of István Bencsik, chief secretary of the Agricultural Research Center, on 1 October 1951, the greatest problem was that they did not know how to extract fiber from the plant. They requested the method from the People's Republic of China but received no reply,[186] and another source hints at the fact that the processing of ramie fiber was a trade secret in China.[187] Perhaps this was the reason why experts classified ramie as weak to frost, and why large-scale production had never begun.[188] In 1953, experiments with ramie stopped altogether.

Experiments to Introduce Citruses to Hungary

The appendix of the regulation issued by the Secretariat of the Hungarian Workers' Party on the state of agricultural research and its tasks states that "by 1957, lemons must be introduced to and produced in Hungary."[189]

Similarly to other non-native crops, experiments involving citruses were managed by the Agricultural Experiment Center of the Ministry of Agriculture. According to the documents, on 31 August 1950, one week after the secretariat session issued the above mentioned regulation, leading bodies of the Hungarian Workers' Party also designated production figures for the introduction of citruses. Experiments involved five research sites, all of them involved in agricultural research (Fertőd, Akali [Balatonakali], Tihany, Keszthely, and Villány), and production was to begin with three thousand saplings. Improvement experiments would be conducted in Fertőd and Akali, with Fertőd as the center. For the year of 1951, Fertőd and Akali were required to produce thirty thousand and ten thousand grafts respectively.[190] The large-scale pilot production center was Keszthely, and Árpád Jeszenszky was entrusted with its direct control. Institutes involved in production were also required to organize citrus meetings quarterly, and Aladár Porpáczy, head researcher of the Fertőd Experimental Farm, was appointed as the national respondent for citruses. This was no coincidence—Fertőd had been experimenting with citruses for well over a year, since lemon trees had been cultivated for several decades in various national arboretums and botanical gardens. When the Fertőd Experimental Farm became the center of citrus improvement experiments, the Agricultural Experiment Center of the Ministry of Agriculture procured 120 three-year-old citrus saplings from the Szarvas Arboretum in the winter of 1948/49. During 1949, seedlings were primarily grown from lemon seeds gathered from national hospitals.

However, despite the above-discussed decision of the party, large-scale lemon planting and grafting began very slowly even though Fertőd had proceeded with scientific experiments at a steady pace. By 5 October 1951, the necessary "planting materials" were ordered, namely thirty thousand uncultivated specimen and fifteen kilograms of lemon seeds from Italy, as well as three thousand grafts from the Soviet Union. By 15 October, the Ministry of Agriculture and the settlements involved had prepared the budgets of these experiments, and by 25 October it was decided that a six-week preparation course would be launched in December, so they assembled the course material and chose ten young experts to be enrolled on the course. They noted that the introduction of lemons and related experiments would cost 1.4 million HUF in 1951 alone; however, László Pálos, referent of the National Planning

Office [Országos Tervhivatal, OT] was also present at the meeting where work schedules and finances were discussed, and stated that they absolutely could not finance the experiments and that the Agricultural Experimental Center would have to deduct the required amount from other experiment budgets because they may only receive 100,000 HUF of credit in advance from the total allocated credit for 1951 to cover the most urgent expenses. Regardless of the complaints, the planning office did not budge until December, when they agreed to issue the total supplementary estimate, which had been reduced to 917,000 HUF. However, it was discovered in January 1951 that the official documents of the Agricultural Experimental Center's credit request had been lost by the planning office—just another obstacle—and by the time the documents were replaced, the originals had been found as well. The planning office began to process the request, but in the meantime, allotted credit and the authority to process the request had been delegated from one department to another, and the supplementary estimates for experiments with citruses had not been included in the allotted credit. A new supplementary estimate request had to be submitted, and due to regulatory changes, different forms were required and other documents had to be acquired to submit the request. The procedure took ages, but on 9 April 1951, the Investment Bank finally issued the allotted credit, a total of 1,054,000 HUF.

It is probable that it was the March 1951 visit of Soviet professor P.A. Baranov, corresponding member of the Soviet Academy of Sciences,[191] that helped Hungarian citrus production and granted the project its supplementary estimates. Baranov, who of course had no idea of the dead ends of Hungarian bureaucracy, came to advise the managers of citrus farms and to assess their achievements thus far; he considered it a notable result that in the Fertőd farms the foliage of tangerine trees was trimmed diagonally to allow the plant to absorb more sunlight. He also found the Fertőd method of seedling replanting promising, as Aladár Porpáczy had managed to use said method to make the hybrid plant yield fruit within two years instead of ten to twelve years.[192] However, these were only the results of the experiments in 1948–1949. Baranov was a guest of the Hungarian Academy of Sciences, and he not only visited Fertőd but also many universities and a lot of lemon orchards in Hungary, and advised producers how to introduce the new methodology.

The winter months were relatively mild, which would have allowed the farms to begin soil preparation for planting citrus seedlings, had they received their funding in time. However, due to the lack of financial resources, they were behind on building the greenhouse in Akali and could not order climate cabinets. Fortunately, they managed to prepay for seeds and wild specimen from Italy from other funds, which then awaited their fate in Fertőd. They also managed to pay for the one thousand grafts of lemon, tangerine, and orange respectively from the Soviet Union, which arrived in Budapest by air on 11 April 1951 and were transferred to Fertőd by trucks, where the plants were placed in pots and shady greenhouses to take root and get accustomed to outside temperatures. These citruses had primarily been purchased for improvement experiments.

The construction of experimental farms for the production of citruses began in April 1951. Each experimental site had devised a plantation system consisting of sloping trenches some eighty centimeters deep in order to protect the plants from winter frost by covering up the trenches. At the beginning of May, as per the expectations of the era, a portion of the work was done through tenders. At the Fertőd experimental site, the improvement of grafts from wild Italian specimen and further experiments with hybrid seedlings was spearheaded by Aladár Porpáczy.[193]

The irrigation system in Akali was finished in the summer of 1951, while the irrigation systems in Keszthely and Villány and the Akali laboratory and greenhouse were finished in the fall; that same year, 1,100 citrus saplings were planted in Fertőd, 800 in Akali, 440 in Tihany, 600 in Keszthely, and 420 in Villány. Most of these were from the shipment of three thousand saplings from the Soviet Union, while the rest had been grown from grafts in Fertőd. Most of the tangerine saplings were grown in Tihany. Production experiments were soon extended to the Alföld as well, where producers' cooperatives in Túrkeve, Kisújszállás, and Mezőtúr agreed to plant forty saplings in two trenches each.[194] Very few plants died upon plantation—a mere 1 percent in some of the sites. Based on experiments in the previous year, experts were hopeful that the Fertőd experiments of grafting seedlings into wild specimen would allow the foliage of these plants to resist up to minus 6–7°C, while saplings growing on their own roots froze at a mere minus 3–4°C.[195]

From the fall of 1951, the Agricultural Research Center began preparing for 1952 since they were instructed by the party decision to produce seventy-five thousand grafts at the experimental site. Thirty-five thousand of these were to be cultivated in Keszthely at the newly constructed greenhouse complex; however, based on the experiences of

1951, these had to be created from wild specimen due to the fact that, according to one of the reports, the seeds imported in 1951 were unable to germinate properly.[196]

Despite all efforts, the effects of production experiments and mass field plantation were delayed and the Agricultural Experiment Center considered their results doubtful, merely more than attempts. The saplings that spent winter in the greenhouses of Fertőd had proven to be particularly fertile and yielded fruit; however, observations had shown that plants that were freshly planted outside became extremely sensitive to the difference between day and night temperatures (dilatation), and consequently lost their flowers. Experts suggested that it would be particularly desirable to conduct meteorological and microclimatic observations and to take measurements of humidity, temperature, dilatation, and the duration of sunshine; however, there was no suitable equipment to do so.[197]

From the fall of 1951, the education of future experts began. Both the Ministry of Agriculture and the Agricultural and Cooperative Department of the Hungarian Workers' Party supported the idea of teaching the introduction of citruses, and other crops, in agricultural higher education.[198]

In 1952, a decision was made to order a collection of species representing the most cold-resistant citruses from the Soviet Union and to graft seventy-five thousand wild specimen, but there are no sources confirming the arrival of these orders.

Hungarian weather remained the greatest enemy of the outdoor plantation of citruses. The cold winter of 1951/52 had destroyed many freshly planted saplings, and the frosty weather in March ensured a sharp drop in temperature (−7 to −15°C), frost lasting for fifteen to twenty days.[199] The winter of 1952/53 was somewhat milder but still took its toll on the citrus plantations. The fact that citruses could not be domesticated to endure the Hungarian climate eventually had to be acknowledged by the upper management of the party,—and so at the 19 March meeting of the Political Committee of the Hungarian Workers' Party on the development of crop production in 1953, nobody was surprised that citruses had been left off the agenda.[200]

Between 1953 and 1955, the introduction of Mediterranean crops was removed from the agenda of the Fertőd Experimental Farm and the improvement of berry plants had become the top priority,[201] although we can still see traces of these unsuccessful experiments at the sites once used for the introduction of citruses, which now grow only in arboretums or the greenhouses of private homes. The politically charged large-scale experiment, however, had disappeared for good.

Rice Production in the 1950s

The Hungarian history of experiments with rice production dates back several centuries. According to national lore, it began under the Turkish occupation when Turks living in Hungary began to grow rice for their national cuisine—the pilaf—in the southern marshlands of the country.[202] The loss of territory after World War I forced Hungarian economic policy to relocate rice fields further north; however, the production methods and hitherto used species of rice proved to be unfavorable. Therefore, in 1933, the Szeged Pedological and Crop Production Research Station was charged with the task of finding the most adaptable species of rice for these northern areas.[203] This research was conducted by Ernő Obermayer and proved successful, leading to a substantial rise in Hungarian rice production during World War II that lasted well beyond 1945 since the 1949 nature transformation program gave a boost to the large-scale production of this highly nutritious crop. By the end of the three-year plan, rice was being produced over 20,000 hectares of land.[204]

One of the most prominent agricultural investments under the first five-year plan was the construction of waterworks necessary for irrigation. (Construction projects such as the dam at Tiszalök and the adjoining Eastern Main Canal, as well as the deficiencies of implementing irrigation, will be discussed later.) The forced development of rice production was key to these endeavors due to the fact that the majority of irrigated land was used for rice, decreasing the efficiency of irrigation. Earlier experiences with rice did show that it also thrives in poor quality soil, leading to the idea that land unsuitable for the cultivation of more delicate crops could be used for rice production instead.[205] This decision led to parts of the Alföld, especially the saline fields of Hortobágy, being used for rice, significantly changing the natural landscape.

In 1950, the Ministry of Agriculture prescribed the production of 3.3 kilograms of rice per person (based on Hungary's population figures) on 14,375 hectares of land.[206] To ensure the implementation of the production plan, "rice committees" were organized to contract state farms, producers' cooperatives and private farmers.[207] However, despite all propaganda, they could only fulfill a meager part of the plan,[208] and one of the main reasons was the loss incurred from rice production in 1949.

The Crop Production Company alone incurred over 10 million HUF of bad debt from producers, while the National Water Management Office incurred 6.6 million HUF of due water fees for the irrigation of rice fields,[209] leading the National Planning Office to suggest that farmers should only be cleared of their debts if they were willing to offer their lands to a state farm.[210]

However, it was not just farmers unwilling to contract due to their experiences in 1949; at that time, the National Water Management Office itself was simply unable to increase the size of flood-irrigated rice land.[211] The party expected it to introduce new rice production sites in the Körös valley; however, due to the relative lack of precipitation, the river's discharges were unable to supply even the already existing production sites that had been established in the previous year. Furthermore, some of the farmland near the Danube was planted with other crops and was thus excluded from rice production.[212] Nevertheless, political pressure proved enough to result in 14,375 hectares of farmland being used for rice production in 1950.

Although official policies emphasized that rice was easy to cultivate due to the fact that it even thrives in saline soil, in reality it required a lot of maintenance, especially in the Hungarian climate. In fact, a large labor force was required not just for crop cultivation but also to prepare the soil and maintain the irrigation system. The local peasantry found themselves overwhelmed with work, and much like in the case of cotton and the construction of the dam at Tiszalök, rice production also required seasonal workers. However, without modern machinery, devising the irrigation system, building floodgates and leveling the plantations required an extensive labor force. Since the Communist dictatorship had no moral objections, people and families who were either deemed enemies of the state or politically untrustworthy were deported to provide the required labor force; and so, simultaneously with the demarcation lines on the southern and western borders of Hungary, the deportation of "class enemies" began in the summer of 1950, and entire families were taken from their homes and transported in railway carriages to the desolate lands of the Tiszántúl, especially to the uninhabited Hortobágy "where they had to work on the rice and cotton plantations."[213] The party decision to employ the deportees living in forced housing was issued on 23 May 1951, providing an official framework for the exploitation of these already abject deportees. "First, kulaks should be gradually removed from the villages, and in order to ensure the institutional use of their labor, forced housing may be designated for them and their family members. Three to five villages must be built on the barren saline fields of the Tiszántúl where the kulaks . . . may work . . . with their family members."[214] Rice fields provided work all year round: the irrigation system had to be prepared in the winter, rice needed to be weeded in the summer, and harvest and thrashing was done in the fall—but all in appalling circumstances, without proper equipment or work clothes, and for wages barely enough for survival. "De-weeding rice on the fields gave plenty of work to the women. However, morning temperatures were around 10–15°C, and they had to work barefoot in the cold, shallow water of the fields. Standing in the cold water caused many illnesses in these women such as articular diseases, rheumatism and gynecological diseases."[215]

The change of production plans in 1953 politics did not affect the introduction and production of rice on account of the fact that, unlike other newly introduced unnative crops, rice production turned out to be successful; therefore, the amount of farmland used for rice was increased in 1954. "Based on the amount of irrigation required and the fulfillment of other necessary conditions for rice production in 1953, the amount of farmland used for rice production is to be increased by 49 percent in 1954 and by 80 percent within the next three years."[216] These plans were then promptly implemented; while in 1953 there were less than 28,000 hectares worth of rice fields, by 1954 rice was being produced on approximately 43,000 hectares of farmland.[217] However, while the chosen species of rice was suitable for production on Hungarian soil, the requirements for rice production were still missing: there was no expertise, rice fields expanded quickly without any attempt at improvement, and monocultural production became the norm, leading to the rapid decline of average crop yields. Weeds, pests, and plant diseases spread quickly as well, and by 1955, the average crop yield of rice was lower than ever before, even with 50,000 hectares of land being used for rice alone.[218]

From our point of view, the history of rice in the 1950s is especially telling since it was an experiment that could have been successful. Unlike the idealistic and completely irrational production of Hungarian oranges or Hungarian cotton, Hungarian rice was actually possible to produce—in fact, after the initial struggles, Hungary still produces and exports rice today. However, Communist Party policies did not aim for economical or efficient production; they sought to prove that socialism was all-powerful and that they could not only triumph over nature but bend it to their will. This "faith" was what ploughed the formerly uncultivated lands of the Hortobágy and prepared them for flood irrigation; as a result, agricultural production could not grow simultaneously with the development of the irrigation system as expected because the irrigation system mainly catered to rice that was produced with unsuitable agricultural technology. By the end of the 1950s, they realized that the saline soil of the Hortobágy was unfavorable for the economical production of the "undemanding" crop, leading to the gradual eradication of these rice fields; however, the few years that passed had left a gaping wound on the Hungarian landscape that is still being remedied today.

Experiments to Improve Wheat

In many parts of the world, grain crops, especially the production of wheat, are one of the most fundamental requirements for feeding a nation; therefore it was only natural that one of the top priorities of Soviet agricultural innovations was to create a species of wheat that would yield an abundant crop even in the worst weather conditions and produce multiple times the average yield of any known species of wheat. To this end, Bolshevik policy expanded wheat production towards the north so that all regions could produce their required amount of wheat; according to party propaganda, the Soviet Union had managed the task by 1952. They not only surpassed 1940 production rates by 48 percent but also introduced the production of wheat in the far north.[219] These results were, of course, attributed to Lysenko, who attempted to advance the development of planting seeds by imitating the vernalization processes of nature. The key to vernalization is that a temporary low temperature is required to induce change in the consecutive development stages of certain plants.[220] In nature, these conditions are given since wheat is planted in the fall, and winter provides enough cold for its development in the spring. This was copied by the glorified Soviet agricultural "innovation" called jarovization, during which the seeds were artificially cooled and then planted. However, the most notable result of this experiment was that after jarovization, autumn wheat could also be planted in the spring—in other words, there was no practical gain. Regardless, the fact that jarovized seeds had managed to grow into wheat in the north was seen as the triumph of socialism, even though production results were low and the production costs of producing wheat in the north far surpassed the transportation costs of wheat grown in more favorable climates.

The "results" discussed above were also to be implemented in Hungarian agriculture. "Wheat is also included in the decision of the Council of Ministries of the People's Republic of Hungary on the development of crop production: while the average crop yield of wheat is 7.5 centners in Hungary, the wheat fields of the Soviet Union yield an average of 12–14 centners of wheat. Therefore, one of the most important tasks of our five-year plan is to develop our lagging agricultural production."[221] The first five-year plan dictated that the average crop yield of wheat must improve by 13 percent in 1954 compared to 1949 averages by introducing new agricultural technologies and methods as well as crop improvement. However, increasing the size of farmland used for wheat had severe consequences; just like in the case of large-scale cotton production, the local environment was not taken into account when choosing new areas for production. In 1951, producers' cooperatives were forced to reorganize their wheat production (which had adhered to soil conditions) and plant wheat into the sandy soil of the Alföld, previously primarily used for rye production.[222] Despite all hardships, the "glorious march" of socialist wheat production continued, and producing spring wheat out of autumn wheat by following Lysenko's methods was regarded as a huge success. The greatest breakthrough, however, was expected of the so-called rivet wheat.

The creation of rivet wheat was considered a very significant feat in the Soviet Union despite the fact that they basically rediscovered an already existing plant long known by the peasantry of the southern Soviet regions. The thick, forked ear of rivet wheat, or cone wheat (Triticum turgidum), is able to yield more seeds than its Hungarian cousin. However, the plant requires warmth and plenty of water, its maturation is slow, and it is also weak to cold; therefore, Hungarian climate is unsuitable for rivet wheat production—not to mention that, while it yields a larger crop, the gluten content of the seeds is extremely low, resulting in low quality flour.[223] Europe had long researched this species of wheat, and by the end of the nineteenth century it was common knowledge that rivet wheat flour is low quality and unsuitable for baking. Regardless, there were great efforts to introduce the crop to Hungary.[224]

In June 1949, the Hungarian Newsreel was proud to announce experiments with the very productive rivet wheat in Eszterháza,[225] which was expected to produce a 200–300 percent greater crop yield than previously produced wheat as soon as it adapted to Hungarian climate.[226] According to their calculations, producing rivet wheat would have allowed them to reduce wheat fields by half, but there was some hesitation once they attempted to estimate the time required to achieve the expected results. The creation of Hungarian rivet wheat was expected by the end of the five-year plan; regarding its production in Hungary, however, all they could say was, "We would only be able to begin production later."[227]

The unusually tentative tone of rivet wheat reports confirms our assumption that even party members who believed in producing cotton and citruses felt that the production of rivet wheat was a hopeless endeavor, and soon enough, all experimentation stopped. It had already been stated in 1950 that jarovization brought mixed results, meaning that wheat production could not be improved by this method,[228] and according to the already referenced newsreel, there were only fifty ears of rivet wheat in 1949 in the whole country, and experiments to cross these with better quality species of wheat were unsuccessful.[229] Therefore, rivet wheat was never introduced to farms in Hungary.

Introducing Viliams' Grass-Crop Rotation in Hungary

One of the key issues of Soviet nature transformation was to make hitherto uncultivable ground into fertile agricultural farmland. This question had already surfaced in the era of Tzarian Russia since the utilization of the extensive Russian steppes would have been very beneficial in terms of alimentation and stock feeding. At the end of the nineteenth century, V.V. Dokuchaev developed a theory on how to make the often arid Russian plains fertile,[230] proposing to do so by river regulation, the establishment of water reservoirs, and assisting the drainage of precipitation through afforestation of certain areas. His theory was further developed by agronomist V.R. Viliams, who declared with the superior self-confidence of the Soviets that, "with the application of the appropriate methods, any land could be transformed into highly cultured, fertile soil."[231] By studying the rotation of cropped and uncropped land, he realized that on uncropped land the fertility of soil is re-established by grass; however, the natural propagation of grass was a slow process, which he intended to shorten through grass- crop rotation. He planted light shrubs and perennial vexillary plants into the unused soil so that their roots would loosen the soil and enrich it with nitrogen.[232] He supported the grass- crop rotation system by the establishment of protection forests that blocked dry winds, provided a more balanced microclimate for crops, and also regulated the distribution of precipitation. In the impoverished Soviet Union torn by civil war, this was a useful agricultural method indeed due to the lack of crop protection chemicals and the scarce availability of agricultural machinery; however, in the long run, this method actually delayed the development of agriculture. Nevertheless, after World War II, Communist leadership made the use of this agricultural method obligatory in all countries within the Soviet sphere of influence.

In Hungary, the first experiments with a grass-crop rotation system began in 1949. Approximately 26,000 hectares of land had been planted with grass, to be doubled by 1951.[233] However, these experiments had dubious results. In 1950, a mixture of grass seeds was planted in twenty-eight different places in small parcels, but the designated sites were badly chosen—or, at least, that was how they explained the failure of the experiment.[234] In reality, these experimental sites were established far from the centers of state farms, and during the middle of agricultural seasons—the grain harvest—there was no available labor force, nor the will to cultivate and mow these seemingly useless grass plantations. However, Hungarian peasants were not alone in their resistance to the idea of planting valuable farmland with grass; agricultural intellectuals also lacked the political "training" to embrace such forced innovations, thus requiring serious propaganda and "education." Therefore, the college of the Ministry of Agriculture instructed all state farms conducting grass-crop rotation system experiments to appoint a referent who would inspect the implementation of necessary measures.[235] Nonetheless, not even referent experts could control the weather, and the summer of 1950 more or less burned the grasslands dry.

The failure of grass experiments prompted the ministry to reorganize experimentation for the year 1951. In September 1950, the Department of Crop Production of the Ministry of Agriculture designated twelve state farms and producers' cooperatives that were required to introduce a grass-crop rotation system in the following year.[236] Next, "brigades" were organized to supervise and control field work. All twelve farms involved were placed under two trained groups of four experts each, and the work of the brigade was supervised by the Department of Crop Production of the Ministry of Agriculture. Before work could begin on the farms, the instructions of these experts (considered "proposals") had to be approved by a supervisory committee, which consisted of delegates of the National Planning Office, the Hungarian Academy of Sciences, the University of Agricultural Sciences, the Agricultural Experiment Center, and the Department of Crop Production of the

Ministry of Agriculture.[237] During the following year, political pressure grew on a legislative level as well when the five-year plan was amended to include that the Soviet-approved grass-crop rotation system was to be implemented by at least five hundred settlements.[238] Every relevant institute was mobilized accordingly to ensure the success of this plan in their respective scientific fields.[239] For example, the Institute of Agricultural Organization was charged with the task of determining agricultural policies based on their data; in short, it had to deduce where, how, and what kinds of grass should be planted. However, as it turned out, the institute was incapable of fulfilling the task.[240]

In defense of the Institute of Agricultural Organization, other scientific institutions were also struggling with the grass-crop rotation system and this was not necessary the fault of their researchers; they were simply trying to force Hungarian agriculture to use a system that was alien to their traditional methods. In short, it was not so much a lack of expertise or experience as the fact that these researchers realized the irrationality of these policies since there were more modern methods of soil improvement available by that time that were already being used among Hungarian peasantry. The idea of planting grass instead of life-sustaining wheat or even stock feed was unsupported, and the method that may have worked perfectly on the enormous Siberian steppes was simply unsuitable for a country with limited territory such as Hungary. Extensive research would have been necessary to compile the most suitable mixture of grass for Hungarian soil, not to mention the fact that importing jarovized planting seeds from the Soviet Union was expensive, and grass production would have required water that not only would have made their production even more expensive but would also have done so to the detriment of industrial crops already struggling for what little irrigated land there was; in fact, the Ministry of Agriculture did complain about the lack of development regarding irrigation.[241] Another unique challenge was posed by the constant organization of state farms and producers' cooperatives, and the near constant reallocation of farmland—in other words, a portion of the experimental parcels may have survived the weather but still died out due to the fact that production plans had to be revised.[242] The most rational decision would have been to introduce a grass-crop rotation system—since it was obligatory—on fields with greater breeding stock, since the mowed grass could have been used as stock feed. However, there was no available data on the density of national stock breeding by district to allow for such an arrangement.[243]

By fall of 1952, the grass-crop rotation system had been introduced to approximately 94,500 hectares of farmland, with more or less success.[244] However, there were still no tangible results, and the new agricultural policy of spring 1953 no longer supported the unsuccessful approach.

There was another important component of the complex production improvement method of Dokuchaev and Viliams, namely the establishment of protection forests. The ability of forests to block drying winds grows exponentially with their height, meaning that they are able to reduce the impact of wind on areas twenty-five to thirty times the size of their height. The humidifying effect of their foliage is beneficial to the crops they protect, and not only do their roots slow down the drainage of precipitation, but they also channel subsoil water into higher layers of soil.[245] Based on these research results, the Soviet Union issued a regulation on protective afforestation in October 1948, which immediately determined the policies of the Hungarian Communist government that tried to imitate the Soviet Union in every area possible.[246] The Ministry of Agriculture promptly instructed the Forestry Center—established in the spring of 1948—to develop plans for the afforestation of Hungary, which was approved by the Forestry Council on 6 October 1949.[247] A month later, the 1950 plans for afforestation were complete, according to which, forest areas in Hungary had to be increased from 12.3 to 20.3 percent. The plan contained the amount and cost of saplings necessary for the afforestation of areas around flood control areas, public roads and railways, pastures, farmlands, farm centers, and settlements.[248] Increasing forest areas was especially necessary on the Alföld plains where forest areas did not even reach 5 percent; therefore, they decided to plant forests on sandy, saline soil unsuitable for agricultural production.[249] The government instructed the Ministry of Agriculture and its Agricultural Science Center to develop the scientific plan and implementation schedule[250] and also to established a new authority called the Institute of Forest Planning (with one center and fifteen county divisions) to ensure the execution of this long-term project.[251] However, bureaucratic (re)organization was far from done since a host of institutes and companies had to be established by 1 April 1950 to implement the project.[252] The Forestry Center was then abolished to give way to the new Forestry Department of the Ministry of Agriculture. The centers of the state forestries were organized on a county level, and the actual forestry units—the National Forestry Companies—were established in 150 districts, boosting the total personnel of forestry administration to 5,310 employees.[253] This reorganization had a fundamental effect on all areas of forestry; however, the issue of forest property had not been entirely resolved yet. Of course by 1950, the majority of forest areas had become state property; however, there were still farms that were in the collective management of farmers or the property of settlements; and there were even private farms, though their numbers were scarce. Nevertheless, the reorganization of forestry administration affected all forest areas equally, since, according to the government, this was in the interest of the national economy.

Although the introduction of the grass-crop rotation system failed completely, the implementation of related afforestation was a significant achievement. For the first time since the Hungarian conquest of the Carpathian Basin, forest areas increased instead of decreasing. Between 1950 and 1960 alone, afforestation was carried out on a total of 248,387 hectares of land.[254] From the end of the 1960s, Hungarian afforestation projects began to dwindle and the resolution of property issues following the system change of 1990 delayed the development of forestry for quite some time. We may therefore say that Hungary's 21 percent forest density is largely due to the implementation of the Stalinist plan to transform nature.

Construction of the Tiszalök Dam and Irrigation System

At the beginning of the 1950s, Eastern Europe was still convinced that nature was not invincible, leading them to believe that the Davydov plan devised to change the course of Siberian rivers was realistic and achievable. It was in this spirit that Hungary had begun to transform the environment of its eastern territories, including the water management and soil quality of areas by the Tisza River.

According to one of Mátyás Rákosi's propaganda speeches at the beginning of the 1950s, "the Soviet Union is now able to begin realizing such colossal projects as establishing water power plants by the Volga, Dnieper, and Amu Darya rivers to change the arid climate of entire countries and transform barren wastelands into fertile farmland."[255] Hungarian newsreels were constantly talking about the "gigantic" construction that produced one of the largest water reservoirs in the world, and plans for the utilization of the Volga River's water power. Hungarian party leaders, swayed by the construction of the dams at Kuibyshev and Stalingrad, had announced plans for the construction of a dam at Tiszalök and the corresponding irrigation system. The Tiszántúl—the eastern part of the Alföld plains—became the site where they attempted to "transform barren wastelands into fertile farmland"; therefore, this also became the site where they sought to introduce the complex production improvement method of Russian scholars Dokuchaev and Viliams discussed above.

Improving the fertility of the eastern regions of the Alföld plains had been an issue long before the plan to transform nature. The soil is suitable for agricultural cultivation and receives plenty of sunshine, but the amount of precipitation greatly varies and a significant portion of the territory had alkalized due to the regulation of the Tisza River and the draining of ancient marshlands. In fact, in droughty seasons, large stretches of farmland had fallen out of production. Demands for the establishment of a Tisza-based irrigation system had surfaced as early as the nineteenth century; however, the region had to wait until the 1930s for the first steps of this project, one of which was the irrigation act that established the Hungarian Royal Irrigation Office in 1937.[256] The new organization was charged with the establishment of a dam on the Tisza and Körös rivers as well as their corresponding irrigation systems. At that time, Tiszalök[257] was chosen as the site of the future dam, and Emil Mosonyi[258] began preparatory research for its construction in 1941.[259] The famous engineer was also charged with designing a dam after the war, and he amended his plans to include a smaller water power plant in 1946. In accordance with the decision of the National Economic Council, the construction of the Tiszalök water management facilities had to begin on 1 January 1950 and were to be finished within the period of the five-year plan.[260] Plans included the construction of more than one dam, and the establishment of another water power plant had been included in the second five-year plan; however, that one was built much later.[261] The National Economic Council charged the successor of the Hungarian Royal Irrigation Office—the National Water Management Office, as it was called from the summer of 1948—to execute deep construction work; the actual facility was to be constructed by the Ministry of Transportation and the Post Office Department, while all work involving machinery and iron structures was supervised by the Ministry of Heavy Industry.[262] The organization of these measures was likely made more difficult by the issue of establishing the authority of the various ministries. Based on certain documents, it is assumed that whenever a problem surfaced, especially regarding finances, these ministries kept deferring responsibility among themselves.[263]

Planning and construction work on the dam were based on the National Water Management Framework Plan and carried out near

Tiszalök, some five kilometers to the west of the settlement. Construction was supervised by the Tiszalök Water Construction Company; however, this major investment required a massive labor force that the settlement of 4,500 inhabitants and its surrounding area simply did not have. Therefore, when major ground work began in the early fall of 1950, most of the workers were prisoners who had either been labeled kulaks or had been sentenced because they could not comply with agricultural produce-surrender requirements; and while work was led by civilian experts, these prisoners were supervised by prison guards.[264] The prisoners were also forced to construct the forced work camp that provided the majority of the labor force for the Tiszalök construction projects between February 1951 and late fall of 1953. From 1951 the majority of inhabitants of this forced labor camp in Tiszalök were prisoners of war who had been released by the Soviet authorities in December 1950 and at the beginning of 1951, but Hungarian examination committees decided to keep them under surveillance.[265]

The dam was placed at the 2.4 kilometer-wide transection of the (Rázompuszta) bend by the border of Tiszalök, and was completed by 1954; in other words, the Tiszalök Dam was one of the few investments of the five-year plan that was actually completed on schedule. However, the power plant only began to produce electricity in 1959.

The establishment of the Tiszalök Dam went hand in hand with the construction of the irrigation system, since the excavation of the riverbed of the Eastern Main Canal began simultaneously with the construction of the dam, and this artificial river was to supply the eastern regions of the Alföld plains with water. Construction was right on schedule, but as we shall see later, the opportunities inherent within were not immediately put to good use. However, before we discuss the details of the plans to transform the Tiszántúl region, it is important to discuss the beginning of experimentation with irrigation.

Measures to expand irrigation in agriculture had begun even before the announcement of the nature transformation program; however, due to financial difficulties, most of these had to be abandoned. For example, one of the tasks of the National Water Management Office was to build the Danube–Tisza Canal, which would have also served as an important transportation route; however, they only managed to construct the section included in the three-year plan.[266] The office had also prepared a five-year plan for irrigation, and the Agricultural Science Center established experimental sites to carry out necessary research and development, with six sites established in 1949 and two more in the following year.²⁶⁷ In the spring of 1950, the reorganized center (see the previous sections) established the institutional framework of water management research; for example, the Institute of Irrigation Experiments in Kisújszállás was in charge of irrigation,²⁶⁸ and was directed by Melanie Frank.²⁶⁹ Irrigation experimental farms were also established in the following settlements: Szarvas, Hódmezővásárhely, Nagytétény, Hortobágy I, Hortobágy II, Kunhegyes, and Szabadszállás.²⁷⁰ However, a reorganization occurred shortly thereafter, and the coordination of all irrigation-related tasks was delegated to the Department of Irrigation and Soil Improvement (established on 27 May 1950) within the Ministry of Agriculture.²⁷¹

Institutionalized experiments with irrigation lead to a rapid increase in the amount of irrigated farmland. While there were only 11,610 hectares of irrigated farmland in 1947, this rose to 50,069 in 1951,²⁷² of which 80 percent was on the Alföld plains, mostly on the banks of the Tisza River. It is important to note that the extensive development of irrigation was closely related to the forced development of rice production, as discussed earlier, which had a great impact on the choice of irrigation methods since surface irrigation in particular was the most favorable method for rice production.

Plans for the nature transformation of the Tiszántúl region were developed in 1952, and the first related document we found was dated 10 April 1951. On the secretariat session of the Hungarian Workers' Party, the Agricultural and Cooperative Department suggested the necessity of a plan that would determine and ensure the soil improvement of farmland located by the Tisza River, the future main canal, the territories by the Körös and Berettyó rivers, and the farmlands on the right bank of the Tisza River between Mezőcsát and Szolnok.²⁷³ Plans revolved around approximately 740,000 hectares of land, and the comprehensive nature transformation plans were clearly developed in order to imitate the Soviet model and adhere to the requirements of the "Big Brother." Choosing the experimental site seemed evident since a dam and corresponding irrigation system were being constructed at Tiszalök. In the preceding years, these farmlands had been irrigated the most, in some shape or form, and in terms of climate, the southeastern regions of Hungary were the most suitable for the introduction of crops that were native to warmer climates, not to mention that, from the perspective of political decision makers, it was favorable that this region had succumbed very quickly to forced collectivization.

The ideas for nature transformation were approved by the secretariat, and it instructed the Agricultural and Cooperative Department to develop a detailed proposal within three weeks,[274] and two weeks later, on 25 April 1951, the submitted detailed proposal was once again discussed by the secretariat.[275] The proposal had listed six main policies for nature transformation, including the need to improve saline soils, the importance of building the Eastern Main Canal, and connecting lakes and reservoirs, thereby expanding the amount of irrigated farmland. To ensure the widespread use of the grass-crop rotation system (which had scarcely been applied thus far), it highlighted the necessity of land plotting, road management, and the organization of state farms focusing on grass seed production, as well as the significance of establishing protection forests. However, all of these tasks required a massive labor force, which raised the issue of mass vocational training as well as the absolute necessity of mechanization. Last but not least, the proposal suggested the establishment of a committee consisting strictly of experts and the directors of involved research institutes to develop a detailed technical plan. The secretariat did approve the majority of these proposals; however, it decided to replace the experts of the committee with party cadres, and the difference between the initial proposal for members of the committee and the final "team" is striking—and telling. For example, the proposal suggested renowned soil chemist Károly Páter, director of the Agrochemical Institute, to be the head of the committee; instead, the secretariat appointed Zoltán Vas, faithful Communist, member of the Political Committee of the Hungarian Workers' Party, and director of the National Planning Office.

Other members included Minister of State Farms and Forestries András Hegedűs, Minister of Agriculture Ferenc Erdei and a few of his subordinate officials, and András Somos, secretary of the Agricultural Science Department of the reorganized Hungarian Academy of Sciences. In the end, the involved scientific fields were represented by five people in the twelve-member committee: Károly Páter, Melanie Frank, Emil Mosonyi, forester Imre Babos, and agriculturalist Ernő Obermayer.[276] The significantly modified proposal was approved by the secretariat on 6 February 1952, which not only diluted the scientific committee but also distorted the original plans. Basically, the previously established objectives had become a reduced hodgepodge of prerequisites to a much more "comprehensive" vision: "All plans to transform nature must be based on long-term production tasks dictated by the characteristics of designated areas and the development interests of the national economy. The development of crop production is necessary, especially that of wheat, rice, and cotton—Hungary must become the most important source of these crops to supply its own necessities."[277] In other words, all greater agricultural innovations were directed to serve the geographically and biologically voluntaristic demagogy of party policies. The propaganda used to influence Hungarian society will be discussed later, but the following half-sentence lurking in the modified plans is worth considering: "[the plan] must be announced to the public at an appropriate time after the completion . . ."[278]—followed by detailed instructions by the secretariat on how this news should be fed to the "working nation."

On 18 April 1952, the Council of Ministries issued a decision on the establishment of the Tiszántúl Nature Transformation Planning Committee,[279] followed by the establishment of work committees and their scopes of activity.[280] Based on the designated deadlines, these teams of experts would have had to devise precise plans by 30 April 1953. However, the death of Stalin brought changes to Hungarian politics, and after the formation of the Imre Nagy government on 5 July 1953, these agricultural plans were significantly reassessed, which meant that the planning committee had probably done little work of any significance, which may be one of the reasons why the examined documents contain no files from the committee itself.

The size of irrigated agricultural farmlands steadily increased in the examined period, but data were particularly high for the year 1953 when Hungary attempted to improve production by introducing irrigation to 94,093 hectares of farmland.[281] At first glance, one would expect significant results behind the data, but it is actually difficult to explain it. By the end of 1952, the five-year plan had been half over and—not surprisingly—Hungarian leaders were faced with several fallbacks compared to the often irrational ideas and expectations in the announced plans, which meant that the experts of the various involved scientific fields had to act fast to comply with their respective tasks. The construction of the Tiszalök Dam and the Eastern Main Canal, mentioned earlier, had proceeded according to plan. However, the amount of irrigated farmland did not increase accordingly, mostly due to the fact that the construction of irrigation equipment and the necessary canal system had not been built yet—in other words, there was water but it was not being put to any use, and the agricultural department of the party reprimanded the Ministry of Agriculture for not using the existing opportunity when political propaganda was desperate to show society what little they actually did achieve.[282] The dam and corresponding canal were outright called a "peace monument," and farmers were recruited en masse to build the irrigation system in the spirit of "peace is yours, you build it for yourself," since labor was still too scarce to execute an investment of this magnitude. The work schedule for the next year instructed all state farms to use every source of water suitable for irrigation (from smaller brooks to high-yielding wells) on their farmlands and to carry out maintenance on all older irrigation systems.[283] However, the ministry only offered professional assistance; the financial background and labor force would have had to be provided by the state farms and producers' cooperatives. These required measures had significant results, even if they only lasted a year, and in 1953 a huge amount of farmland was irrigated on command.

Despite the increasing amount of irrigated farmland, opportunities remained unexplored and unused. We have already discussed that the canal had been built, but there was no simultaneous development of the water system; above all, irrigation had primarily been introduced on the lowest quality farmlands, further reducing its efficiency. In short, the actual amount of irrigated farmland did not even reach 50 percent of the farmland suitable for irrigation.[284]

After the turning point of 1953, the distorted agricultural development of the previous years was re-evaluated and the construction of the next dam on the Tiszar River, originally scheduled for the second five-year plan, was delayed in favor of the utilization of existing opportunities, supporting the expansion of irrigation with smaller investments. From 1954, the government was planning the construction of floodgates and compression pumps for producers' cooperatives, and sought to enable private farmers to purchase irrigation pumps as well.[285]

The Eastern Main Canal was presented in its entirety in 1956. The artificial river between Tiszalök and the settlement of Bakonszeg on the bank of the Berettyó River cuts across ninety-eight kilometers of land in the region of the Tiszántúl, and it still provides water for the surrounding farmlands along with its various side canals. It is also a significant drinking water supply for the surrounding area and its high quality water provides for an abundance of wildlife as well. We may therefore say that, despite its mostly irrational ideas, nature transformation did have some positive results for Hungary, especially when we consider that in that era, forced industrialization was a priority; but nature transformation enabled the development of agriculture as well—and to the national economy of a country lacking in industrial resources, agriculture is definitely one of the key means of survival.

The Transformation of Thinking through Propaganda

Besides widespread terror, the dictatorship of the Rákosi era also sought a "gentler" way of regulating people's lives and ideas, which entailed the top–down regulation and uniformization of the private sphere to the smallest detail, using every possible method of propaganda available.[286] The popularization of party politics was, of course, a top priority in manipulating everyday life, and was presented as the only road to happiness and peace. It was not surprising then that all reports of the three- and five-year plans and the exaggeration of achieved "results" was frequent in the media as well as the press and contemporary film, and "national education" documentaries on the transformation of nature were also included.

"Instead of taking a passive approach to the development of living nature, Soviet science based on the theories of Marx and Lenin had consciously and systematically changed nature for the benefit of man. In his well-known work, Michurin says, "Man can triumph and must triumph over nature . . . we cannot expect nature to take pity on us, our job is to take what we want."[287] Faithfully imitating the Soviet Union, the Hungarian Workers' Party did indeed take what it wanted, not just from nature but also from the peasantry directly dependent on nature, all the while trying to convince those involved with propaganda that it was necessary to give back the distributed fields and to turn in all the produce, and explain how the mostly unviable "innovations" would become the key to a better future. The quote above shows that scientific journals, too, had published papers on the concept of the Soviet transformation of nature. The journal *Természettudományi Közlöny* and its successor *Természet és Technika*, as well as *Erdészeti Lapok* and *Élet és Tudomány*—to name but a few journals from the era—were full of "scientific" support of the current political policies, and articles attempting to persuade scientific intellectuals. However, it was not only journals that published faux-scientific papers; propaganda and pseudo-scientific works written by Soviet scholars and translated from Russian to Hungarian also flooded the market, not to mention smaller essays written by Hungarian researchers. In the current section, however, we forego the analysis of texts written for the scientific public to instead discuss a few examples of popular scientific education and the "dumbing down" aimed at the public at large.

The hardest task for propaganda was to bring the "voice" of the party to the villages; therefore, nature transformation went hand in hand with the distribution of the "popular radio"—a simple device that only operated on medium-wave frequencies, ensuring that only two Hungarian stations were available and there was no access to Western broadcasts or any foreign propaganda that could have gone against the interests of the party.[288] At that time, there were approximately 525,000 "foreign-tuned" devices in Hungary, which meant that the distribution of "simplified" radio devices was an ill-fated party decision, not to mention that people were cunning enough to modify these popular radios so as to tune in to the "prime enemy stations" such as *Radio Free Europe* and *Voice of America* on short-wave frequencies.

Another form of public information was the installment of speaker systems on the streets of various settlements; however, in villages that had no speaker systems installed at the beginning of the 1950s, all official regulations and local public service announcements were publicized in the traditional way by drumming. Until the introduction of the council system in 1950, it was the village judge and then an employee of the local council apparatus, usually an office clerk, who would use drums to announce news in various parts of the village.[289] "Councils should do more than drumming and fliers: members of the council need to take the lead in persuasion and organization," reads the column of *Szabad Nép* (the daily newspaper of the Hungarian Workers' Party).[290] To this end, reideologized "documentary" lectures were organized in local culture houses. These lectures had to be "factual, scientifically sound, and militant,"[291] and they were not unsuccessful: the event series *Szabad Föld Téli Esték* [Free Land Winter Evenings] was so successful that they held 50,000 lectures in 1950 and 123,000 in 1953.[292] The party seized every opportunity to reach the ears of the Hungarian population; for instance, they instructed the Hungarian Society of Natural Sciences [Magyar Természettudományi Társulat] to organize scientific exhibitions in the entrance hall of several cinemas and in the waiting rooms of railway stations.[293] By 1952, several counties had begun screening films and slide-strips[294] in the waiting rooms of railway stations, helping propaganda to reach new sites.[295] During these screenings, speakers kept repeating the clichés and inspiring work chants of communism. Slide-strips were considered relatively simple and inexpensive instruments of propaganda; therefore, they were used extensively in every area of political agitation, but especially in relation to agricultural issues since persuading the Hungarian peasantry was a key point of Communist policies. The Stalinist Plan for the Transformation of Nature made its appearance in slide-strips soon enough: we have knowledge of nearly twenty slide-strips produced between 1949 and 1953, all of them popularizing various agricultural innovations.[296] This number is confirmed in a decision by the Council of Ministries from 1953 in which they dedicated a separate chapter to methods of agricultural propaganda.[297] The document contains instructions for the production of at least twenty strips of film on agricultural topics by 1954, and the documentaries on the agricultural technologies of the most important industrial crops were also to be finished by the same year.

Besides documentaries, weekly newsreels also served to broadcast the "success" of Soviet agricultural technologies. The first news relevant to our current topic appeared at the end of the 1940s, accompanied by current Hungarian political events. In January 1949, the former estate of Duke Pál Eszterházy was filmed to showcase experiments with new non-native industrial crops in the spirit of Lysenko's and Michurin's teachings.[298] However, the scientific aspect of the report was largely abandoned in favor of denigrating the duke and the aristocracy in general, as well as the recently incarcerated Cardinal József Mindszenty. It is also apparent that newsreels gradually began to focus on results and, consequently, on glorifying the Communist Party. Sztálinváros and the Stalin Iron Factory became the most spectacular backdrop for the construction of the country of iron and steel, and from the beginning of 1951, cinemagoers were faced with repeated screenings of happily smiling workers. The Tiszalök Dam also received a lot of screen time, as did the produce-laden industrial crops on irrigated farmland.

Daily newspapers and journals were supplied with information by the Hungarian News Agency which also determined what topics were currently relevant and required to be published. On 2 September 1948, the news agency informed the public that "cotton is thriving in Borsod."[299] This was one of the first instances of news broadcasts on the transformation of nature, which is especially shocking since it pinpoints the most northern region of Hungary as the production site of tropical cotton. We found no evidence in contemporary documents that Borsod had produced cotton in 1948, so we may assume that this was a prime example of the conscious distortion of reality: if cotton was said to grow even in the mountainous climate of a northern county, then plans for the plantation of cotton on farmland the following year would seem completely justified. By 2 July 1949, news informed that the quality of the last year's crops had been on a par with cotton from India, and there was an "uplifting" report on former textile worker and current minister Anna Ratkó arranging chunks of cotton—the "white gold"—on an assembly line.[300] The event was, of course, filmed and photographed, and footage was distributed all over the country.

The political importance of cotton production is evident in the frequency of reports that followed production every year from planting the seeds until harvesting crops. The current state of cotton production was almost constantly featured in the weekly newspaper *Szabad Föld*, and during 1951 and 1952 the population was informed weekly on the state of cotton production. Political agitation soon appeared in the titles of articles as well; "Produce more cotton this year than you have last year!"[301] was followed by words of wisdom on how these expectations should be accomplished: "Successful cotton production depends on hard work, not the weather."[302] The topics discussed in the tabloids did not differ significantly from the obligatory schemas of political papers, though the approach and presentation of the topics was perhaps a little fresher. For example, the weekly paper, *Nők Lapja*, sought to persuade its female readers by saying "More Cotton, More Clothes".[303] However, it is striking that cotton production was mostly favored by the written press; it would have obviously been harder to mask the sad truth on film. Nevertheless, they did shoot a propaganda film on cotton, and in May 1953, spectators of the cinema newsreel were greeted on the silver screen by cotton-picking expert Katalin Pávkovits.[304] The 22-year-old woman had become a national success story when she was the first to master the Soviet method of picking cotton with both hands on both sides, thereby quadrupling her previous record in picking cotton.

Besides cotton, the introduction of other non-native industrial crops was also featured in Hungarian news. The newspaper *Szabad Föld* welcomed the introduction of the sweet lupins that were not only considered one of the best stock feeds, but also as one of the pioneers of soil improvement.[305] There were also reports of the Agricultural Science Council on 15–16 December 1950, where "Aladár Porpáczi, leading researcher of the Fertőd Experimental Farm, stated that thanks to the application of Soviet agricultural technologies and methods, lemon saplings that otherwise take 8–10 years to bear fruit had blossomed and grown fruit within six months."[306] This "miracle" was also achieved in the Ukraine, Soviet member state and Hungary's northeastern neighbor, where they established tea plantations and produced citruses, figs, and pomegranates as well—at least according to the official newspaper of the party.[307] When we see reports of how farmers were producing juicy watermelons near Leningrad (as it was called in that period),[308] one cannot help but think of the tall tales of Baron Munchhausen, but the answer is quite simply, "Man is in charge, not the weather!"[309]

In 1951, a picture album titled *Magyarország új arca* [The New Face of Hungary] was published in honor of the second congress of the Hungarian Workers' Party.[310] The volume served to show the changes of the previous few years by praising the party and painting a much more favorable image of reality than was warranted. The section on agriculture features a two-meter-tall specimen of Uzbekistanian cotton, but two pages later, we see Hungarian peasant women bending low as they struggle to pick the Hungarian cotton. Did the striking difference elude the editors of the volume? Nonetheless, this period produced the chant called *Gyapot éneke* [The Song of Cotton], which sounds more satirical in light of the pictures than had been intended: "I am rather small,/ But I will grow tall,/ If I manage to learn / Soviet agricultural modes."[311]

Everything from changing the course of rivers to the production of exotic industrial crops in extreme conditions, experiments that would make even science fiction authors uncomfortable, were presented in the contemporary press as Stalin's great "monument to peace" and as a devastating strike against imperialist states. "We will keep striking down on imperialist war adventurers and their cronies, the bloodthirsty Tito gang. Participating in this fight is the greatest honor and foremost duty of every worker," reads the article of a national paper on the significance of picking cotton.[312]

Upon the eruption of the Korean War, the transformation of nature became associated with peace even more, as it presented a great opportunity to compare the war and destruction of the United States with the Soviet Union's struggles for the future. The satirical weekly paper *Ludas Matyi* created to entertain—or rather, to uneducate the working nation through entertainment—reflected on contemporary events with the following "joke", killing two birds with one stone: "Soviet Scientists: This land [pointing at the deserts of Siberia] is a desert and we are working on making it thrive in five years! American scientists: This land [pointing at the Korean peninsula] is thriving and we are working on turning it into a desert within five minutes!"[313] "Behold two different ways of transforming nature, two different policies. On the one hand, new power plants and gigantic river regulations, and on the other, a ruthless and insane arms race, plans for genocide, and preparation for a new world war."³¹⁴

In the 1950s, the "heroic" plan of building dams on the Volga received a lot of media attention alongside the popularization of water management investments along the Tisza River. "The Volga is the course of communism,"³¹⁵ reads the conclusion of a Soviet novel, and the constant parroting of the omnipotence of Stalin, god of communism or paradise on Earth, is borderline blasphemy in the following lines: "If you say, let there be roads, then roads shall be built in the mountains; if you say let there be no mountains, the mountains shall be gone. Rivers will be born if you say, let there be rivers, and we shall block their path if you say, stop!"³¹⁶ The simplistic approach of news writers often shows in how they attempted to show the significance of constructions with incredibly high numbers, such as the amount of earth that had to be extracted to lay down foundations for the dam or the construction of the canals, or how much concrete and cement, how many bricks and who knows what else would be used. The results, of course, would speak for themselves. "Teacher: Show us the river Don, boy! Student: Which one, today's or yesterday's?"³¹⁷ "Tell me, father, why are we building these irrigation canals? For two reasons: one, to irrigate Soviet lands, and two, to cool down American war aggressors."³¹⁸ Another telling joke was the "wisecrack" of a stork that met Old Lady Drought on the endless plains of the Alföld and said, "You hear me, Drought? You'll be going under water soon!"³¹⁹

Heavy industrial investments were naturally more publicized in "advertisements" of the five-year plan than new agricultural achievements, so it is not surprising that the construction of Sztálinváros was a popular topic in the press. "You do not build cities by building a house and then another one and another one. Only amateurs would build cities like that. Cities are built in one go. Entire rows of streets are built together and all the walls are erected simultaneously. In the end it feels like the whole city was made in half an hour."³²⁰ And indeed the rising chimneys had built a new world in the regions formerly living from agriculture, significantly changing the characteristics of the landscape. However, these seemingly frightening changes were considered huge achievements, and there were satirical comics of well-sweeps, the symbol of the Alföld plains, looking sad upon realizing that they are dwarfed by the skyscrapers and chimneys of developing cities.³²¹ The only thing sadder was the return of migrating birds that only recognized their homes by the road sign of Dunapentele.³²²

The above excerpts and quotes are merely a handful of examples of the flood of political propaganda aiming to rule and transform Hungarian thinking. It also shows how the Communist thinking not only ignored the idea of protecting nature and pursuing sustainable development, but was outright proud of forcefully interfering with its environment and transforming living nature to its own image, as exemplified by the following lines of a contemporary movement song: "Smokestacks are pouring out smoke all over the country,/Factories, furnaces and machines are blaring."[323]

After 1953, the above topics still appeared in the news, but did so less and less, and were significantly toned down. News concentrated first and foremost on irrigation and rice fields, and then in the second half of the 1950s began paying greater attention to the afforestation of the Alföld plains and the establishment of protection forests. The concept of nature transformation was still used, but was now used within reasonable boundaries as man and nature arrived at a compromise.

Life After the Plan

In the sciences, the approach that relies on a unilateral interpretation of the relationship of society and nature, disregarding the role of the geographical environment, is called geographic nihilism, and this perspective is based on the idea that the change and development of society is a faster process than the change of the geographical environment, which further decreases the role assigned to natural factors. Representatives of these views ignored the role of natural factors and considered social order to be the determining factor in socioeconomic development, while they viewed nature as completely and wantonly customizable. Therefore, economic policy decisions had largely disregarded the impact of natural factors on production and the practical and efficient placement of productive forces.[324] In the 1950s, this approach dominated the development of plans for industrial investments in Hungary, the construction of new Socialist cities to serve as industrial sites, and the idea of becoming "the country of iron and steel," with the Soviet model as their guide.

With the death of Stalin, the Stalinist Plan for the Transformation of Nature was mostly abandoned in Hungary, although changes had only become tangible upon the government change in July 1953. At the 19 March 1953 session of the Political Committee of the Central

Committee of the Hungarian Workers' Party, criticism regarding industrial crop production was centered on how changing the ratio of farmland and decreasing the farmland used for grains was having a negative impact on national industrial crop production and how low crop yields were also decreasing the average crop yield. However, besides the droughts, the low quality of manure and fertilizer, and inefficient irrigation, their reasons included neglecting the use of Soviet scientific achievements and superior production knowledge.[325] Nevertheless, the things left out of these documents are also telling: from that time, party decisions no longer contained the encouragement of experimentation with the new industrial crops that they had previously attempted to introduce and produce.[326]

On 6 April 1953, Minister of State Farms and Forestries András Hegedűs prepared a report for the Secretariat of the Hungarian Workers' Party, informing them of the state of Hungarian agricultural scientific work. Among his negative observations, Hegedűs listed that "there is still no sign of an ideological struggle in favor of materialistic biology, and a portion of agricultural researchers are working with Morganist theories."[327] His suggestions included that the Hungarian Academy of Sciences and the Ministry of Agriculture hold a public debate on the most significant practical issues of Michurinist biology as well as Lysenko's new theory of species formation.[328] However, he did not mention anything about the future of non-native industrial crop experiments.

The political committee had only touched upon agriculture on 22 July, after the establishment of the Imre Nagy government, and regarding our topic, the title of their agenda speaks volumes: "Javaslat a mezőgazdaság terén elkövetett hibák kijavítására" [Proposal for the Correction of Mistakes Committed in Agriculture]. The document contains only minimal hints at the mistakes of nature transformation policies; however, we do find the intention to revise the production structures of agriculture by reading between the lines: for example, they intended to significantly increase the farmland used for wheat and corn while deciding to reduce the farmland used for cotton by 40,250 hectares.[329]

At the 19 December 1953 session of the Central Committee of the Hungarian Workers' Party, the decision of the Central Committee and the Council of Ministries on the development of agricultural production was approved. This decision was obviously critical of the agricultural strategies of previous years, stating that: "We need to stop the planning and implementation of agricultural schemes disregarding local environmental characteristics, and must support the production of crops and breeding of livestock that are compatible with local environmental conditions."[330] Criticism eventually turned into self-criticism as well: "The forced production of industrial crops in regions unsuitable for their production and with no regard to the natural environmental conditions of Hungary is faulty practice and must be stopped."[331] Judging by these quotes, we could assume that this decision was the end of cotton production in Hungary, but we would be wrong: of all non-native crops, cotton was the only one that was still prescribed in agricultural agendas, and while the amount of farmland used for cotton was reduced, an increased crop yield was requested. The reasons behind this decision are hard to determine. One factor may have been that, in 1950, the Soviet Union made cotton production obligatory in all COMECON member states,[332] and this decision was not withdrawn even after the death of Stalin. Another reason could have been that cotton supplied not only the textile industry but also the military industry, and Cold War politics relied on the production of explosive cotton.

The decision of 19 December 1953 on the development of agricultural scientific work no longer mentioned Michurin or Lysenko; it did, however, mention that "in order to restore the fertility of Hungarian soil, we must develop a system for the crop rotation, fertilization and soil cultivation of Hungary's top agricultural production sites"—meaning that the grass-crop rotation system and jarovization were no longer listed as priorities.[333]

As a result of this decision, pressure on Hungarian scholars to study and apply Michurinist and Lysenkoist biology gradually decreased, largely due to the fact that the miraculous production results promised by Lysenko had been proven false and the Soviet Union was forced to keep relying on food imports. However, Hungarian scholars could only truly return to ideologically independent research after the Revolution of 1956. Eventually, the concept of chromosomes was reintroduced into Hungarian scientific discussion along with the determining role of genes and mutation in inheritance, and the rules of Mendel. Lysenko received criticism in the Soviet Union as well, and thus the slow rehabilitation of scientific genetics began. The process only stopped temporarily around 1957 when it was discovered that Khrushchev, head secretary of the Soviet Communist Party, endeavored to restore the reputation of his comrade.[334]

Lysenko had only travelled abroad once when he visited Hungary in January 1960 to attend a two-day national corn production meeting. He also visited several universities and research institutes during his stay, but the highlight of his visit was holding a lecture titled "A micsurini biológia időszerű kérdései" [Relevant Issues of Michurinist Biology] at the Hungarian Academy of Sciences on 23 January 1960. The lecture was attended by party and government leaders, and the assembly room and adjacent rooms of the academy were completely full.[335] His audience anxiously awaited the lecture; however, Lysenko failed to answer the over two hundred questions previously addressed to him, leaving his audience stunned.[336] Had there been any faithful disciples of Lysenko in Hungary left by that time, he had definitely lost their support now.

It is difficult for historians to assess the amount of damage—and for that matter, if there were any benefits—arising from direct interference with nature in the first half of the 1950s. We have no means of assessing how much damage was incurred by taking farmland away from traditional industrial crop production to plant cotton, especially at the beginning of the 1950s. Experiments with citruses, kenaf, ramie and the *Taraxacum kok-saghyz* have long since been forgotten—politicians were simply forced to realize that no matter how much they craved success, the temperate climate of Hungary was simply unsuitable for the natural introduction of tropical industrial crops. Cotton persisted the longest, its farm production forced until the very end of the 1950s, only to be gradually left out of annual plantation plans. Rice, on the other hand, was considered a more successful investment since it is still produced in Hungary—now with appropriate production technologies, reasonable costs, and new methods of production.[337] However, erasing the traces of former rice production experiments is still an ongoing process in the Hortobágy National Park, where a gene bank has been established in order to reintroduce once native greenery and, essentially, to revert the land to its former state.[338] Nevertheless, afforestation and the establishment of protection forests in the 1950s have proven to be valuable endeavors, as have the dams on the Tisza River and the finished main canals. The real problem is that these have not been developed since, the issue of drainage has not yet been resolved, there are not enough reservoirs to store flood water for droughty seasons, and the irrigation system is still insufficient.

The former Socialist cities brought forth by voluntarism were struggling for survival and their shifting identities. After the system change of

1990, most of the industrial compounds that supported the inhabitants were privatized; the majority of them changed owners several times and a good number of them closed down, increasing unemployment rates not only in the involved settlements but also their surrounding area.

Zsuzsanna Borvendég is a PhD student and researcher at the Historical Archives of the Hungarian State Security in Budapest. She specializes in contemporary Hungarian history. She published *The International Organization of Journalists in Hungary* (2015) and co-authored, together with Mária Palasik, *Vadhajtások – A sztálini természetátalakítási terv átültetése Magyarországon 1948–1956* (2015, in Hungarian) about the transformation of nature in Hungary.

Mária Palasik is a Jean Monnet Professor and head of the research department at the Historical Archives of the Hungarian State Security. She specializes in contemporary Hungarian history and the history of technology. Her book *Chess-Game for Democracy: Hungary between East and West in 1944–47* (Montreal & Kingston, McGill-Queen's University Press, 2011) discusses the dramatic struggle for power in Hungary after the Second World War. Her most recent book, co-authored with Zsuzsanna Borvendég, is *Vadhajtások – A sztálini természetátalakítási terv átültetése Magyarországon 1948–1956* (2015, in Hungarian) about the transformation of nature in Hungary.

Notes

1. M. Palasik, *Chess Game for Democracy: Hungary between East and West 1944–1947*, Montreal and Kingston, London, and Ithaca, 2011, 101–35.
2. Mátyás Rákosi (1892–1971) earned his trading certificate at the Eastern Academy of Trade. At the beginning of 1915, he was ordered to the Eastern Front as a cadet candidate, and captured by the Russians in April. However, he escaped at the beginning of 1918, returning to Hungary in May of the same year. In November 1918, he became a founding member of the Hungarian Party of Communists. When the Hungarian Republic of Councils was declared in 1919, he was appointed Deputy Commissary of Trade, and became a member of the Revolutionary Governing Council as well. Upon the fall of the Republic of Councils in August 1919, he escaped to Vienna where he was interned along with other communist emigrants. He then traveled to Soviet Russia in 1920 and returned to Hungary illegally in 1924, where he was arrested in September 1925 on charges of rebellion, and received an

eight-and-a-half-year prison sentence from the Criminal Court of Budapest in August 1926. On completion of his sentence, he was persecuted for his activities under the Hungarian Republic of Council and received a life sentence in February 1935; however, following the Hungarian–Soviet treaty, he was released in 1940 on the condition that he leave the country and relocate to the Soviet Union immediately. He returned to Hungary once again at the end of January 1945, where he not only became the leader of the party, but was also appointed Vice President of the Council of Ministers (September 1949 – August 1952), eventually becoming the President of the Council of Ministers until 4 July 1953. After he was relieved from his position as State Secretary and was not re-elected onto the political committee—on account of his state of health—at the Central Committee session of the Hungarian Workers' Party between 18–21 July 1956, Mátyás Rákosi went to the Soviet Union to 'receive treatment' and eventually passed away in the USSR. Besides Rákosi, another prominent figure of the small political group was Ernő Gerő (1898–1980), who joined the Communist Party in 1918. He escaped to Vienna after the fall of the Hungarian Soviet Union and was sent back to Hungary in 1922 to lead Hungarian Communist operations. He was arrested in September and sentenced to fifteen years in May 1923, but was transferred to the Soviet Union in 1924. He continued to work for the party in France, Belgium, Spain, and Portugal, and represented the Communist Party of Hungary at the Communist Internationale between 1939 and 1941. After 1945, he received key positions in government and economic matters, becoming Minister of Trade and Transportation, Minister of Transportation, Minister of Finance, Minister of State and Minister of the Interior as well as Deputy Prime Minister. Following the dismissal of Mátyás Rákosi, he became General Secretary of the Central Committee of the Party from July 1956 until 25 October 1956. Discharged from his position during the revolution of 1956, he escaped to the Soviet Union, and upon his return in 1960 he was persecuted for his illegal activities and political maneuvers between 1948 and 1956, and dismissed from the party.

3 Act XX of 1949. *Törvények és rendeletek hivatalos gyűjteménye I. Törvények és törvényerejű rendeletek 1949* [Official Compilation of Acts and Regulations, Volume I. Acts and Statutory Regulations of 1949], Presidential Bureau of the Council of Ministers, Budapest, 1950, 3–9.

4 M. Palasik, *A jogállam megteremtésének kísérlete és kudarca Magyarországon, 1944–1949* [Attempt and Failure: Hungarian Rule of Law between 1944–1949], Budapest, 2000, 311–14.

5 Ibid., 319 and 322.

6 Hungarian and French teacher László Rajk (1909–1949) joined the Communist Party in 1931 and was arrested several times for illegal political activities. He participated in the Spanish Civil War and was a political commissary and party secretary of the Hungarian battalion of the International Brigade. Upon his return in 1941 he was arrested and interned, only to

become one of the leaders of the resistance after his release in September 1944. He was then arrested by members of the Arrow-Cross in December 1944 and deported to Germany. He returned in May 1945, joining the party's work and national politics, and eventually became a member of all governing bodies of the Hungarian Communist Party: he was secretary of the party committee in Budapest from May to November 1945, then deputy party secretary from November 1945 to March 1946, was Minister of the Interior until August 1948 and then Minister of Foreign Affairs until 20 May 1949. On 30 May he was arrested on false charges, sentenced to death and executed. He was rehabilitated in 1955 and given a ceremonial reburial in the cemetery of Kerepesi on 6 October 1956.

7   This involved such as former Minister of Justice István Ries as well as Árpád Szakasits and György Marosán, key social democrat allies in merging the two parties together. The purge continued with the leaders of the illegal Communist movement in Hungary: János Kádár, Dr. Ferenc Donáth and Gyula Kállai. Not even the state security apparatus was spared: from the end of 1949, Rákosi's leadership had tried and sentenced a host of collaborators, leading to the arrest of Gábor Péter, head of the secret services, in January 1953. Gy. Gyarmati and T. Valuch, *Hungary under Soviet domination 1944–1989*, New York, 2009, 166; Gy. Gyarmati, *A Rákosi-korszak. Rendszerváltó fordulatok évtizede Magyarországon, 1945–1956* [The Rákosi Era: The Decade of Regime Changes in Hungary, 1945–1956], Budapest, 2011, 249–77.

8   M. Palasik, "A *Rákosi-rendszer* hétköznapjai" [Daily Life in the *Rákosi System*], *Hitel*, no. 27 (March 1990), XXI–XXV.

9   I. Pető and S. Szakács, *A hazai gazdaság négy évtizedének története 1945–1985* [The History of Four Decades of Hungarian Economy, 1945–1985], Budapest, 1985, 95–104.

10 T. Valuch, *Magyarország társadalomtörténete a XX. század második felében* [The History of Hungaran Society in the Second Half of Twentieth Century], Budapest, 2001, 94–118.

11 The number of affected families was 75,000–80,000. T. Stark, "Magyarország háborús embervesztesége" [Hungary and Its Casualties of War], *Rubicon*, no. 9 (2000); T. Valuch, *Magyarország társadalomtörténete*, 121–22.

12 See B. Bank, Gy. Gyarmati and M. Palasik, '*Állami titok'. Internáló- és kényszermunkatáborok Magyarországon, 1945–1953* ["State Secret." Internment and Forced Labor Camps in Hungary, 1945–1953], Budapest, 2012.

13 I.T. Berend, *Terelőúton. Közép- és Kelet-Európa 1944–1990.* [Detours. Central and Eastern Europe, 1944–1990], Budapest, 1999, 113; P. Germuska, "Szocialista csoda? Magyar iparfejlesztési politika és gazdasági növekedés, 1950–1975" [Socialist Miracle? Hungarian Politics of Industrial Development and Economic Growth, 1950–1975], *Századok*, no. 1 (2012), 47–78.

14 See I.T. Berend, *Gazdaságpolitika az első ötéves terv megindításakor 1948–1950* [Economic Policy at the Initiation of the First Five-Year Plan, 1948–1950], Budapest, 1964, 95; Pető and Szakács, *A hazai gazdaság négy évtizedének*

története, 95–104, 154; I. Birta, "A szocialista iparosítási politika néhány kérdése az első ötéves terv időszakában" [Issues of the Socialist Industrialization Policy during the First Five-Year Plan], *Párttörténeti Közlemények*, no. 3 (1970), 113–50; I. Okváth, *Bástya a béke frontján. Magyar haderő és katonapolitika 1945–1956* [Bastion at the Front of Peace. The Hungarian Armed Forces and Army Politics Between 1945–1956], Budapest, 1998, 79, 110; P. Germuska, "A szocialista iparosítás Magyarországon 1947–1953 között" [Socialist Industrialization in Hungary Between 1947–1953], *Évkönyv 2001* [Year Book 2001], Budapest, 2001, 160; Gy. Gyarmati, *A Rákosi-korszak. Rendszerváltó évtizedek története Magyarországon 1945–1956* [The Rákosi Era: The Decade of Regime Changes in Hungary, 1945–1956], Budapest, 2011.
15 In 1949: 1 USD = 11.74 HUF.
16 I.T. Berend, *A szocialista gazdaság fejlődése Magyarországon 1945–1968* [Socialist Economic Development in Hungary Between 1945–1968], 1974, 78.
17 See Berend, *Gazdaságpolitika az első ötéves terv megindításakor*, 95; Pető and Szakács, *A hazai gazdaság négy évtizedének története*, 154; Birta, "A szocialista iparosítási politika néhány kérdése az első ötéves terv időszakában," 113–50; Okváth, *Bástya a béke frontján*, 79, 110; Germuska, "A szocialista iparosítás Magyarországon 1947–1953 között," 160.
18 K. Kaplan, *Dans les Archives du comité central: Trente ans de secrets du bloc soviétique*, Paris, 1978, 165–66.
19 T. Toranska, *"Them": Stalin's Polish Puppets*, New York, 1987, 46.
20 C. Cristescu, "Ianuarie 1951: Stalin decide înarmarea Romanei," *Magazin Istoric*, no. 10 (1995), 15–23. Čepička, Ochab and Bodnăraş are referenced in V. Mastny, *NATO in the Beholder's Eye: Soviet Perceptions and Policies, 1949–1956*, CWIHP Working Paper No. 35, Washington, DC, no. 3 (2002), Woodrow Wilson International Center for Scholars, Cold War International History Project.
21 M. Rákosi, *Visszaemlékezések, 1940–1956* [Memoirs, 1940–1956], Vol. 2, Budapest, 1997, 860–61.
22 Cristescu, "Ianuarie 1951: Stalin decide înarmarea Romanei."
23 Rákosi, *Visszaemlékezések*, 861. For the meeting in Moscow, see M. Palasik, "A nők tömeges munkába állítása az iparban az 1950-es évek elején" [Mass Employment of Women in the Industrial Sector at the Beginning of the 1950's], M. Palasik, and B. Sipos, eds., *Házastárs? Munkatárs? Vetélytárs? A női szerepek változása a 20. századi Magyarországon* [Spouse, Colleague, Rival? Changing Female Roles in 20th Century Hungary], Budapest, 2005, 82–84.
24 Body of three that operated between the fall of 1949 and the end of 1952 without the knowledge of the Central Committee of the Hungarian Workers' Party—comprised of Mátyás Rákosi, Ernő Gerő, and Mihály Farkas.
25 Rákosi, *Visszaemlékezések*, 862.

26 P. Germuska, *Indusztria bűvöletében. Fejlesztéspolitika és a szocialista városok* [Under the Spell of Industria: Development Policies and Socialist Cities], Budapest, 2004, 109.
27 The number of citizens had nearly quadrupled from 21 million to 83 million. See *Strana sovetov za 50 let* (Sbornik schtatisticheskih materialov) [Страна советов за 50 лет. (Сборник статистических материалов) – Fifty Years of the Soviet State (Collection of Statistical Data)], eds. S.J. Genin and O.K. Makarova, et al., Moscow, 1967, 22.
28 Ibid.
29 Germuska, *Indusztria bűvöletében*, 48–51.
30 Ibid., 49.
31 Following the system change, Berente separated from Kazincbarcika in 1999 and has been a separate municipal entity since.
32 Most of the labor for the future factory was supplied by the forced labor camp opened on 6 October 1951 and managed by the State Security Authority, populated by former prisoners of war, most of them soldiers and gendarmes who had been in Soviet *lager*s and were released by the Soviet Union in December 1950 and at the beginning of 1951, only to be placed under surveillance by an examination committee. There were also people of German nationality among their ranks whose families had been interned from Germany and had "no place" to be liberated in contemporary Hungary. See Bank, Gyarmati, and Palasik, "Állami titok," 46–49.
33 Zs. Borvendég, "A Former Socialist Model City Tatabánya on the Edge of Environmental Disaster." Environmental Histories of the Visegrad Countries: Cold War and the Environmental Sciences. Workshop, Institute for Contemporary History of the Academy of Sciences of the Czech Republic, Prague, 23–25 March, 2012.
34 "1949. december 14-i ülés" [Session of 14 December 1949], National Archives of Hungary (Magyar Nemzeti Levéltár Országos Levéltára, henceforth: MNL OL) M-KS 276. f. 54/77. ő. e.
35 A. Horváth, *A Dunai Vasmű felépítésének rövid története* [A Brief History of the Construction of the Danube Iron Factory], 1972, Fejér County Archives of the National Archives of Hungary (Magyar Nemzeti Levéltár Fejér Megyei Levéltára, henceforth: MNL FML), Library, Manuscript Archive, 864, 6–7.
36 MNL OL M-KS 276. f. 53/42. ő. e.
37 S. Marosi and J. Szilárd, eds, *A dunai Alföld. Magyarország tájföldrajza* [The Danube-Alföld Region. Hungarian Regional Landscape Geography], Vol. I, Budapest, 1967, 255.
38 Horváth, *A Dunai Vasmű felépítésének rövid története*, MNL FML, Library, Manuscript Archive, 864, 8.
39 Ibid., 9–11.
40 I. Lázár, *Életrajzi beszélgetések* [Biographical Interviews], 1978, 8. http://orkenyistvan.hu/elertajzi_beszelgetesek (accessed 13 December 2012). Literal translation by Éva Misits.

41 See M. Miskolczi, *Az első évtized. Dunapentelétől – Dunaújvárosig* [The First Decade. From Dunapentele to Dunaújváros], Dunaújváros, 1975; S. Horváth, *A kapu és a határ: Mindennapi Sztálinváros* [The Gate and the Border: Daily Sztálinváros], Budapest, 2004.
42 On this subject, see M. Palasik, "A New Industrial Town and Environment in Hungary in the 1950s and its Heritage." Environmental Histories of the Visegrad Countries: Cold War and the Environmental Sciences. Workshop, Institute for Contemporary History of the Academy of Sciences of the Czech Republic, Prague, 23–25 March 2012.
43 B. Éberhardt, *A dunai partfal* [The Danube Embankment]. http://eberhardt.rpotor.com/16ADunaiPartfal.pdf (accessed 29 February 2012).
44 T. Bereczkei, "Tudás a semmiből: a Liszenko-ügy" [Knowledge from Nowhere: The Case of Lysenko], *Természet Világa* 3 (1998). http://www.kfki.hu/chemonet/hun/teazo/liszenko.html (accessed 23 December 2012).
45 T.D. Lysenko (1898–1976) was a Soviet agronomist and plant-breeder who created the Theory of Michurinist Biology claiming as predecessor the autodidact horticulturist Ivan Vladimirovich Michurin (1855–1939). Lysenko was a member of the Soviet Academy of Sciences and director of the Genetic Institute of the Academy, as well as president of the Lenin Academy of Agricultural Sciences. Referring to Michurin's work, Lysenko concluded that inherited characteristics are determined by environmental circumstances rather than by genes. He rejected the principles of modern biology and modern genetics in particular, deeming the inheritance of acquired characteristics a proven fact. See S. Igali, "A liszenkoizmus Magyarországon" [Lysenkoism in Hungary], *Valóság* 2 (2002), 41–43; M. Müller, "Liszenko emlékezetes előadása a Magyar Tudományos Akadémián 1960-ban" [Lysenko's Memorable Presentation at the Hungarian Academy of Sciences in 1960]. *Magyar Tudomány* [Hungarian Sciences] II (2011), 1356.
46 Bereczkei, "Tudás a semmiből: a Liszenko-ügy."
47 Z. Hajdú, "Környezet és politika: a természetátalakítás 'zseniális' sztálini terve" [Environment and Politics: The 'Brilliant' Stalinist Plan to Transform Nature], *Változó környezetünk* [Our Changing Environment], ed. J. Tóth and Z. Wilhelm, Pécs, 1999, 131–45.
48 Bereczkei, "Tudás a semmiből: a Liszenko-ügy."
49 Ibid. According to Professor Bereczki: "These manipulations were, of course, unsuccessful since—as we all know—the characteristics acquired during individual life cycles are not transferred to the genetic pool."
50 Z. Hajdú, "A szocialista természetátalakítás kérdései Magyarországon, 1948–1956" [The Issues of Socialist Nature Transformation in Hungary, 1948–1956], in *Táj, környezet és társadalom. Ünnepi tanulmányok Keveiné Bárány Ilona professzor asszony tiszteletére* [Landscape, Environment, and Society. Jubilee Papers in Honor of Professor Ilona Keveiné Bárány], ed. A. Kiss, G. Mezősy, and Z. Sümeghy, Szeged, 2006, 246.
51 *Agrártudomány* 1, no. 1 (1949).

52 T.D. Lysenko, *A biológiai tudomány állásáról* [On the State of the Biological Sciences], Budapest, 1949.
53 *Micsurin, I.V. Válogatott tanulmányai* [Selected Papers of I.V. Michurin], Budapest, 1950; T.D. Lysenko, *Agrobiológia* [Agrobiology], Budapest, 1950.
54 E.J. Gluschenko, *Micsurin tanítása az idealista biológia elleni küzdelemben* [Michurin's Teachings in the Struggle against Idealist Biology], Természettudományos kiskönyvtár [Little Library of Natural Sciences], no. 12, Budapest, 1949; A.I. Molodšikov, *A szovjet tudomány – a természet átalakítója* [Soviet Science – Transformer of Nature], Természettudományos kiskönyvtár [Little Library of Natural Sciences], no. 15, Budapest, 1949; K.A. Timirjazev, I.V. Michurin, and T.D. Lysenko, *Hogyan alakítja át az ember a növényeket* [How Plants Are Transformed by Man], Természettudományos kiskönyvtár [Little Library of Natural Sciences], no. 16, Budapest, 1949.
55 Igali, "A liszenkoizmus Magyarországon," 43.
56 *A micsurini biológia tanítása a falusi iskolákban* [Teaching Michurinist Biology in Village Schools], A szocialista mezőgazdaság [Socialist Agriculture], no. 15, Budapest, 1950; B. Faludi, Gy. Kiszely, and H. Tangl, *Az ember szervezete. Gimnáziumok számára. Ideiglenes tankönyv* [The Human Organism. For High Schools. Temporary Textbook], Budapest, 1950; M. Vaszkó, *Természetismeret. Ideiglenes tankönyv* [Natural History. Temporary Textbook], Budapest, 1950.
57 MNL OL M-KS 276. f. 93/621. ő. e., 104.
58 Igali, "A liszenkoizmus Magyarországon," 43.
59 G. Iby, "I.E. Gluscsenko professzor látogatásához" [To the Visit of Professor Gluschenko], *Erdészeti Lapok*, March 1949, 49–50. http://erdeszetilapok.oszk.hu/00845/pdf/EL_1949_03_49.pdf (accessed 26 December 2012).
60 MNL OL M-KS 276. f. 54/92. ő. e., MNL OL M-KS 276. f. 54/102. ő. e.
61 MNL OL M-KS 276. f. 54/136. ő. e., MNL OL M-KS 276. f. 54/149. ő. e.
62 MNL OL MK-S 276. f. 93/167. ő. e., 415. Issues included the application of grass-crop rotation, farmland protection by the establishment of protection forests and improving the average production of cotton.
63 "In memoriam Dr. Igali Sándor" [In Memory of Dr. Sándor Igali], in *Az Országos Frédéric Joliot-Curie Sugárbiológiai és Sugáregészségügyi Kutató Intézet ötven éve* [Fifty Years of the Frédéric Joliot-Curie National Research Institute for Radiobiology and Radiohygiene], ed. I. Turai, Budapest, 2007, 145–46.
64 Igali, "A liszenkoizmus Magyarországon," 45.
65 Zoltán Kodály (1882–1967) was a Hungarian composer, musicologist, folk music researcher, member of the Hungarian Academy of Sciences and developer of the "Kodály method." He was president of the Hungarian Academy of Sciences between 1946 and 1949.
66 F. Glatz, "Akadémia és tudománypolitika a volt szocialista országokban 1922–1999" [The Academy and Scientific Policies in Former Socialist Countries between 1922–1999], *Magyar Tudomány*, no. 4 (2002).

67 "Jegyzőkönyv a Magyar Dolgozók Pártja Titkársági üléséről" [Minutes of the Secretariat Session of the Hungarian Workers' Party], 1 July 1948, MNL OL M-KS 276. f. 54/3. ő. e.
68 Act XXXVIII of 1948 on the Establishment of the Hungarian Scientific Council. http://www.1000ev.hu/index.php?a=3&param=8320 (accessed 20 December 2012).
69 On the operation and role of the Hungarian Science Council, see T. Huszár, *A hatalom rejtett dimenziói. Magyar Tudományos Tanács 1948–1949* [Secret Dimensions of Power. The Hungarian Scientific Council 1948–1949], Budapest, 1995.
70 "A Magyar Tudományos Tanács 1949. évi munkaterve súlyponti kérdéseinek részleges részletezése" [Detailed Discussion of the Key Issues of the 1949 Work Plan of the Hungarian Science Council], 24 February 1949, MNL OL M-KS 276. f. 54/31. ö. e., 266.
71 Ibid., 256.
72 Government Decree 1590/1949 (II. 19) on the Establishment of the Agricultural Science Center.
73 "A Magyar Tudományos Tanács 1949. évi munkaterve súlyponti kérdéseinek részleges részletezése," 24 February 1949, MNL OL M-KS 276. f. 54/31. ö. e., 256.
74 During its eight months of operation, the Hungarian Science Council established twenty-six individual institutes; however, most of these were merely reorganized former institutions, and many were phantom institutions that only existed on paper. J. Pótó, "Harmadik nekifutásra. A Magyar Tudományos Akadémia 'átszervezése', 1948–1949" [On the Third Try: The "Reorganization" of the Hungarian Academy of Sciences Between 1948–1949], *Történeti Szemle* 94, no. 1–2, 89.
75 Huszár, *A hatalom rejtett dimenziói*, 105–6.
76 Huszár, *A hatalom rejtett dimenziói*, 108. Ernő Gerő's letter "Rákosi, Farkas, Kádár elvtársaknak!" [To Comrades Rákosi, Farkas, Kádár!] of 1 June 1949.
77 "Javaslat a Magyar Tudományos Akadémia és a Magyar Tudományos Tanács átszervezése tárgyában" [Proposition Regarding the Reorganization of the Hungarian Academy of Sciences and the Hungarian Science Council], 14 September 1949, MNL OL M-KS 276. f. 54/62. ö. e., 86.
78 S. Kónya, "A Magyar Tudományos Tanács és a Magyar Tudományos Akadémia egyesítése (1949)" [Uniting the Hungarian Scientific Council and the Hungarian Academy of Sciences (1949)], *Magyar Tudomány*, no. 2 (1990), 222.
79 Pótó, "Harmadik nekifutásra," 97.
80 Kónya, "A Magyar Tudományos Tanács," 222.
81 S. Kónya, 'Az akadémiai tagság összetételének változásai 1945–1949' [Changes in the Membership Composition of the Academy Between 1945–1949], *Magyar Tudomány*, no. 6 (1989), 503–504.

82 "Feljegyzés a Magyar Tudományos Akadémia IV. Osztálya és a Mezőgazdasági Kísérletügyi Központ Kísérletügyi Osztályának feladatai szabályozásával kapcsolatban" [Note Regarding the Regulation of the Tasks of Department IV of the Hungarian Academy of Science and the Experiment Department of the Agricultural Experiment Center], undated, MNL OL 276. f. 93/168. ö. e., 255.
83 Decree 145/1950 (V. 20.) of the Council of Ministers on the Establishment of the Agricultural Experiment Center.
84 "Kísérleti Tangazdaságok szervezésének bejelentése" [Announcing the Organization of Experiment Farms], 10 December 1949, MNL OL OT XIX-A-16-a 8060/19/1949.
85 The experimental farms were as following: Mosonmagyaróvár, Keszthely, Debrecen, Herceghalom, Kecskemét, Szarvas, Karcag, Mohács, Martonvásár, and a Nagytétény. "Előterjesztés a Népgazdasági Tanácshoz" [Memorandum to the National Agricultural Council], 27 March 1950, MNL OL 276. f. 93/223. ö. e., 20–21.
86 "A Mezőgazdasági Tudományos Központhoz Tartozó intézetek és telepek" [Institutions and Sites of the Agricultural Science Center], 9 February 1950, MNL OL 276. f. 93/168, ö. e., 167–75.
87 "Feljegyzés a Magyar Tudományos Akadémia Elnöksége részére" [Note to the Presidency of the Hungarian Academy of Sciences], 14 September 1950, MNL OL 276. f. 93/168. ö. e., 259.
88 "Mezőgazdasági Szervezési Intézet" [Agricultural Organization Institute], undated, MNL OL 276. f. 93/224. ő. e.
89 Ibid.
90 "Előterjesztés a Népgazdasági Tanácshoz" [Memorandum to the National Economic Council], undated, MNL OL 276. f. 73/223 ö. e., 278–82.
91 "Javaslat a Földművelésügyi Minisztérium Tudományos Tanácsának felállításáról" [Proposition for the Establishment of the Scientific Council of the Ministry of Agriculture], 10 October 1950, MNL OL 276. f. 93/223. ö. e.
92 Ernő Gerő's speech in Parliament, 9 December 1949, 279, *Országgyűlési Napló* [Parliament Journal] Session 15 of Parliament, 9 December 1949. http://www3.arcanum.hu/onap/pics/a.pdf?v=pdf&a=pdf&p=PDF&id=KN-1949_1/KN-1949_1%20144&no=0 (accessed 8 September 2012).
93 Act XXV of 1949. Article 32. http://www.1000ev.hu/index.php?a=3&param=8370 (accessed 10 December 2012).
94 Act II of 1951. Article 28 (3). http://www.1000ev.hu/index.php?a=3&param=8380 (accessed 10 December 2012).
95 The proposal was made by the Agricultural and Cooperative Department of the Hungarian Workers' Party on 22 August 1950. It was approved by the Secretariat of the Hungarian Workers' Party with the addendum that the necessity of disseminating information on materialist biology had to be emphasized further. MNL OL M-KS 276. f. 54/114. ő. e.

96 The label 'reactionary' had become the most widespread concept at this time and was used often to stigmatize known or presumed enemies of the system.
97 All references cite the following document: MNL OL M-KS 276. f. 54/114. ő. e.
98 Ibid. It is important to note that very often only lemons are mentioned in documents, but it is obvious that, in the majority of cases, this refers to every type of citrus to be introduced to Hungary, including oranges and tangerines.
99 MTI, 18 June 1949.
100 M.A. Nagy, "Magyar gyapot" [Hungarian Cotton], *Búvár*, no. 11 (1937), 872.
101 Ibid., 873.
102 Ibid.
103 E. Obermayer, "A rizs és a gyapot meghonosítása" [Introducing Rice and Cotton] *Természettudományi Közlöny*, no. 4 (1949), 212.
104 Ibid., 216.
105 "A Gyapottermeltető Nemzeti Vállalat alapító levele" [Founding document of the National Cotton Production Company], 30 December 1948, MNL OL XIX-K-1-ee 2. d.
106 "A Gyapottermeltetési Nemzeti Vállalat szervezete" [Organization of the National Cotton Production Company], 3 May 1950, MNL OL 276. f. 93/251. ő. e., 98. Offices were located in Agárd, Katalinpuszta, Siklós, Baja, Kisújszállás, Szentes, Békéscsaba, and Nyíregyháza, with factory farms on Székkutas and Középhídvégpuszta. The two plant improvement farms were established in Szentes and Gyulatanya, with experimental farms on Siklós, Pusztaecseg, Iregszemcse, and Mosonmagyaróvár.
107 Ibid. In 1949, Schüller was also charged with directing the Crop Improvement and Crop Production Research Institute by Minister of Agriculture Ferenc Erdei; however, due to being heavily preoccupied with cotton production, he was not in office for long.
108 "Feljegyzés a szerződéses termelésbonyolításával foglalkozó nemzeti vállalatokról" [Note Regarding National Companies Engaging in Contractual Production], 13 March 1950, MNL OL 276. f. 93/219. ő. e., 58.
109 Ibid.
110 Ibid.
111 "Előterjesztés a Titkárság részére. A KGST legközelebbi ülésszakán napirendre kerülő kérdések" [Memorandum to the Secretariat. Issues to Be Discussed at the Next Session of the COMECON], 19 April 1950, MNL OL M-KS 276. f. 54/95, ő. e., 129.
112 Ibid.
113 "Javaslat a Gyapottermelési Kutató Intézet szervezésére" [Proposal for the Organization of the Cotton Production Research Institute], 25 October 1950, MNL OL 276. f. 93/184. ő. e.
114 Ibid.

115 "Előterjesztés a Népgazdasági Tanácshoz a Gyapottermelési Kutató Intézet megszervezésére" [Memorandum to the National Economic Council on the Organization of the Cotton Production Research Institute], 27 December 1950, MNL OL 276, f. 93/223, ő. e., 106-8. Not only did the National Cotton Production Company lose its right to cotton-related research, but also the right to experiments introducing other exotic crops to Hungary. All of its experimental farms and their entire personnel had been placed under appropriate institutions of the Agricultural Experiment Center; for example, the experimental Russian dandelion farms in Püski and Mosonmagyaróvár were assigned to the Agricultural Research Institute of Mosonmagyaróvár, while the division in Igerszemcse was attached to the research farm at Igerszemcse.

116 The anti-company lobby would probably not have been entirely satisfied since Ferenc Erdei suggested at the 22 November 1950 secretariat session of the Hungarian Workers' Party that cotton production and cotton harvesting should be done by county cotton production companies under the Ministry of Agriculture, and that the National Cotton Production Company should be liquidated. The secretariat returned the proposal for modification, and the version approved two weeks later had reinforced the company's authority that had been drastically reduced compared to its original state. "Határozati javaslat a titkárság számára a szerződéses termelés szervezetéről" [Decision Proposal to the Secretariat on the Organization of Contractual Production], 22 November 1950, MNL OL M-KS 276. f. 54/119 ő. e., 298; and ibid., 6 December 1950, MNL OL M-KS 276. f. 54/121, ő. e., 31.

117 Ancient unit of measurement in Hungary. 1 cadastral yoke = 1,600 square fathoms, 5,755 square meters, or 0.575464 hectares.

118 "FM Mezőgazdasági Főosztályának feljegyzése" [Note by the Agricultural Department of the Ministry of Agriculture], 24 May 1950, MNL OL 276. f. 93/252, ő. e., 211.

119 Ibid., 210.

120 The size of cotton fields in 1950 by county were as follows: 1,465 hectares in Békés, 1,305 hectares in Bács-Kiskun, 834 hectares in Tolna, 738 hectares in Baranya, 547 hectares in Szolnok, 465 hectares in Csongrád, 295 hectares in Somogy, 287.5 hectares in Fejér, 226 hectares in Hajdú-Bihar, 29 hectares in Pest, 5 hectares in Veszprém and 0.5 hectares in Szabolcs County, for a total of 6,200 hectares of land used for cotton production. 'Vázlat a Gyapottermeltetési N. V. 1950. évi gyapottermő területéről' [Draft of cotton producing sites of the National Cotton Production Company in 1950], MNL OL 276. f. 93/251. ő. e.

121 "A Gyapottermeltető Nemzeti Vállalat jelentése az állami gazdaságoknál tartott szemle után" [Report of the National Cotton Production Company after Holding Audits at State Farms], 15 February 1950, MNL OL XIX-K-1-ee 17. d. 2764/1950.

122 "Gyapottermeltetési előirányzat módosítása" [Modification of Cotton Production Plans], undated, MNL OL XIX-K-1-ee 17. d. 8.214-Gy-3/1950.

123 "Szelevényi gyapottermelők panasza a miniszterhez" [Complaint Sent to the Minister by Cotton Producers in Szelevény], 20 July 1950, MNL OL XIX-K-1-ee 17. d.
124 Report of the National Cotton Production Company after Holding Audits at State Farms, 15 February 1950, MNL OL XIX-K-1-ee 17. d. 2764/1950.
125 "Összefoglaló jelentés a gyapottermelés helyzetéről" [Report on the State of Cotton Production], 23 May 1950, MNL OL 276, f. 93/251, ő. e.
126 Cotton plantations formed a homogenous stretch of land except for fields belonging to the Szeged district, home of the world famous Hungarian paprika. However, there had been cases where cotton occupied fields more suitable for paprika production. L. Görög, *Magyarország mezőgazdasági földrajza* [The Agricultural Geography of Hungary], Budapest, 1954, 99.
127 I. Orgoványi, "Határvédelem Magyarország déli határán 1948 és 1953 között" [Border Security at the Southern Hungarian Border between 1948 and 1953], in *A Nagy Testvér szatócsboltja. Tanulmányok a magyar titkosszolgálatok 1945 utáni történetéből* [Big Brother's Miserable Little Grocery Store. Studies on the History of the Hungarian Secret Services after World War II], ed. Gy. Gyarmati and M. Palasik, Budapest, 2012, 156.
128 Ibid., 158.
129 MNL OL XIX-K-1-ee 17. d. 11696/1950. Border Approach Permit Issued to Agents of the National Cotton Production Company. Undated. According to the letter, the quarantined territory had 1,725 hectares of cotton plantations, or 30 percent of Hungary's borderline cotton plantations.
130 "Éghajlati adatokhoz hozzájárulás" [Contribution to Climate Data], undated, MNL OL XIX-K-1-ee 17. d. 8214-Gy-51/1950.
131 Ibid.
132 "Feljegyzés az 1950. évi gyapottermelés begyűjtéséről és az 1951. évi gyapotterületekkel kapcsolatban végzett munkáról" [Note on the 1950 Cotton Harvest and Work Done on Cotton Plantations in 1951], 13 September 1950, MNL OL 276, f. 93/219, ő. e., 100.
133 "Gyapottermelési jelentés" [Cotton Production Report], 16 October 1950, MNL OL 276. f. 93/184. ő. e.
134 "Then these scoundrels come with this shitty cotton!" MNL OL XIX-K-1-ee 19. d. "Lődi Béla iváncsai tanácselnök felelősségre vonása" [Interrogation of Béla Lődi, Chairman of the Council of Invácsa], undated.
135 "A Gyulai Tanács VB mezőgazdasági osztálya által engedélyezett gyapotterületek csökkentése" [Reduction of Cotton Production Areas Approved by the Agricultural Department of the Central Committee of the Council of Gyula], 22 March 1951, MNL OL XIX-K-1-ee 18. d.
136 "A Gyulai Tanács VB mezőgazdasági osztálya által engedélyezett gyapotterületek csökkentése" [Reduction of Cotton Production Areas Approved by the Agricultural Department of the Central Committee of the Council of Gyula], 29 March 1951, MNL OL XIX-K-1-ee 18. d.

137 "A gyapot vetésterületének csökkentése" [Reduction of Cotton Production Areas], 12 April 1951, MNL OL XIX-K-1-ee 18. d.
138 "Gyapotvetés beállítása az északi megyékben" [Configuration of Cotton Planting in the Northern Counties of Hungary], 23 March 1951, MNL OL XIX-K-1-ee 18. d. FM 8214-Gy-307/1951.
139 "Nógrádi termelők kérése" [Request Made by Producers in Nógrád], 14 December 1951, MNL OL XIX-K-1-ee 19. d.
140 Everything is relative. If we take into account the fact that, in October 1950, Hungary sold 500 kilograms of cotton planting seeds to Czechoslovakia for experimentation, Nógrád no longer seems like a northern territory in comparison. "Vetőmag Csehszlovákiának" [Planting Seeds for Czechoslovakia], 25 October 1950, MNL OL XIX-K-1-ee 17. d. 8214-Gy-122/1950.
141 Ibid.
142 MNL OL 276. f. 93/354. ő. e. Phoned-in report of 7 July 1951, 84–85.
143 University and college party committees argued in favor of involving Hungarian youth in cotton harvests by saying that "universities notoriously divorce theory from practice, especially in the liberal arts, and picking cotton would allow Hungarian youth to engage in practical issues." MNL OL XIX-K-1-ee 19. d. "Egyetemi és főiskolai pártbizottság 1951. október 25-i ülése" [25 October Session of the University and College Party Committee in 1951].
144 Youth organization of the Hungarian Workers' Party.
145 It was established in 1950 to train industrial workers for the forced industrialization of Hungary. The act excluded the employment of school-age children as industrial trainees, but it was not uncommon that 7th and 8th graders (13–14 year olds) were recruited to meet designated requirements, meaning that these children could not even acquire their elementary school diplomas.
146 "Kiértékelő jelentés a gyapotszedési munkáról" [Assessment Report on Cotton Picking], 31 October 1951, MNL OL 276, f. 93/261, ő. e.
147 "Hivatalos feljegyzés Sándi Ottó elvtársnak" [Official Note to Comrade Ottó Sándi], 6 March 1952, MNL OL 276, f. 93/411, ő. e., 88–89.
148 A. Pünkösti, Kiválasztottak [The Chosen Ones], Budapest, 1988, 20–21.
149 Ibid., 21.
150 "Utasítás az öntözött gyapot termesztésére az 1952. Évben" [Instructions for Irrigated Cotton Production in the Year 1952], undated, MNL OL XIX-K-1-ee 85. d.
151 Görög, Magyarország mezőgazdasági földrajza, 99.
152 I. Romsics, Magyarország története a XX. században [20th Century Hungarian History], Budapest, 1999, 350.
153 "Jelentés a gyapot betakarításáról" [Report on the Cotton Harvest], 28 October 1953, MNL OL XIX-K-1-ee 116. d.
154 "Kérelem Nagy Imre elvtárshoz a gyapot betakarításához szükséges külső munkaerő biztosítására" [Request for Comrade Imre Nagy to Provide the

Labor Force Necessary to Harvest Cotton], 10 October 1953, MNL OL XIX-K-1-ee 116. d.
155   The word *Taraxacum* means "dandelion" and *saghyz* denotes plants that contain rubber, while *kok* is a Khirgiz term meaning "blue," referring to the blue or sage green color of the plant's foliage, setting it apart from other dandelions. László Kovátsits, *A hamvas gumipitypang termesztése és nemesítése* [Production and Improvement of the *Taraxacum kok-saghyz*], Mosonmagyaróvár, 1944, 7.
156   In the search for plants containing rubber, over one thousand plants were examined and some six hundred contained rubber in various quantities, but only a few of them were eligible for industrial processing. The *Taraxacum kok-saghyz* was deemed the most important of these plants.
157   A.U. Buranov and B.J. Elmuradov, "Extraction and Characterization of Latex and Natural Rubber from Rubber-Bearing Plants," *Journal of Agricultural and Food Chemistry* 58, no. 2 (2009), 734–43. http://pubs.acs.org/action/doSearch?action=search&searchText=Taraxacum+kok\-saghyz+\%28TKS\%29&qsSearchArea=searchText (accessed 7 January 2013).
158   K. Mayhood, "Roots to Rubber," *The Columbus Dispatch*, 1 July 2008. http://www.dispatch.com/content/stories/science/2008/07/01/sci_latex.ART_ART_07-01-08_B4_HNAJBMK.html (accessed 7 January 2013). There is still extensive research being done in the United States and the European Union on sources of natural rubber, reviving experimentation with the *Taraxacum kok-saghyz*. B. Jan van Beilen, and Y. Poirier, "Guayule and Russian Dandelion as Alternative Sources of Natural Rubber," *Critical Reviews in Biotechnology* 27, no. 4 (2007), 217–31; *EU-based Production and Exploitation of Alternative Rubber and Latex Sources*. http://www.eu-pearls.eu/UK/http://informahealthcare.com/doi/abs/10.1080/07388550701775927 (accessed 7 January 2013).
159   Minutes made in Budapest on 4 January 1950 at the conference held by Department IV of the Hungarian Academy of Sciences, MNL OL M-KS 276. f. 93/168. ő. e., 317; "Jelentés a gumipitypang üzemszerű termeléséről" [Report on the Large-Scale Production of the *Taraxacum kok-saghyz*], Budapest, 26 June 1950, MNL OL M-KS 276, f. 93/167, ő. e., 288. Lajos Herz, party secretary of the Agricultural Experiment Center and head of the Inspection Center.
160   MNL OL M-KS 276. f. 93/167. ő. e., 289.
161   László Kovátsits (1908–1988) acquired his diploma in 1929 at the Agricultural Academy of Magyaróvár. He worked as a researcher at the National Plant Improvement Institute and later at the National Plant Production Experiment Station. He is responsible for corn silage production and the cold induction method, about which he wrote a book and published it himself in 1934. The book has been reprinted several times. See P. Kimlei, *Kovátsits László* [László Kovátsits]. http://mmel.nansoft.hu/?o=szc&c=296 (accessed 7 January 2013).

162 "Határozati javaslat a gumipitypang nagyüzemi termesztésének megkezdésére" [Proposal on the Initiation of Large-Scale *Taraxacum kok-saghyz* Production], 1 August 1950, MNL OL M-KS 276, f. 93/224, ő. e. Barna Zoltán Sólyom (1888–1960) was a chemical engineer and university lecturer. He began his industrial career in 1923 at the Hungária Guttapercha Rubber Product Factory, later becoming director of engineering at the same factory. From 1937, he became a private lecturer and later a regular public lecturer at the Budapest University of Engineering. Between 1950 and 1952 he was head of department at the Organic Chemical Industry Research Insitute and head of the Chemistry and Physics laboratories of the Technological Rubber Factory. http://www.kfki.hu/physics/historia/historia/egyen.php?namenev=solyom&nev5=S%F3lyom+Barna+Zolt%E1n (accessed 8 January 2013).

163 "Jelentés a gumipitypang üzemszerű termeléséről" [Report on the Large-Scale Production of the *Taraxacum kok-saghyz*], Budapest, 26 June 1950, MNL OL M-KS 276, f. 93/167, ő. e., 288–89.

164 Ernő Kurnik (1913–2008) was an agricultural engineer, plant improver and a standard member of the Hungarian Academy of Sciences. He directed the institute in Iregszemcse (renamed the Feed Production Research Institute in 1955) until 1983.

165 Although the Hanság farm carried on cultivating 37 hectares of land during the war, and ten kilograms of *Taraxacum kok-saghyz* planting seeds were saved in 1944 and transferred to the agricultural cooperative of Kapuvár, these had become ungerminable by that time. MNL OL M-KS 276, f. 93/167, ő. e., 292.

166 Ancient unit of measurement in Hungary: 1 cadastral yoke = 1,600 square fathoms, 5,755 square meters or 0.575464 hectares.

167 MTI, 20 April 1949. http://archiv1945-1949.mti.hu/Pages/PDFSearch.aspx?Pmd=1 (accessed 8 January 2013).

168 MTI, 12 December 1949. http://archiv1945-1949.mti.hu/Pages/PDFSearch.aspx?Pmd=1 (accessed 8 January 2013).

169 "Dr. Balázs Ferencnek, a mosonmagyaróvári Mezőgazdasági Kísérleti Intézet intézetvezetőjének bizalmas jelentése" [Confidential Report of Dr. Ferenc Balázs, Director of the Agricultural Experimental Institute of Mosonmagyaróvár], 30 July 1951, MNL OL M-KS 276, f. 93/364, ő. e., 117.

170 Ibid., 117. Based on the univocal opinion of the two experts (Kovátsits and Kurnik), it was determined that "planting pieces of *Taraxacum kok-saghyz* root for large-scale production does not yield reliable results and is therefore an unreliable method of the production and propagation of the *Taraxacum kok-saghyz*." "Dr. Kurnik Ernő levele a Földművelésügyi Minisztérium Mezőgazdasági Kísérleti Központjába" [Letter from Dr. Ernő Kurnik to the Agricultural Experiment Center of the Ministry of Agriculture], 30 July 1951, MNL OL M-KS 276, f. 93/364, ő. e., 120.

171 MNL OL M-KS 276. f. 93/168. ő. e., 269.

172 MNL OL M-KS 276. f. 93/184. ő. e. No page number. Note by deputy head of department László S. Hegedűs, 20 October 1950.
173 Pünkösti, *Kiválasztottak*, 75. Károly Sarlai was originally a miller who joined the Hungarian Workers' Party and became exemplary in "plan subscription," and was thus rewarded with a six-week party course. Upon graduation, he immediately became the director of the National Rubber Crop Production Company and its seventy employees.
174 Ibid.
175 MNL OL M-KS 276. f. 93/364. ő. e., 119.
176 Ibid. Note on the completion of planned tasks for the production of the *Taraxacum kok-saghyz*. 4 September 1951, 121. Signed by Bedők.
177 Ibid., 121–22.
178 Ibid. Note regarding the state of *Taraxacum kok-saghyz* production, 18 September 1951, 123. (Three handwritten signatures.)
179 MNL OL M-KS 276. f. 93/168. ő. e., 317–18.
180 MNL OL M-KS 276. f. 53/117. ő. e.
181 "Különleges rostnövények termelése" [The Production of Special Fiber Plants], Report by Klára Sándiné Faggyas, 23 July 1951, MNL OL M-KS 276, f. 93/362, ő. e., 84.
182 Ibid.
183 "Előzetes jelentés a kenaf honosítási és nemesítési kísérletekről" [Preliminary Report on Experiments with Introducing and Improving Kenaf], 15 February 1951, MNL OL M-KS 276, f. 93/364, ő. e., 21.
184 "Feljegyzés új növényfajták, új növények és új termelési módszerek bevezetésével kapcsolatban" [Note Regarding the Introduction of New Plant Species, Crops and Production Methods], 1 October 1951, MNL OL M-KS 276, f. 93/364, ő. e., 129.
185 "Jelentés a ramienemesítési, termesztéstechnikai és rostkinyerési kísérletek állásáról" [Report on the Status of Experiments Regarding the Improvement, Production Technologies, and Fiber Extraction of the Boehmeria Nivea], 13 October 1951, MNL OL M-KS 276, f. 93/364, ő. e., 140.
186 MNL OL M-KS 276, f. 93/364, ő. e., 129–30.
187 "Jelentés" [Report], 15 October 1951, MNL OL M-KS 276, f. 93/364, ő. e., 137.
188 MNL OL M-KS 276, f. 93/362, ő. e., 84/2.
189 MNL OL M-KS 276, f. 93/167, ő. e., 288–89.
190 "Jelentés a citromnemesítési kísérletek állásáról" [Report on the Status of Lemon Improvement Experiments], 24 April 1951, MNL OL M-KS 276, f. 93/364, ő. e., 162–64.
191 Baranov, Pavel Aleksandrovich (1892–1962) was a Soviet botanist, the deputy director of the Main Botanic Garden of the Soviet Academy of Sciences (1944–1952; in 1952 he became director); from 1949 to 1954 he was the president of the Presidium of the Moldavian Branch of the Academy of Sciences of the Union of Soviet Socialist Republics. Baranov's basic works

deal with the problems of plant ontogeny and morphology, the complex botanical study of cotton and grape plants, agricultural appropriation of high mountain territory, and the history of botany. From the end of 1950s he played an important role in the introduction of hybrid maize in the USSR.

192 Report of 25 September 1951, 176. MNL OL M-KS 276, f. 93/364, ő. e. Basically, the sprout was grafted into saplings of *Citrus trifoliata* (wild lemon) that were decorative deciduous native plants to Hungary.
193 Reports of 24 and 27 April 1951, MNL OL M-KS 276, f. 93/364, ő. e., 162–64 and 166–68, respectively.
194 Ibid. Report of 8 May 180.
195 Report of 18 September 1951, 91–96.
196 Ibid., 96.
197 Ibid., 178.
198 Ibid., 179.
199 Görög, *Magyarország mezőgazdasági földrajza*, 171–73.
200 MNL OL M-KS 276. f. 53/117. ő. e. 2.
201 G. Kollányi, *A Fertődi Kutatóintézet története az alapítástól napjainkig* [The History of the Fertőd Research Institute from Its Foundation to Today], undated. http://www.gykut.hu/index.php?option=com_content&view=article&id=69&Itemid=115 (accessed 9 September 2012). The spearhead of citrus experiments, Aladár Porpáczy sr. (1903–1965), received the Kossuth award in 1952 and was appointed as a correspondent member of the Hungarian Academy of Sciences in 1954.
202 F. Somorjai and G. Járányi, *Rizstermesztés* [Rice Production], Budapest, 1954, 9.
203 Obermayer, "A rizs és a gyapot meghonosítása," 212.
204 Hajdú, "A szocialista természetátalakítás kérdései," 255.
205 Obermayer, "A rizs és a gyapot meghonosítása," 216.
206 "Előterjesztés a Népgazdasági Tanácshoz az 1949/50. évi növénytermelési terv tárgyában" [Memorandum to the National Economic Council Regarding Plant Production Plans for the Years 1949/50], undated, MNL OL XIX-K-1-ee 2. d. 8202-7/1949.
207 "A termelőszövetkezetek lemaradása a rizstermelés megszervezése vonalán" [Producers' Cooperatives are Falling Back in the Organization of Rice Production], 25 February 1950, MNL OL 276, f. 93/252, ő. e., 288.
208 Ibid.
209 "Jegyzőkönyv a rizstermelésből eredő tartozásokról" [Records of Debts Arising from Rice Production], 16 January 1950, MNL OL XIX-K-1-ee 2. d.
210 Ibid.
211 "Feljegyzés Adamkó János elvtárs részére az 1950. évi rizstermesztéssel kapcsolatban" [Memorandum for Comrade János Adamkó Regarding Rice Production in 1950], 30 March 1950, MNL OL 276, f. 93/220, ő. e., 146.
212 Ibid.

213 I. Orgoványi, "A déli határsáv 1948 és 1956 között" [The Southern Border between 1948 and 1956], in *Bács-Kiskun megye múltjából. Évkönyv 17* [The Past of Bács-Kiskun County. Yearbook 17], ed. Á. Tóth, Kecskemét, 2001, 264.

214 "Előterjesztés a letartóztatottak, rendőrhatósági őrizet alatt állók, munkahelyre bocsátandók foglalkoztatására és kényszerlakóhelyek létesítésére" [Memorandum on the Establishment of Forced Housing and the Employment of Arrested Persons, Persons under Police Surveillance and Persons Eligible for Labor], 23 May 1951, MNL OL M-KS 276, f. 54/145, ő. e., 267.

215 B. Jolsvai, "Munkavégzés" [Labor], ed. Z. Hantó, *Kitaszítottak III* [Outcasts III], Budapest, 2007, 152.

216 "A Magyar Dolgozók Pártja Központi Vezetőségének és a Magyar Népköztársaság Minisztertanácsának határozata a mezőgazdasági termelés fejlesztéséről" [Decision of the Central Committee of the Hungarian Workers' Party and the Council of Ministries of the People's Republic of Hungary on the Development of Agricultural Production], 19 December 1953, MNL OL M-KS 276, f. 52/26, ő. e., 10.

217 "Feljegyzés a rizstermelés helyzetéről" [Notes on the State of Rice Production], 2 June 1954, MNL OL 276, f. 93/491, ő. e.

218 *A rizstermesztés – kutatás és nemesítés – története hazánkban* [The History of Rice Production, Research and Improvement in Hungary]. Website of the Research Institute for Fisheries, Aquaculture and Irrigation, http://www.haki.hu/index.cgi?rx=&nyelv=hu&item=&searchwords2=&menuparam4=38&menuparam_4=47&type_=4 (accessed 3 February 2013).

219 "A Szovjetunió Kommunista Pártja XIX. kongresszusa anyagának legfontosabb kérdései a mezőgazdaság területén" [Key Agricultural Issues from the Prepared Material of the 19th Congress of the Communist Party of the Soviet Union], undated, MNL OL 276, f. 93/621, ő. e., 170.

220 Cs. Gallyas and I. Sárossy, eds, *Mezőgazdasági kislexikon* [The Small Encyclopedia of Agriculture], Budapest, 1989, 249.

221 "Haladó tudománnyal, jobb búzatermések felé" [Towards Better Wheat Production with Increasing Knowledge], undated, MNL OL 276, f. 93/167, ő. e., 460.

222 G. Belényi, "A termelőszövetkezeti mozgalom első kísérlete Szegeden és vidékén (1948–1953)" [The First Experiment of the Producers' Cooperative Movement in Szeged and the Szeged Region (1948–1953)], in *Agrárpolitika és agrárátalakulás Magyarországon (1944–1962)* [Agricultural Politics and Agricultural Transformation in Hungary (1944–1962)], ed. S. Balogh and F. Pölöskei, Budapest, 1979, 160.

223 Gallyas and Sárossy, *Mezőgazdasági kislexikon*, 32.

224 "A gabonafélék nemesítése" [The Improvement of Grains], *Mezőgazdasági Szemle* [Agricultural Review], Magyar-óvár, 1885, 475. http://lib.szie.hu/sites/default/files/mgszemle/1885/1885_10009.pdf (accessed 9 March 2013).

225 Hungarian Newsreel, 1949/27. http://www.filmintezet.hu/uj/hirado/heti/index.php?y=hir&id=1621&eid=10047#go (accessed 9 March 2013).
226 "A Micsurin–Liszenko-féle agrobiológia alkalmazása hazánkban" [The Application of the Michurin and Lysenko School of Agrobiology in Hungary], undated, MNL OL 276, f. 93/168, ő. e., 183.
227 Ibid., 192.
228 "Növénynemesítés" [Plant Improvement], undated, MNL OL 276, f. 93/167, ő. e., 329.
229 Ibid.
230 L.T. Zemljanickij, *Dokucsajev és Viljamsz tanítása a fásításos talajjavításról* [Dokuchaev and Viliams on Soil Improvement by Afforestation], Budapest, 1950, 4–7.
231 V. Smirnov, *V.R. Viljamsz élete és munkája* [The Life and Work of V.R. Viliams], Budapest, 1949, 8.
232 V.M. Slobodin, *A Viljamsz-féle földművelési rendszer* [The Viliams Agricultural System], Budapest, 1950.
233 "Javaslat a füves vetésforgó bevezetésére 1950–1951-ben" [Proposal on the Introduction of Grass-Crop Rotation in 1950–1951], 6 September 1950, MNL OL 276, f. 93/224, ő. e.
234 "Jelentés a földművelés füves vetésforgós rendszere bevezetésének előkészítéséről szóló határozat végrehajtásának munkáiról" [Report on the Implementation of the Decision on the Introduction of Grass-Crop Rotation in Agriculture], 5 August 1950, MNL OL 276, f. 93/167, ő. e., 315–18.
235 Ibid.
236 "Javaslat a füves vetésforgó bevezetésére 1950–1951-ben."
237 Ibid.
238 Act II of 1951.
239 "Jelentés a füves vetésforgós rendszer bevezetésének előkészítéséről szóló 1950 áprilisában hozott párthatározat végrehajtásáról" [Report on the Implementation of the Party Decision of April 1950 on the Preparation of the Introduction of Grass-Crop Rotation], 27 April 1951, MNL OL 276, f. 93/353, ő. e., 48–54.
240 Ibid.
241 "Előterjesztés az állami gazdaságokban és a termelőszövetkezetekben a füves vetésforgó bevezetése, mezővédő erdősávok telepítésére és az ezekkel összefüggő üzemen belüli földrendezési feladatok végrehajtására" [Memorandum on the Introduction of Grass-Crop Rotation in State Farms and Producers' Cooperatives, on the Plantation of Protection Forests and the Execution of Land Consolidation Tasks on the Involved Sites], 17 October 1951, MNL OL 276, f. 93/353, ő. e.
242 "A földművelési miniszter előterjesztése a füves vetésforgó rendszer bevezetéséről" [Memorandum of the Minister of Agriculture on the Introduction of the Grass-Crop Rotation System], 15 November 1951, MNL OL 276, f. 93/353, ő. e., 240–44.

243 Ibid.
244 "Jelentés a füves vetésforgó és üzemen belüli földrendezési tervek bevezetéséről és meghonosításáról" [Report on the Introduction of Grass-Crop Rotation and the Dissemination of On-Site Land Consolidation Plans], 12 November 1952, MNL OL 276, f. 93/411, ő. e., 106–8.
245 L.T. Zemlyanickij, *Dokucsajev és Viljamsz tanítása a fásításos talajjavításról* [Dokuchaev and Viliams on Soil Improvement by Afforestation], Budapest, 1950.
246 G. Fekete, "A fásítás tíz éve" [Ten Years of Afforestation], *Erdészeti Lapok*, no. 12 (1962), 547.
247 Ibid.
248 "Országfásítás 1950. évi részletezése az erdőgazdasági üzemi területeken kívül eső fásítások 5 éves terve keretében" [Detailed Plans for National Afforestation in the Year 1950 Within the Framework of the Five-year Plan of Afforestation Occurring outside Industrial Forest Holdings], 2 November 1949, MNL OL OT XIX-A-16-a 8710/5/1949.
249 Illegible title, 19 January 1950, MNL OL 276. f. 93/252, ő. e., 3.
250 Ibid., 5.
251 Ibid., 7.
252 No title, 21 February 1950, MNL OL 276, f. 93/174, ő. e., 193–97.
253 Ibid., 194.
254 Fekete, "A fásítás tíz éve," 552.
255 E. Mosonyi, *A kujbisevi és sztálingrádi vízlépcsők jelentősége a Szovjetunió természetátalakítási tervében* [The Importance of the Dams at Kujbishev and Stalingrad in the Soviet Plan to Transform Nature], (excerpt from) *Magyar Technika*, no. 2 (1951), 39.
256 Article XX of 1937: *Az öntözőgazdálkodás előmozdításához szükséges intézkedésekről* [Required Measures for the Promotion of Irrigation]. http://www.1000ev.hu/index.php?a=3&param=8047 (accessed 14 January 2013).
257 Tiszalök is located near the bank of the Tisza River in Szabolcs-Szatmár County, thirty kilometers from Nyíregyháza, the administrative center of the county.
258 Dr. Emil Mosonyi (1910–2009) was an internationally renowned water engineer whose expertise had been applied on every continent to plan and implement water management investments. He proved indispensable for the Communist Party and had designed several water management structures for Hungary including the dam at Tiszalök. He was dismissed for his participation in the revolution of 1956, and soon left Hungary. He could only return in 1990.
259 E. Mosonyi, I. Pados, and P. Ötvös, *A vízlépcső és erőmű tervezési, építési és üzemelés, ökológiai és társadalmi előnyei, tapasztalatai 50 év tükrében* [The Planning, Construction, Operational, Ecological, and Social Advantages and Experiences of the Dam and Powerplant in the Last 50 Years]. http://www.

hidrologia.hu/vandorgyules/26/7szekcio/Mosonyi_EmilOK.htm (accessed 14 January 2013).

260 "Határozati előterjesztés a Népgazdasági Tanácshoz a tiszalöki vízlépcső létesítése tárgyában" [Decision Memorandum to the National Economic Council on the Establishment of the Tiszalök Dam], 24 August 1949, MNL OL XIX-A-16-a 8754/83/1949.

261 The Kisköre power plant was finished in 1974 and is the most efficient water power plant in Hungary today.

262 "Határozati előterjesztés a Népgazdasági Tanácshoz a tiszalöki vízlépcső létesítése tárgyában" [Decision Memorandum to the National Economic Council on the Establishment of the Tiszalök Dam], 24 August 1949, MNL OL XIX-A-16-a 8754/83/1949.

263 A good example would be the letter of Minister of Agriculture Ferenc Erdei to 'Comrade Ottó Sándi' on 9 January 1953 (Ottó Sándi was the political referent of the Agricultural and Cooperative Department of the Central Committee of the Hungarian Workers' Party, and deputy head of department from 1954). With the approach of the opening deadline, Erdei expressed his concern over how slowly construction was going compared to the original plans. The Ministry of Agriculture had been granted 43 million HUF but would have needed an extra 47 million HUF and a significant amount of building material, as well as for the Ministry of Transportation to "build the structures of the half-sectioned canal, whether it is included in their plans or not. It is also necessary that the Ministry of Transportation provide the appropriate machinery and sufficient labor force for construction. This is currently the most problematic issue, and the National Planning Office insists that the Ministry of Agriculture should finance this additional investment from its approved budget, which is absolutely impossible." MNL OL 276. f. 93/411. ő. e., 31.

264 Bank, Gyarmati, and Palasik, *"Állami titok"*, 43.

265 Ibid.

266 The necessity of constructing a Danube-Tisza Canal is still discussed by experts today, since establishing an irrigation system on the western region of the Alföld plains would significantly improve agricultural production.

267 "Összefoglaló jelentés a kísérleti öntözési telepekről" [Assessment Report on the Expermental Irrigation Sites], 26 June 1950, MNL OL 276, f. 93/167, ő. e., 373.

268 "A Mezőgazdasági Kísérletügyi Központ létszámának alakulása a racionalizálással kapcsolatban" [Changes in the Number of Personnel at the Agricultural Experiment Center due to Rationalization], 16 May 1950, MNL OL 276, f. 93/167, ő. e., 426-28.

269 M. Palasik, "Frank Melanie" [Melanie Frank], in *Nők a magyar tudományban* [Women in the Hungarian Sciences], ed. M. Balogh and M. Palasik, Budapest, 2010, 256-57.

270 "A Mezőgazdasági Tudományos Központ szervezete" [The Organization of the Agricultural Science Center], 9 February 1950, MNL OL 276, f. 93/168, ő. e., 167–68.
271 "A mezőgazdasági vízügyek mai állása és 1951. évi terve, 1950. december 5" [The Current State of Agricultural Water Management and the 1951 Plan], 5 December 1950, MNL OL 276, f. 93/229, ő. e., 268.
272 *A magyarországi öntözési kultúra kialakulása* [The Establishment of Hungarian Irrigation Culture] Website of the Szarvas Research Institute for Fishery and Irrigation. http://www.haki.hu/index.cgi?menuparam4=38&menuparam_4=44&nyelv (accessed 18 January 2012).
273 "Határozati javaslat a titkárság számára" [Decision Proposal for the Secretariat], 10 April 1951, MNL OL M-KS 276, f. 54/138, ő. e.
274 Ibid., 63.
275 "Javaslat a titkárság számára az Észak-tiszántúli természetátalakító terv kidolgozására" [Proposal for the Secretariat on the Development of Nature Transformation Plans for the Northern-Tiszántúl Region], 24 April 1951, MNL OL M-KS 276, f. 54/140, ő. e., 259–61.
276 "Határozati javaslat a titkárság számára a Tiszántúl egy része természetátalakító tervének kidolgozására" [Decision Proposal for the Secretariat on the Development of Nature Transformation Plans for One Portion of the Tiszántúl Region], 28 January 1952, MNL OL M-KS 276, f. 54/179, ő. e.
277 Ibid.
278 Ibid.
279 Decision 2037/1952 of the Council of Ministries.
280 "A Tiszántúli Természetátalakító Tervbizottság feladatköre, szervezete és ügyrendje" [Scope of Activities, Organization, and Agenda of the Tiszántúl Nature Transformation Planning Committee], 16 May 1952, MNL OL 276, f. 93/411, ő. e., 12–19.
281 *A magyarországi öntözési kultúra kialakulása* [The Establishment of Hungarian Irrigation Culture]. Website of the Szarvas Research Institute for Fishery and Irrigation. http://www.haki.hu/index.cgi?menuparam4=38&menuparam_4=44&nyelv (accessed 18 January 2012). Data for the year 1953 is especially high compared to the years before and after. In 1952, the size of irrigated farmland was a mere 56,508 hectares, while in the second half of the 1950s, it rose to some 70–80,000 hectares.
282 "Határozati javaslat az öntözéses termelés fejlesztésére az 1953–54. Években" [Decision Proposal on the Development of Irrigation in the Years 1953–154], 22 December 1952, MNL OL 276, f. 93/411, ő. e., 27.
283 Ibid., 27–30.
284 Berend, *A szocialista gazdaság fejlődése*, 148.
285 "Jegyzőkönyv a Politikai Bizottság üléséről" [Minutes of the Meeting of the Political Committee], 22 July 1953, MNL OL M-KS 276, f. 53/127, ő. e., 30.
286 For details, see Gy. Gyarmati, "Nők, játékfilmek, hatalom" [Women, Feature Films, and Power], in *Az ötvenes évek Magyarországa játékfilmeken* [1950s

Hungary in Feature Films], ed. Gy. Gyarmati, M. Schadt, and J. Vonyó, Pécs, 2004, 41–67.
287 I. Karcag, "A természetátalakítás kérdései" [Nature Transformation Issues], *Magyar–Szovjet Közgazdasági Szemle*, no. 2 (1951), 152.
288 Z. Ólmosi, "A néprádió és a vezetékes rádió az ötvenes években" [Popular Radio and Wire Radio in the 1950s], *Archívnet*, no. 6 (2007). http://www.archivnet.hu/politika/a_nepradio_es_a_vezetekes_radio_az_otvenes_evekben.html (accessed 20 February 2013).
289 E. Gaálné Barcs, *Dobolások a propaganda szolgálatában az 1950-es években* [Drumming in the Service of Propaganda in the 1950s], Presentation at the Conference of the Pest County Archives on 25 November 2010. http://www.pestmlev.hu/data/files/202705480.pdf (accessed 20 February 2013).
290 *Szabad Nép*, 13 June 1951.
291 Cs. Halász, "A diafilm mint a kultúragitáció eszköze a Rákosi-korszakban" [Film Strips as Instruments of Cultural Agitation in the Rákosi Era], *Archívnet*, no. 1 (2008). http://www.archivnet.hu/hetkoznapok/a_diafilm_mint_a_kulturagitacio_eszkoze_a_rakosikorszakban.html (accessed 20 February 2013).
292 Romsics, *Magyarország története a XX. században*, 372.
293 *A Magyar Természettudományi Társulat zsebkönyve* [Pocketbook of the Hungarian Society of Natural Sciences], Budapest, 1952, 40.
294 Slide-strip: a roll of photos where the pictures are visible in positive film and they can be projected with a slide projector.
295 Halász, "A diafilm mint a kultúragitáció eszköze a Rákosi-korszakban."
296 We would like to thank Ferenc Bíró, founder and owner of the Hungarian Museum of Slide-Strips, for introducing us to the referenced slide-strips and allowing us to use them.
297 "Tervezet. A Magyar Népköztársaság Minisztertanácsának határozata a növénytermelés fejlesztéséről" [Draft. Decision of the Council of Ministries of the People's Republic of Hungary on the Development of Plant Production], 8 June 1953, MNL OL 276, f. 93/455, ő. e.
298 Hungarian Newsreel 45, January 1949. http://www.filmintezet.hu/uj/hirado/heti/index.php?y=hir&id=1597&eid=9933#go (accessed 16 February 2013).
299 MTI, 2 September 1948.
300 MTI, 2 July 1949.
301 *Szabad Föld*, 24 February 1952.
302 Ibid.
303 *Nők Lapja*, 1 May 1952.
304 Hungarian Newsreel, no. 19 (1953). http://www.filmintezet.hu/uj/hirado/heti/index.php?y=hir&id=1292&eid=7937 (accessed 28 February 2013).
305 *Szabad Föld*, 13 January 1952.
306 *Szabad Nép*, 17 December 1950.

307 *Szabad Nép*, 5 September 1952. We would like to thank Róbert Takács for calling our attention to this article.
308 "Ahol megváltoztatják a természet arculatát" [Where Nature is Changed], *Kincses Kalendárium*, 1949, 92.
309 "Nem az idő: az ember a gazda! Újfajta agrotechnikai módszerekkel a magasabb termésért" [Man Is in Charge, Not the Weather: New Agricultural Technologies and Methods for Better Crop Yield], *Kincses Kalendárium*, 1953, 153.
310 Z. Ólmosi, "Az ötvenes évek propaganda képei" [Propaganda Images of the 1950s], *Archívnet*, no. 3 (2006). http://www.archivnet.hu/politika/az_otvenes_evek_propaganda_kepei.html (accessed 20 February 2013).
311 *Ludas Matyi*, 4 October 1951. Literal translation by Éva Misits. The original text is: "Kicsi vagyok én/ Majd megnövök én/ Ha a szovjet agrármódszert/ Eltanulom én."
312 M. Braunné Szendi, "Biztonsági fehér arannyal az imperializmus ellen" [White Gold Insurance against Imperialism], *Archívnet*, no. 2 (2004). http://www.archivnet.hu/hetkoznapok/biztonsagi_feher_arannyal_az_imperializmus_ellen.html (accessed 22 January 2013).
313 *Ludas Matyi*, 22 September 1950.
314 *Szabad Nép*, 30 August 1950, 4.
315 V. Galaktionov and A. Agranovsky, *Egy nagy építkezés hajnala* [The Dawn of a Great Construction Project], Budapest, 1952, 194.
316 *Szabad Nép*, 13 April 1952. We would like to thank Róbert Takács for calling our attention to this article.
317 *Ludas Matyi*, 4 October 1951.
318 *Ludas Matyi*, 29 September 1950.
319 *Ludas Matyi*, 13 February 1952.
320 *Ludas Matyi*, 14 June 1951.
321 *Ludas Matyi*, 27 February 1952.
322 *Ludas Matyi*, 23 November 1950.
323 "Magyarország köröskörül füstölög/gyárak, kohók, gépek zaja dübörög."
324 J. Dömsödi, *Természeti erőforrás és környezetgazdálkodás, 1. A természeti erőforrások szerepe a társadalom és a gazdaság fejlődésében* [Natural Power Sources and Environmental Management I: The Role of Natural Power Sources in Social and Economic Development], Nyugat-magyarországi Egyetem Geoinformatikai Kar, 2010, 6.
325 "Javaslat a Politikai Bizottság számára" [Proposal for the Political Committee], 18 March 1953, MNL OL M-KS 276, f. 53/117, ő. e., 2–3.
326 MNL OL M-KS 276, f. 53/117, ő. e., 2.
327 "Jelentés a Titkárság számára" [Report for the Secretariat], 6 April 1953, MNL OL M-KS 276, f. 54/239, ő. e., 2–3.
328 Ibid.
329 "Jegyzőkönyv a Politikai Bizottság 1953. július 22-i üléséről" [Minutes of the 22 July 1953 Session of the Political Committee], MNL OL M-KS 276, f. 53/127, ő. e., 2.

330 "A Magyar Dolgozók Pártja Központi Vezetőségének és a Magyar Népköztársaság Minisztertanácsának határozata a mezőgazdasági termelés fejlesztéséről" [Decision of the Central Committee of the Hungarian Workers' Party and the Council of Ministries of the People's Republic of Hungary on the Development of Agricultural Production], 19 December 1953, MNL OL M-KS 276, f. 52/26, ő. e., 2.
331 Ibid., 14.
332 "Előterjesztés a Titkárság részére. A KGST legközelebbi ülésszakán napirendre kerülő kérdések" [Memorandum for the Secretariat. Issues on the Agenda of the Next Session of the COMECON], 19 April 1950, MNL OL M-KS 276, f. 54/95, ő. e., 129.
333 MNL OL M-KS 276. f. 52/26. ő. e., 36.
334 Igali, "A liszenkoizmus Magyarországon," 48.
335 Müller, "Liszenko emlékezetes előadása a Magyar Tudományos Akadémián 1960-ban," 1355–56.
336 Ibid., 1358.
337 In the 1960s and 1970s, rice was produced on 27–30,000 hectares of farmland in Hungary. Rice was still being produced on some 27,000 hectares of land in the 1980s—primarily in large-scale agricultural farms, but after the system change, the amount of farmland used for rice decreased significantly. By 2012, rice production had reduced to a mere 2,800 hectares of farmland.
338 I. Újszász, *Visszavadítják a tájat – A szikesedés régebbi, mint a folyószabályozás* [Reverting the Land: It Had Been Saline Long Before River Regulation]. http://www.u-szeged.hu/egyetemrol/tudomany-innovacio/visszavaditjak-tajat (accessed 1 March 2013).

Bibliography

Balogh, M., and M. Palasik, eds, *Nők a magyar tudományban*, Budapest, 2010.
Balogh, S., and F. Pölöskei, eds, *Agrárpolitika és agrárátalakulás Magyarországon (1944–1962)*, Budapest, 1979.
Bank, B., Gy. Gyarmati, and M. Palasik, *"Állami titok." Internáló- és kényszermunkatáborok Magyarországon, 1945–1953*, Budapest, 2012.
Beilen, Jan B. van, and Y. Poirier, "Guayule and Russian Dandelion as Alternative Sources of Natural Rubber," *Critical Reviews in Biotechnology* 27, no. 4 (2007), 217–31.
Bereczkei, T., "Tudás a semmiből: a Liszenko-ügy," *Természet Világa* 3 (1998).
Berend, I.T., *A szocialista gazdaság fejlődése Magyarországon 1945–1968*, Budapest, 1974.
———, *Gazdaságpolitika az első ötéves terv megindításakor 1948–1950*, Budapest, 1964.
———, *Terelőúton. Közép- és Kelet-Európa 1944–1990*, Budapest, 1999.
Birta, I., "A szocialista iparosítási politika néhány kérdése az első ötéves terv időszakában," *Párttörténeti Közlemények*, no. 3 (1970), 113–50.

Braunné Szendi, M., "Biztonsági fehér arannyal az imperializmus ellen," *Archívnet*, no. 2 (2004).
Buranov, A.U., and B.J. Elmuradov, "Extraction and Characterization of Latex and Natural Rubber from Rubber-Bearing Plants," *Journal of Agricultural and Food Chemistry* 58, no. 2 (2009), 734–43.
Cristescu, C., "Ianuarie 1951: Stalin decide înarmarea Romanei," *Magazin Istoric*, no. 10 (1995), 15–23.
Dömsödi, J., *Természeti erőforrás és környezetgazdálkodás, 1. A természeti erőforrások szerepe a társadalom és a gazdaság fejlődésében*, Nyugat-magyarországi Egyetem Geoinformatikai Kar, 2010.
Faludi, B., Gy. Kiszely, and H. Tangl, *Az ember szervezete. Gimnáziumok számára. Ideiglenes tankönyv*, Budapest, 1950.
Gallyas, Cs., and I. Sárossy, eds, *Mezőgazdasági kislexikon*, Budapest, 1989.
Genin, S.J., and O.K. Makarova, Yu.S. Chuprova, K.K. Senchilo, I.S. Khoroshilova, eds, *Strana sovetov za 50 let* (Sbornik schtatisticheskih materialov), Moscow, 1967.
Germuska, P., *Indusztria bűvöletében. Fejlesztéspolitika és a szocialista városok*, Budapest, 2004.
———, "Szocialista csoda? Magyar iparfejlesztési politika és gazdasági növekedés, 1950–1975," *Századok*, no. 1 (2012), 47–78.
———, "A szocialista iparosítás Magyarországon 1947–1953 között," *Évkönyv 2001* Budapest, 2001. 147–172.
Glatz, F., "Akadémia és tudománypolitika a volt szocialista országokban 1922–1999," *Magyar Tudomány*, no. 4 (2002). 494–506.
Gluschenko, E.J., *Micsurin tanítása az idealista biológia elleni küzdelemben*, Budapest, 1949.
Görög, L., *Magyarország mezőgazdasági földrajza*, Budapest, 1954.
Gyarmati, Gy., *A Rákosi-korszak. Rendszerváltó évtizedek története Magyarországon, 1945–1956*, Budapest, 2011.
Gyarmati, Gy., and M. Palasik, eds, *A Nagy Testvér szatócsboltja. Tanulmányok a magyar titkosszolgálatok 1945 utáni történetéből*, Budapest, 2012.
Gyarmati, Gy., M. Schadt, and J. Vonyó, eds, *Az ötvenes évek Magyarországa játékfilmeken*, Pécs, 2004.
Gyarmati, Gy., and T. Valuch, *Hungary under Soviet domination 1944–1989*, New York, 2009.
Halász, Cs., "A diafilm mint a kultúragitáció eszköze a Rákosi-korszakban," *Archívnet*, no. 1 (2008).
Hantó, Zs., *Kitaszítottak III*, Budapest, 2007.
Horváth, S., *A kapu és a határ: Mindennapi Sztálinváros*, Budapest, 2004.
Huszár, T., *A hatalom rejtett dimenziói. Magyar Tudományos Tanács 1948–1949*, Budapest, 1995.
Igali, S., "A liszenkoizmus Magyarországon," *Valóság* 2 (2002).
Kaplan, K., *Dans les Archives du comité central: Trente ans de secrets du bloc soviétique*, Paris, 1978.

Kiss, A., G. Mezősy, and Z. Sümegh, eds, *Táj, környezet és társadalom. Ünnepi tanulmányok Keveiné Bárány Ilona professzor asszony tiszteletére*, Szeged, 2006.

Kónya, S., "A Magyar Tudományos Tanács és a Magyar Tudományos Akadémia egyesítése (1949)," *Magyar Tudomány* 2 (1990).

Kovátsits, L., *A hamvas gumipitypang termesztése és nemesítése*, Mosonmagyaróvár, 1944.

Lysenko, T.D., *Agrobiológia*, Budapest, 1950.

Lysenko, T.D., *A biológiai tudomány állásáról*, Budapest, 1949.

Marosi, S., and J. Szilárd, eds, *A dunai Alföld. Magyarország tájföldrajza*, Vol. I, Budapest, 1967.

Mastny, V., *NATO in the Beholder's Eye: Soviet Perceptions and Policies, 1949–1956*, CWIHP Working Paper No. 35. Washington, DC, no. 3 (2002), Woodrow Wilson International Center for Scholars, Cold War International History Project.

Micsurin, I.V., *Válogatott tanulmányai*, Budapest, 1950.

Miskolczi, M., *Az első évtized. Dunapentelétől – Dunaújvárosig*, Dunaújváros, 1975.

Molodšikov, A.I., *A szovjet tudomány – a természet átalakítója*, Budapest, 1949.

Müller, Miklós, "Liszenko emlékezetes előadása a Magyar Tudományos Akadémián 1960-ban," *Magyar Tudomány* 2 (2011).

Okváth, I., *Bástya a béke frontján. Magyar haderő és katonapolitika 1945–1956*, Budapest, 1998.

Ólmosi, Z., "A néprádió és a vezetékes rádió az ötvenes években," *Archívnet*, no. 6 (2007).

Palasik, M., *A jogállam megteremtésének kísérlete és kudarca Magyarországon, 1944–1949*, Budapest, 2000.

———, and B. Sipos, eds., *Házastárs? Munkatárs? Vetélytárs? A női szerepek változása a 20. századi Magyarországon*, Budapest, 2005.

———, "A Rákosi-rendszer hétköznapjai," *Hitel*, no. 27 (March 1990), XXI–XXV.

———, *Chess Game for Democracy: Hungary between East and West 1944–1947*, Montreal and Kingston, London, and Ithaca, 2011.

Pető, I., and S. Szakács, *A hazai gazdaság négy évtizedének története 1945–1985*, Budapest, 1985.

Pótó, J., "Harmadik nekifutásra. A Magyar Tudományos Akadémia 'átszervezése', 1948–1949," *Történeti Szemle* 94, no. 1–2.

Pünkösti, A., *Kiválasztottak*, Budapest, 1988.

Rákosi, M., *Visszaemlékezések, 1940–1956*, Vol. 2, I. Feitl, M. Gellériné Lázár, and L. Sipos eds., noted by M. Baráth, I. Feitl, Gy. Gyarmati, M. Palasik, L. Sipos, L. Szűcs, and Gy. T. Varga, Budapest, 1997.

Romsics, I., *Magyarország története a XX. században*, Budapest, 1999.

Sibanov, A. A., *A micsurini biológia tanítása a falusi iskolákban*, Budapest, 1950.

Slobodin, V.M., *A Viljamsz-féle földművelési rendszer*, Budapest, 1950.

Smirnov, V., *V.R. Viljamsz élete és munkája*, Budapest, 1949.

Somorjai, F., and G. Járányi, *Rizstermesztés*, Budapest, 1954.

Stark, T., "Magyarország háborús embervesztesége," *Rubicon*, no. 9 (2000).
Timirjazev, K.A., I.V. Michurin, and T.D. Lysenko, *Hogyan alakítja át az ember a növényeket*, Budapest, 1949.
Toranska, T., *"Them": Stalin's Polish Puppets*, New York, 1987.
Tóth, Á., ed., *Bács-Kiskun megye múltjából. Évkönyv 17*, Kecskemét, 2001.
Tóth, J., and Z. Wilhelm, eds, *Változó környezetünk*, Pécs, 1999.
Valuch, T., *Magyarország társadalomtörténete a XX. század második felében*, Budapest, 2001.
Vaszkó, M., *Természetismeret. Ideiglenes tankönyv*, Budapest, 1950.
Zemljanickij, L.T., *Dokucsajev és Viljamsz tanítása a fásításos talajjavításról*, Budapest, 1950.

 CHAPTER 3

The Conspiracy of Silence
The Stalinist Plan for the Transformation of Nature in Poland

Beata Wysokińska

Stalin's Plan for the Transformation of Nature, launched in the USSR in October 1948, and partly implemented throughout Eastern Europe, has been largely forgotten nowadays, obscured by the various levels of the collective memories of each of the Eastern Bloc nations, which over the decades have emphasized different facets of the Stalinist Communist period.

The satellite Stalinist Eastern Bloc governments and their scientific elites were nevertheless fascinated by the Soviet plans, including Stalin's Plan and its underpinnings by the postwar belief in the endless possibilities of modern science to orchestrate large-scale transformations of society as well as the natural environment. These new Communist Party "elites" in Eastern Europe, within the first of the Soviet six-year plans, overlapped the same area as the Stalin Plan for the Transformation of Nature, such as the constructions of new dams, irrigation canals, and shelterbelts, as well as the introduction of exotic plants, at the beginning of the 1950s.

In Poland, with its traditions in natural farming methods coupled with its determination to implement different strategies for the protection of nature, the Stalin Plan, when analyzed within the wider societal context, helped to ensure that the plan would never be fully implemented in Poland as it was in other East European countries.

Origins of the Communist Regime in Poland

The presence of the Red Army on Polish soil, a result of the offensive against the Third Reich, was a significant reason behind the formation of the Communist regime in Poland. The case of Poland is an example of the political decision made by Stalin to introduce a Communist regime. The position of Communists and their proponents was further strengthened by demonstrations of Stalin's power, interference by the People's Commissariat for Internal Affairs (NKVD), as well as terror in the country itself used by the Polish security services.[1] The general pattern and chronology were similar to the process taking place in other states of Central and Eastern Europe, which seems to suggest that these political transformations were somehow consistently orchestrated from above, directly from Moscow.

Before the war ended, in July 1944, the Polish Committee of National Liberation [Polski Komitet Wyzwolenia Narodowego, PKWN] was formed in Moscow, and later, in December 1944, transformed into the Provisional Government of the Republic of Poland [Tymczasowy Rząd Jedności Narodowej, TRJN]. This was done through a resolution taken by the National People's Council [Krajowa Rada Narodowa, KRN],[2] chaired by a loyal executor of Stalin's orders, and supporter of fast Sovietization, Bolesław Bierut (1892–1956). The decisive factor was the Red Army's offensive, marching into Warsaw's left-bank in January 1945, and finally defeating the Germans in May 1945.

The demonstrations of Stalin's power and his ability to decide on matters concerning Poland persuaded some politicians on political emigration, mainly with Stanisław Mikołajczyk (1901–1966) in the lead, to go back home and join Stalin's game. Mikołajczyk became the leader of the Polish People's Party, created in August 1945, and then he was appointed deputy prime minister and minister of agriculture in the provisional Government of the Republic of Poland, formed on 28 June 1945. The government, appointed by Bierut's National People's Council in Warsaw, was dominated by Communist Party members and their proponents. Using the strategy of fait accompli, Stalin created institutions and structures subordinated to him as germs of future Communist authorities. It was also supposed to be an element of pressure on the Allies in the rivalry about Poland's postwar political system and borders.[3]

New Borders

The matter of Poland's territory and political system was one of the most important issues to be resolved at meetings of The Big Three: Churchill, Roosevelt, and Stalin. The latter, using his military victories in Central and Eastern Europe and making promises to the United States that he would join the war in the Far East, was very successful in this respect. At the Teheran Conferences in 1943, and then in Yalta and Potsdam in 1945, it was settled that Poland's border would move at the expense of Germany, and the USSR would take over a piece of Polish territory in the East.[4] Decisions taken in Yalta, rejected by the Polish government in exile in London, were of utmost importance. In July 1945 the Allies withdrew their support for the Polish government in exile, citing the decisions taken in Yalta. Poland, along with other Central European countries, was to be included in the Soviet sphere of influence, and its eastern borders were to run along the so-called Curzon Line,[5] on the Bug River to the first stretch of the San River in the Eastern Carpathians. As a result, the majority of the Polish prewar eastern and southeastern territories—the so-called Eastern Borderlands, which included such centers of Polish culture and science as Lviv and Vilnius—were excluded by the new border. The Provisional Government of National Unity agreed to it on 16 August 1945. Poland was to be compensated by receiving German land in the West and North. At the Potsdam Conference (17 July – 2 August 1945) it was decided that territories to the east of the line drawn from the Baltic Sea, along the Oder River to the estuary of the Lusatian Neisse, then along the Lusatian Neisse to the border with Czechoslovakia and part of East Prussia and the former Free City of Danzig, would be included in the territory of Poland, and the German population living there would be relocated. In the postwar propaganda, this piece of land was called "the Regained Lands," referring to the fact that they had belonged to Poland in the Middle Ages.

The conferences in Yalta and in Potsdam drastically changed not only Poland's political situation, but also its territory—in fact, Poland was shifted to the West. The change of borders left Poland with 76,000 sq km (or 20 percent) less compared to the period before the war. It had huge consequences, both social and organizational, as millions of people were resettled (the so-called repatriation) from the East to the Regained Lands, and Germans were expelled.[6]

Postwar Poland, with its new altered borders, had urgent economic needs, primarily the reconstruction of the country shattered by the war and the provision of fairly normal living conditions to the citizens by the new Communist authorities. Due to the destruction of almost 40 percent of the national wealth caused by the war, the authorities faced an immense challenge: the task of reconstruction and development. The death toll and migration had decreased Poland's population from the prewar number of 35 million to about 24 million in 1946.[7] It took a long time, until the end of the 1970s, for Poland's population to recover to its prewar numbers. Nevertheless, the dynamic of population growth was very impressive: in 1950 Poland's population was over 25 million, in 1960 over 29 million and by the end of the People's Republic of Poland era it had reached approximately 38 million (38.53 million in 2013). Such growth inevitably meant a rise in food demand, and therefore necessitated development of most areas of the economy, first and foremost agriculture. The exploitation of natural resources, also taking place in other Eastern European countries, was mainly caused by those postwar needs. However, Western Europe received aid in postwar reconstruction in the form of the Marshall Plan, but the newly formed Eastern Bloc rejected the aid, under pressure from Moscow. Economic problems of the Eastern Bloc were to be solved by another instrument—the Council for Mutual Economic Assistance (COMECON), which was Eastern Europe's response to the political and economic integration in Western Europe. The COMECON was formed in January 1949, on Stalin's initiative.

The Communist Government

Decisions taken at the conference in Yalta obliged the Provisional Government of National Unity to prepare and organize quick elections in Poland. After long preparations (including a referendum in June 1946), there were parliamentary elections held in January 1947, but the results were falsified. It enabled the Communists to take over all political institutions and to pass the so-called Small Constitution (19 February 1947). The next step was taking control of the main opposition party—the Polish People's Party (its leader Mikołajczyk left the country out of fear of arrest)—and then one by one they took control of all the coalition parties and the domestic Communist party,[8] which proposed gradual changes. W. Gomułka (1905–1982), who was a representative of the Polish Communist Party, was charged along with his followers of "a right-wing deviation," and was subsequently arrested.

The political situation in Poland in the late 1940s was complicated, and dominated by Stalin's constant interference. Some define it as a civil war, paving the way for creating a new Communist nation. Under these circumstances, a disciplined Communist Party under an unified leadership could be formed—the Polish United Workers' Party [Polska Zjednoczona Partia Robotnicza, PZPR].[9] As the president of the Republic of Poland (between 1947 and 1952) and the leader of the Communist Party, B. Bierut, who still loyally followed Stalin's orders, came to power. In the years 1945–1952, the Polish state operated under the name "Republic of Poland." The formal domination of Sovietization was marked by the passing of a constitution on 22 July 1952, changing the country's name to the People's Republic of Poland [Polska Rzeczpospolita Ludowa, PRL]. The authorities exerted themselves to keep up the illusion that institutions such as the Parliament, the Cabinet, the State Council, the Supreme Chamber of State Control, and the Supreme Court remained democratic. In practice, all chances for democracy were wiped away by the party's "leadership role." However, as N. Davies points out, in spite of it all, Stalinism in Poland was less ruthless than in other neighboring countries of the Eastern Bloc.[10]

Political and Economic Decisions

Economic and Social Plans

The most urgent need after the war was the reconstruction of the country. In the period 1947–1949, a Three-Year National Reconstruction Plan of recovery from the destruction caused by the war was implemented. It was carried out by the Central Planning Office [Centralny Urząd Planowania, CUP] consisting of pre-war economists led by Professor Czesław Bobrowski (1904–1996), and was developed based on economic measures adopted in Western Europe rather than in the USSR.[11] It aimed at rebuilding the biological strength of the nation. The plan was completed ahead of time, in two years and ten months, as Bierut announced in his speech at the third plenary meeting of the Polish United Workers' Party Central Committee on 11 November 1949.[12] In retrospect, that plan is perceived as the only successful economic plan completed under the Communist rule. It recovered the economy from the aftermath of the war and literally "fed the nation," which gave citizens hope for a better future.

After the Central Planning Office was dissolved, economic development management was taken over by the State Commission for Economic Planning [Państwowa Komisja Planowania Gospodarczego, PKPG]. The new plan included the goal of further industrialization and facing new economic challenges. Between 1950 and 1955, Poland was following a Six-Year Plan drafted by the State Commission for Economic Planning, which at the time was headed by economist Hilary

Minc (1905–1974), deputy prime minister 1949–1957. The main objectives of the plan were presented by Minc much earlier, at the Unification Congress of the Polish Workers' Party in December 1948. Objectives of the plan, developed in the form of an act passed by the parliament,[13] and then extended in the late 1950s due to the Korean conflict, included: fast industrialization of the country, the expansion of heavy and chemical industry, mining, agriculture, the food industry and light industry—based on Lenin's doctrine that capital goods were supposed to come before consumption goods. Collectivization in agriculture, with its production cooperatives and State Agricultural Farms [Państwowe Gospodarstwa Rolne, PGR], were planned.

The process of compulsory collectivization and scheduled nationalization of individual farms, commenced in 1949, was significantly accelerated in the following few years. In those times the Soviet agriculture was indiscriminately emulated. It was said that progress was possible only in collective agriculture, and that the Western agriculture was decadent and doomed.[14] The plan aimed at increasing agricultural output by 35 percent, possibly even 45 percent, within six years. The main factor increasing agricultural production and farming culture was supposed to be widely applicable achievements of modern science—among others, the rational crop rotation and zoning of agricultural production, depending on the natural and economic conditions.[15] The Six-Year Plan ascribed big tasks to agriculture: increasing grain, vegetable, and fruit production, as well as boosting the output of livestock and fodder. The cultivation of basic Polish grains, such as wheat, rye, barley, and oats, was supposed to increase significantly by 1955 compared with 1949. Large growth in the production of potatoes, white beets, and oilseeds was also expected.

As far as forestry was concerned, the Six-Year Plan draft guidelines set new tasks—among others, for forest farms. They were supposed to "protect, and even enhance, by afforestation and the rationalization of forest farming, the significantly depleted tree numbers, but on the other hand to meet the rising national demand for wood, and get ready for further increases in demand in the following periods."[16]

The main industrial investments of the Six-Year Plan were the following: constructing the metallurgical plant near Krakow named the V. Lenin Steelworks, the chemical factory in Oświęcim, the mining and steel plant "Bolesław," car factories in Warsaw (Żerań area) and Lublin, and the power plant in Jaworzno.[17] At the same time, the Upper Silesian Industrial Region [Górnośląski Okręg Przemysłowy, GOP]—an area dominated by a coal-based economy during the previous two hundred years—also underwent significant rationalization.

When evaluating the realization of the Six-Year Plan, it is worth noting that the plan's goals were exceeded in the field of heavy industrial production; however, at the same time lighter industries such as food production, textiles and other consumer goods fell behind.[18]

System Transformation: State Land Fund

Immediately after the war, the Communist authorities in Central and Eastern Europe began implementing system reforms, aiming to eliminate private property and the market economy system. In Poland, political decisions and economic activity took into account the actual needs of the country undergoing reconstruction, but to a large extent their direction was subjected to ideology and Communist propaganda.

The basic program and future reform trend was outlined in the Manifesto of the Polish Committee of National Liberation dated 22 July 1944. It stated that agriculture reforms were to be based on, among others, the return of land to peasants. On the basis of the agriculture reform decree dated 6 September 1944, the State Land Fund [Państwowy Fundusz Ziemi, PFZ] was constituted and subsequently it took over farms from traitors, and landowners whose property exceeded 50 ha (hectares), and in the case of the Regained Lands, landholdings exceeding 100 ha. Thus acquired land was then reallocated by the fund among the former farm servants, peasants, and sharecroppers—they were given up to 5 ha of arable land per household. The authorities created small farms (2–5 ha) on purpose—in the future it was supposed to facilitate agriculture collectivization, modeled upon the Soviet Union.[19] Those interventions were primarily dictated by Communist ideology and led to extensive farming and semi-subsistence agriculture, which from a long-term perspective can be viewed negatively. Woodland was also involved in nationalization. As a result, the majority of forests, or about 85 percent of the entire forest surface, went into state hands. This happened by virtue of the Polish Committee of National Liberation decree of 6 September 1944 on agriculture reform, mentioned above, as well as the Polish Committee of National Liberation decree of 12 December 1944 on the nationalization of forests exceeding 25 ha. The remaining woodland belonged to farmers.[20]

General nationalization of the basic segments of the economy was legitimized by an act of the National People's Council signed in January

1946. Reforms stipulated therein consisted in nationalizing industry and other sectors of the economy. Big manufacturers became state owned. In the early 1950s Poland was still undergoing system transformation, which was supposed to make the Polish economy similar to the Soviet model. The central planning system and forced industrialization were being enhanced.[21]

Transformation in Science

In the first years after the war, society needed to make up for the losses inflicted on science. Many research institutes were destroyed, and additionally Poland lost two important academic centers—in Vilnius and Lviv—due to the postwar border shift. The universities from these cities were moved, along with their staff, to Wrocław, Toruń, and Łódź. A large number of the academic staff, including prewar professors, died during the war or were relegated from the academic milieu for political reasons. Those who stayed were given the task of rebuilding the academia under new conditions. They succeeded fairly quickly, as the new regime helped by setting up academies and universities that educated specialists. Some scholars strived to rebuild academic structures to their previous form, but this turned out to be impossible. Any development of science was only possible if it followed the dialectical materialistic ideology and guidelines coming from Moscow.[22]

The reorientation of research policies and the ideological offensive were fully reflected at a sitting of the Central Council of Science and Higher Education in January 1949, when the academia was accused of unfounded concern about limited scientific freedom and underestimation of the achievements of Soviet science. The ideological offensive was visible in all aspects of scientific life, and in the promotion of science.[23] In the late 1940s and early 1950s, the education and science system was centralized, imitating the Soviet model. The new system was supposed to have three divisions: the Science and Education Division (academies and universities), the Science and Technology Division (institutes and research institutions of ministries), and the Science and Development Division.

The Communist authorities assumed that Polish science, just as was the case in other Central and East European countries, would be significantly influenced by Soviet science, and would remain in conscious, politically imposed opposition to Western science. They supported new, ideologically "correct" scientific concepts, based on the achievements of Soviet science. Promoting science in the society was subject to the cult of Stalin and the only acceptable ideology.[24] Those publishing houses that popularized mathematics and natural sciences, agricultural science, and engineering were favored, while in the period between 1944 and 1951 the number of popular publishers in the field of human sciences (philosophy, history) gradually decreased.[25] Therefore, drastic changes were inevitable, both in terms of the structure of science and in the mentality of its representatives, as well as the way the scientific achievements were popularized. Generally, the ideology's influence on science and the economy in Poland was significant and outweighed rational arguments. It resulted not only in appointing supporters of the ideology to chief positions in the academia and economy, but also in many other irrational decisions in both spheres.

As Polish economist R. Manteuffel wrote in retrospect, the postwar period in Poland

> was a time of the implementation, in the Polish economy (including agriculture) and social life, all sorts of theories that—sometimes right and based on serious research—were vulgarized in the Polish version. It was recommended that they be used at all times and everywhere, irrespective of whether conditions were favorable or not. At the same time, Western achievements were rejected. It all resulted in considerable economic losses, not to mention the moral ones caused by the elimination of those people who had different views from their professions.[26]

Until 1951, the largest scientific organization was the Polish Academy of Arts and Sciences [Polska Akademia Umiejętności, PAU], with a great tradition dating back to the time of the partitions and the interwar period. It was created in 1872 in Krakow as the Academy of Arts and Sciences, transformed from the Krakow Learned Society functioning since 1815. The Polish Academy of Arts and Sciences was an independent association of scholars, which had its own material and financial resources, collected thanks to various foundations. In organizational terms, it was independent from the regime. Due to political pressure, PAU as well as another leading academic association, the Warsaw Scientific Society [Towarzystwo Naukowe Warszawskie] created in 1907, had to cave in and support the idea of appointing one central state-controlled institution. They also had to end their functioning by the end of 1952, handing over all of their academic records and property to the new supreme academic institution in Poland, which the Polish Academy of Sciences [Polska Akademia Nauk, PAN] was to become.[27]

The Polish Academy of Sciences was created at the First Congress of Polish Science, which took place (after a year-long delay) between 29 June and 2 July 1951.[28] The main goals of the congress, according to a draft issued by the Political Bureau of the Polish United Workers' Party Central Committee, were the following: resistance to cosmopolitan tendencies in the Polish academia, drawing on the tradition of Polish progressive science, and referring to the experience and achievements of Soviet science.[29] It is worth noting that there were a significant number of natural science representatives at the First Congress of Polish Science, as well as in the organizational structures of the newly founded Polish Academy of Sciences.

The founding principles of the Polish Academy of Sciences were finally determined by a law passed on 30 October 1951.[30] The new institution was supposed to follow the example of the Soviet Academy of Sciences, perceived by the Communist decision makers as the leading institution. By virtue of the law, the Polish Academy of Sciences was divided into departments in the following manner: Division One—Social Sciences and Humanities (philosophy, history, philology, literature, art, economics, law); Division Two—Biological Sciences (biology, agricultural sciences and forestry, medicine, veterinary medicine); Division Three—Mathematics, Physics, Chemistry, Geology and Geography (mathematics, astronomy, physics, chemistry and Earth sciences); Division Four—Engineering.[31] The first president of the Polish Academy of Sciences was biologist Jan Dembowski (1889–1963), who played a significant role in promoting Soviet science in Poland by, among others, supporting natural science theories originating from the USSR.

Agriculture and Forestry Science in Poland

In the 1950s, a new academic structure was developed, based on one central science institution, the Polish Academy of Sciences. It was supposed to be provided with academic support by the staff of the numerous newly-created science institutes. They were, for the most part, institutions subordinated to the ministries for a given area. For the subject at hand, that is the transformation of nature in the Stalinist era, the most relevant were the agriculture and forestry institutes, therefore they will be covered more extensively.

Agriculture and forestry science in Poland was traditionally associated with the town of Puławy, where there had been a renowned agricultural science center since the mid-nineteenth century. It is also worth noting that at that time Poland was under Russian rule, as a result of an education system reform, the Agronomy Institute was moved from Warsaw-Marymont to Nowa Aleksandria (as Puławy was then called). In 1862 a teaching and research institute was set up there, the Polytechnic Institute of Agriculture and Forestry, which was an important agronomy institute in that part of Europe. The Russian authorities discontinued the teaching activity of the establishment a few months after its creation, following the defeat of the January Uprising. In 1869 the institute was replaced with a higher education establishment, with Russian as the teaching language, called the Institute of Rural Husbandry and Forestry in Nowa Aleksandria. In 1895, thanks to the efforts of Russian scholar Vasily Dokuchaev (1846–1903), the first chair of soil science in the world was established here. Professor Dokuchaev, as the institute's head in 1892–1895, created foundations for further studies in the fields of soil and agricultural sciences. Paradoxically enough, his theories and the results of his research became the theoretical basis for the scientific agenda of the future Stalin Plan for the Transformation of Nature.

Theoretical bases for the plan were the achievements of Russian scholars of such prominence as V. Dokuchaev, geologist and creator of soil science; Pavel Kostychev (1845–1895), one of the founders of modern soil science; and Vasily R. Viliams (1863–1939), creator of the ley farming system. The Plan for the Transformation of Nature primarily included measures aimed at counteracting drought and crop failures. Its basic directions involved the introduction of shelterbelts as protection from dry winds, irrigation of fields and other hydro-technical investments, and V.R. Viliams's ley farming (a grassland amelioration system).

Time and nature itself quickly verified to what extent these were rational assumptions. Projects such as adequate wasteland afforestation were scientifically founded and served nature well. Others, such as the construction of huge canals, turned out to be destructive to nature and to the people working on these projects.

After the war, the institute's management was moved to Warsaw and a far-reaching reorganization of the State Research Institute of Rural Husbandry was pursued.[32] Eventually, this institute, along with its Bydgoszcz department, was dissolved by a law.[33] It was replaced by specialist farming institutes located in various places all over the country. They were coordinated by the Central Agriculture Institute [Centralny Instytut Rolniczy, CIR], functioning in the 1951-1954 period.

One of the new institutes was the Institute of Soil Science and Plant Cultivation (Instytut Uprawy, Nawożenia i Gleboznawstwa, IUNG), formed in 1950 and functioning to this day in Puławy. The institute initially was a state science institution under the authority of the Ministry of Agriculture. It was first headed by biologist Anatol Listowski (1904-1987), who held the position until 1968, while former director L. Kaznowski (1890-1955) was deputy director. One of the institute's departments was the Department of Soil Science run by Michał Strzemski (1910-1992). They created a laboratory, which boasted introducing thermophiles to the south hills of the town of Kazimierz Dolny. The remaining institutes under the Ministry of Agriculture, among others, were: (1) The Plant Breeding and Acclimatization Institute [Instytut Hodowli i Aklimatyzacji Roślin, IHAR]—set up in 1951 in Puławy, which was moved two years later to Radzików near Warsaw, where it functions to this day, with a few departments and numerous establishments located all around the country. In the early period (1951-1969) it was run by biologist Jadwiga Lekczyńska (1899-1983); (2) The Institute for Land Reclamation and Grassland Farming [Instytut Melioracji i Użytków Zielonych, IMUZ] in Falenty—functioning since 1953 until 2009, initially headed by agroecologist Józef Prończuk (born in 1911); (3) The Research Institute of Pomology in Skierniewice [Instytut Sadownictwa]—created in 1951, on the initiative of famous cultivator and horticulturist Szczepan A. Pieniążek (1913-2008), who was its director in the 1951-1984 period, as well as a member of the newly created Polish Academy of Sciences.

These institutes were involved in the introduction of new theories from the Soviet Union, and their managers had to create conditions for work on the theoretical and practical aspects of Stalin's Plan. The situation looked similar in forestry. Specialized institutes were created by the state to undertake research in accordance with Stalinist Plan.

The Ministry of Forestry supervised the Forestry Research Institute, established before the war in 1930 under the name of State Woodland Experimental Department. In 1934 it was transformed into the State Woodland Research Institute, and since 1945 it has functioned under the name of the Forest Research Institute [Instytut Badawczy Leśnictwa, IBL],[34] currently under the jurisdiction of the Ministry of Environment.

The institute conducted studies on natural bases of forestry. After 1948 it had an afforestation department that did research on land cultivation, forest amelioration, and afforestation.

There were also noteworthy natural sciences institutions under the jurisdiction of the Polish Academy of Sciences, including the Institute of Ecology (1952) and the Department of Botany (created in 1953, since 1956 called the Institute of Botany) in Krakow. The Polish Academy of Sciences also took over the Department of Dendrology and Pomology, created as the Department of Dendrology in 1933 in Kórnik as a part of the Kórnik Gardens owned by the Foundation "Zakłady Kórnickie." Based on this establishment, in 1952 the Department of Dendrology and Pomology was created.[35]

There were also numerous natural sciences societies and organizations in Poland, of which the Polish Forest Society [Polskie Towarzystwo Leśne] should be mentioned. It originated from the Galician Forest Society, set up in 1882 in Lviv. From 1873 until now the Society has published a monthly magazine about forests—*Sylwan*. Also, many biologists involved in the reconstruction of postwar science were members of the Polish Botanical Society [Polskie Towarzystwo Botaniczne], which had been founded in 1922.

Polish Science and the Plan for the Transformation of Nature

The Stalin Plan affected Central and East European states to different extents, which is why it is worthwhile asking how it was received in Poland. It was promoted in the country in the late 1940s and early 1950s under the name The Great Plan for the Transformation of Nature. It was supposed to be an inspiration for transformations of nature similar to those that had been conducted in the USSR for economic purposes, mostly with the aim of increasing crop yields.

The Great Plan, initiated by Stalin in 1948, was implemented in the USSR primarily until his death—that is, until 1953. It was supposed to have been a response to the great drought in 1946 and famine in 1947, which had led to a death toll of approximately one million people in the USSR. The plan's goal was agriculture amelioration by appropriate land development, new farming practices, and water management projects. Designed in the tradition of Stalin's cult, it was presented in the regulation of the USSR Cabinet and the Central Committee of the All-Union Communist Party (Bolsheviks), issued on 23 October 1948: "About the plan of protective afforestation, introduction of grassland crop rotation, establishment of ponds and water reservoirs in order to obtain large and stable crops in steppe and mixed forest and steppe regions in the European part of the Soviet Union."[36]

In retrospect, the Stalinist Plan has been assessed negatively, as a plan potentially causing climate change around the entire world, and leading to a global environmental disaster. One of the failed projects of Soviet hydro-engineering, based on ideas predating the Stalin Plan, is the Rybinsk Reservoir, an artificial water reservoir on the Volga River. Its creation in the 1940s cost many people's lives and it turned the neighboring area into swamps.[37] Another example is the destruction of the Aral Sea (currently on the territory of Kazakhstan and Uzbekistan), which almost completely dried out due to misguided amelioration measures—clear evidence of how dangerous irrational nature transformation can be. In that period the laws of nature were disregarded, as Stalin was assumed to be a genius who could control them.

A plan combining science with economics, supported by scientific theories, must have inspired confidence and attracted the interest of the Polish authorities and society. The plan and its supporting theories resonated mostly with scholars. For the more than five years of war, the country was basically completely disconnected from all scientific developments. The subsequent inclusion of Poland in the Eastern Bloc made it follow the Soviet trends in science. Any scientific novelties that reached the country came from the USSR or other Eastern Bloc countries. In many respects, Western science was rejected altogether and criticized, as was the case with genetics. Poland actually stopped teaching it for a few years, manuals were withdrawn, and curious students and scientists were reprimanded. Fortunately, in Poland there were no repressions for those interested in genetics, whereas in the USSR repressions in an extreme case led to the arrest and death of one of the most prominent Russian geneticists—Nikolai Vavilov (1887–1943).

The Basic Trends of the Stalin Plan

The Stalin Plan primarily required vast stretches of land to be afforested in the South of the country in order to prevent the dry wind (called *sukhovey*), coming from the steppes of Kazakhstan and the Central Asian deserts, from penetrating the fields and causing damage.[38] Additionally, it was planned that eight stretches of forest, of a total length of 5,000 km, would be planted, running from north to south on the dry steppes in order to retain water and fence off the area from winds (four belts on drainage divides and four on the river banks of the Donets, Don, Volga, and Ural). The stage described above and subsequent stages of the plan—among others, the new plan announced in 1950 to build large canals and water power plants—aimed at a complete transformation of climate conditions and agriculture in the USSR. The main objectives of the plan were also announced in Poland.[39] In the further stages it was planned to reverse the flow of Siberian rivers in order to irrigate Central Asia.

The plan was supposed to be further supported by results of studies carried out in other fields of science—for example, the latest Soviet achievements and models in hydro-engineering and nuclear power engineering. It was also mentioned enthusiastically in Poland, such as in the party's periodical *Nowe Drogi*: "In the Soviet Union, studies on nuclear energy progress fast. They were used among others to blow up the Turgai Gate in order to implement the Stalin Plan for the Transformation of Nature in the Asian part of the USSR, to reverse rivers and irrigate deserts."[40]

With overestimated objectives and an unrealistic pace of nature transformation, social and environmental costs were ignored. Immediate afforestation of USSR's steppes was forced, as well as indiscriminate induction and acclimatization of plants coming from other geographical regions. The chief mistakes were a willingness to change nature as fast as possible, and a faith in Stalin and his plan's executors' unlimited capabilities.

Examples of the Stalin Plan Implementation in Poland

When studying the practical aspect of the Stalin Plan reception in Poland, one may assume that its implementation was an attempt to adapt its best elements to the specific geographic and climatic conditions of the country. Such an approach is represented by Adam Czermiński in his book with the meaningful title *The History of the Great Plan* published in 1951 by publishing house Wiedza Powszechna, as part of the series 'Library for Everyone.'[41] In his book Czermiński introduced the theme of the subjugation of nature in the USSR and discussed it, often irrationally with the knowledge we have today—for example: afforestation, irrigation, and introducing grassland crop rotation in the steppe areas of the European part of the USSR, building canals (e.g., the Volga–Don Canal and the South Ukrainian Canal), and reversing the flow of Siberian rivers. The author enumerated the possibilities for nature transformation in Poland in line with the Six-Year Plan in the chapter "From Green Spruce Trees to White Coal." He also gave a few examples and feasible plans that could be accomplished in Poland in the following domains: hydro-engineering—the construction of canals, dams, and hydroelectric power stations that would change the water system in Poland. One example was the Danube-Oder Canal, and another the East–West Canal—neither were accomplished. The implementation of several dams on Polish rivers and hydroelectric plants became more advanced. In forestry, the most important task was increasing the area of forests by the afforestation of infertile land and fallow land. Land amelioration was also important. The construction of water outflow canals, for example irrigation works in Żuławy Gdańskie, was realized. Żuławy is the alluvial delta area of the Vistula River, in large part reclaimed artificially by means of dykes, pumps, channels, and a drainage system. This complex system of channels was destroyed during World War II. In the 1950s in Żuławy, according to Czermiński, "grains rustle again, and trees in orchards bend down under the weight of fruit."[42] In mining, the essential project mentioned by the author was a trunk line for the transportation of sand from the Błędów Desert to fill no longer active adits in Silesian coal mines.

These examples show that the popular and practical reception of the Stalin Plan was limited to selected issues and ideas, which were compatible with the country's conditions. But ideas such as river flow reversal had no followers in Poland.

Artificial Reservoirs and Irrigation

It first of all needs to be noted that water resources are scarce in Poland. As industrial and agricultural demand was growing, the postwar Poland had to tackle this problem. Therefore, it cannot be stated that the Stalinist Plan was the reason for the implementation of large hydro-engineering investments or irrigation projects in agriculture. There was neither demand nor interest among Polish water engineers in extreme Soviet ideas, like reversing the flow of rivers. Undoubtedly, however, the plan could have given encouragement to build artificial reservoirs in the early People's Republic of Poland period, but such reservoirs were supposed to solve Poland's own energy and water problems. These problems had been known and tackled for a long time. Creating artificial reservoirs has a long tradition in Poland. It is worth noting that the Czorsztyn Reservoir, the Rożnów Reservoir, and the Goczałkowice Reservoir were built mainly in the People's Republic of Poland period, but these plans dated back to previous periods. The idea to create a reservoir in the Czorsztyn region appeared as early as 1905, although it was not acted on at the time. Studies and conceptualization work for the Czorsztyn–Niedzica reservoir and the Sromowce Wyżne reservoir on the Dunajec River began in the 1950–1964 period, and works started in the 1970s.

As for the Rożnów Reservoir and the Czchów Reservoir, there was a terrible flood in 1934 that led to an immediate decision to build these reservoirs on the Dunajec River. However, the works were interrupted by World War II. The first dam was finished in 1941 and the second one in 1949. Reservoirs with a capacity of 240 million m^3 and a power plant generating 140 million KW annually were created. The devices were designed and later on their construction was managed by prominent Polish hydro-engineers, including Edward Czetwertyński (1901–1987).

In order to supply water to the Silesian Industrial Region, in the 1950–1955 period Poland built a dam on the Mała Wisła River (Little Vistula River) and a reservoir in Goczałkowice with a capacity of 170 million m^3. A barrage on the Vistula River and a port in Przewóz (1948–1954) were to supply water to a great investment of the Six-Year Plan, Nowa Huta. In 1947, the Ministry of Communication ordered the Regional Directorate of Waterways in Krakow to conduct a feasibility study for the construction of a reservoir. Work on the design was completed and approved at the end of 1949. On 1 January 1950 it was constituted, and the construction of one of the biggest investments of the Six-Year Plan began. The project was completed by 1955. It should, however, be noted that the first mention of the need for a reservoir to reduce flood waves on the Mała Wisła River dates back to the Partitions period.[43]

There is, however, one example from the field of water management indicating that the idea of the Plan for the Transformation of Nature had penetrated into the field of water amelioration, namely the Wieprz–Krzna Canal. The Six-Year Plan aimed primarily at completing older amelioration projects, but with the current trend to pursue fast improvement of crop quality through irrigation, a decision to build this canal was made. The idea to construct the Wieprz–Krzna Canal appeared in 1952, and it was supposed to improve the hydrographic conditions in the region of Lublin. It turned out to be one of the largest agricultural investments in Poland.[44] The first part of the canal was opened in 1958.[45] However, it has been determined in hindsight that the impact of the Wieprz–Krzna Canal on the environment in the Lublin Polesia was negative.[46]

Reception of the Stalin Plan and the New Scientific Theories

Studies on the influence of the Stalin Plan on Polish science and on the economy were limited in scope. Few researchers have so far taken an interest in the subject. The area that has been explored to the greatest extent has been the process of introduction of one of the main trends of the plan, Lysenkoism, into Eastern Bloc science. In the 1980s the subject was more widely presented by science historian Stefan Amsterdamski (1929–2005).[47] Research on various aspects of Lysenkoism was subsequently done in the 1990s by theologian Zbigniew Kępa,[48] and presently by botanist and historian of botany Piotr Köhler.[49] As for foreign authors conducting research on the reception of Lysenkoism in Poland nowadays, American historian William deJong-Lambert[50] should be mentioned. Information on the Stalin Plan can also be found in the works and memoirs of scholars who remember the People's Republic of Poland period, such as L. Kuźnicki[51] and R. Manteuffel. The latter wrote in the 1980s:

> Personally, I got to know Viliams's ley farming system closely, very useful at times, but only in extensive economy. Its indiscriminate application in inadequate conditions and circumstances led to a decline in production per unit area. The application of Lysenko's ideas was a complete humiliation. Unfortunately, many of our scientists and practitioners were "seduced" by the charm of his "science." Prior to 1956, there was almost no contact with the Western countries. All novelties came to us from the USSR and other countries of the Socialist community.[52]

In this context, Polish scholars inevitably had to express "interest" in the new theories and take a stance in the matter of the Plan for the Transformation of Nature. Its supporting theories in the domain of biology, forestry, agronomy, and particularly Michurin-Lysenkoism, found their followers in many countries of the Eastern Bloc. They were referred to as Michurinists. There were, however, scholars in Poland such as Tadeusz Dominik (1909–1980) who suggested that in our conditions such extreme nature transformations were not necessary. Some of the studies were conducted by scientists following the pre-war schools.

Although admittedly there were not many of them, they were skeptical about the new natural sciences theories. Natural science concepts resulted, to a certain extent, from independent research. Polish scholars were keener to devise theories on nature transformations than to actually engage in putting them into practice. In some cases they deftly presented the results of their work in order to make them fit in with the obligatory ideology. Many introductions and endings of publications were enthusiastic about the new theories, but the main body was independent. Papers by A. Listowski, the director of the Institute of Soil Science and Plant Cultivation, are a good example of such occurrences. He adopted this strategy to be able to satisfy decision makers and retain his scientific independence in the study of biology and physiology of agricultural plants. Obviously, there were some researchers who carried out studies in line with the Stalinist Plan guidelines. Their activity was to a large extent subject to political pressure (the implementation of the general idea of a Socialist planned economy) and ideology. This was clearly supported by the regime and people responsible for science organization in postwar Poland. Among important organizers of science in Poland there were a few representatives of the natural sciences. Biologists of different specialties made a great contribution to the First Congress of Polish Science: previously mentioned general biologist Jan Dembowski, parasitologist Włodzimierz Michajłow (1905–1994),[53] and zoologist Kazimierz Petrusewicz (1908–1982).[54] They held key functions in the academic structures: W. Michajłow, as a decision maker in education who held many positions as a manager, director and vice minister of High Education; Jan Dembowski, as the first president of the newly created Polish Academy of Sciences; and K. Petrusewicz, as head of the Science and Higher Education Department at the Polish United Workers' Party Central Committee in the 1948–1955 period, and the secretary of Division Two of the Polish Academy of Sciences. Petrusewicz was invited on many occasions to sittings of the Polish United Workers' Party Central Committee Politburo, and was ascribed the task of drafting program guidelines for the First Congress of Polish Science.

The largest response to the Plan for the Transformation of Nature came from natural science representatives: biologists, botanists, foresters, and agronomists. Those representatives of natural sciences who were also party activists and science organizers in the postwar reality clearly made key contributions to promoting the "new biology" and the Stalin Plan. The most active were the three mentioned above: Michajłow, Dembowski, and Petrusewicz.

At the initial stage of Stalinism, the new biology was promoted by W. Michajłow, a high-ranking functionary in the Ministry of Education and a recognized authority in biology, especially in parasitology. His account of a session at the Lenin All-Union Academy of Agronomy held in August in 1948, during which Lysenkoism found support, was published in the daily *Głos Ludu* as part of a series of papers under the title "Discussion on Biological Science in the USSR."[55] Michajłow also presented the new biology in his paper "What Michurin and Lysieńko [sic] Gave the Soviet Countryside."[56] A more complete view of the new theory was presented by Michajłow in the party's periodical *Nowe drogi* in 1949 in the paper "At the First Stage of Biology Development in Poland."[57] In that paper he outlined the concepts that were elaborated during a deliberation organized by the Marxist Naturalists' Society, which functioned from 1948 and was affiliated with *Nowe Drogi*. The main ideas were the following:

1. It is necessary to keep promoting the Michurin–Lysenko theory, publishing as full a literature on the matter in Polish as possible.
2. Education of young biologists and agrobiologists should be based on the rules of the new theory, and it should be included in curricula of schools and universities.
3. Higher schools of agriculture should conduct research and provide teaching in the spirit of the new theory. Young academics on the territory of the USSR should be acquainted with the new theory and the practice of the Soviet agrobiology.
4. Work of agriculture institutes, experiment stations and other agricultural organizations should be based on the Soviet biology and agrobiology.
5. State farms and cooperatives should become the site of extensive experiments aimed at the creative mastery of the theory and practice of the new agrobiology.
6. There is a necessity for the new biology findings to be adopted by other sciences.[58]

Michajłow had the opportunity to implement these ideas personally. He was famous as an authority in biological sciences, a party activist and a high-ranking official working for education organizations. As a functionary of the Ministry of Education he could recommend his own manuals, explaining the new biology to schools and academies.[59] At this time, it was believed that "implementing a curriculum in secondary schools, pedagogy schools and schools educating future kindergarten teachers, meant that teachers needed not only to have the knowledge of the new biology, but also the ability to combine theory with experiments conducted by students."[60] School biology curricula were supposed to take into account the issue of plant variability, influence of external conditions on plants and nature transformation based on works by Michurin and Lysenko. The new biology, also at schools, was supposed to be taught in combination with practical agrobiology based on experiments carried out on school plots.[61] Indoctrination did not even spare kindergartens, where teachers were to talk about the following issues: "Familiarizing children with the Soviet Union; Stalin; Stalin's Five-Year Plan—reconstruction of the USSR after the war; the Plan for the Transformation of Nature; and Stalin leading millions fighting for peace."[62]

There was certain caution, masked by enthusiasm, in the promotion of the Stalin Plan in Poland. It is perceivable in the words of authors writing manuals under W. Michajłow's supervision. The Stalin Plan was described as incredibly courageous, and Lysenko as a genius agronomist, nevertheless the obvious differences were pointed out to students: "In our climate, shelterbelts in fields are less useful, and fallow lands are afforested by the State Forests with forest tree species."[63]

Preparations for the introduction of the new "scientific theories" in the natural sciences predated even the First Congress of Polish Science in summer 1951. From 27 December 1950 through to 13 January 1951, there was a conference held in Kuźnice on theories in biology, agrobiology, and medicine. The content of the Kuźnice conference was included in the materials of the First Congress and published.[64]

There is another conference that is worth noting—the agrobiological conference for forestry specialists, held 9–13 September 1952 in Rogów. A. Makarewicz (1905–1990), S. Białobok (1909–1992), and W. Krajski were among the scholars who presented their papers on the new agrobiological theories, and their application in forestry. Presentations were followed by a debate, and not everyone agreed with what had been said. For instance, T. Dominik stated: "I do not agree with Viliams's claims in their entirety; Viliams, for example, maintains that trees can develop only if they have mycorrhiza. That's not true."[65] We may add here a comment based on today's state of knowledge—Viliams was right to a large extent; mycorrhiza is the occurrence of plant and fungi symbiosis,[66] which probably appears in most plants. However, Dominik's statement shows that it was possible to disagree with Soviet science, and that constructive criticism was, within some limitation, accepted by decision makers. During the debate Dominik was critical of Lysenkoism as well. The conference was concluded by Petrusewicz, who said enthusiastically: "The new biology sprouts in forestry, it has shoots and they are getting stronger." He also noted that despite constructive criticism and academic debates, contributions of forestry practitioners were too little. He underlined that it was "a tradition among practitioners–field workers, inherited after the interwar period, when science was not very useful and isolated from practice." Petrusewicz recommended that conference participants take a pragmatic approach, and as an example to follow he indicated the great prospects of Soviet forestry, based on dialectical materialism and Lysenko's theories.[67] During another biology conference, which took place in Kortowo, a district of Olsztyn, 18–28 August 1953, the papers and debates indicated an increasingly clear division between Lysenkoism supporters and its opponents.[68]

Scholars were familiar with the Stalin Plan due to numerous conferences and scientific as well as popular science publications on the matter in the 1940s and early 1950s.[69] Practical implementation of the Stalinist Plan and Lysenkoism in Poland was probably limited compared to the USSR and other Communist countries. The end of the Lysenkoism era came earlier in Poland.[70] Scholars realized this and talked about it during a special conference for biologists in April 1956 at the editorial office of the periodical *Po prostu*.[71] It was summarized by contemporary American historian W. deJong-Lambert: "The Lysenko era in Poland ended relatively earlier than in neighboring, Soviet-allied states such as Hungary, East Germany and Czechoslovakia, as well as nine years before Lysenko was forced from power in the USSR."[72] The main reasons for the end of the fascination with Lysenkoism in 1956 were political, as deJong-Lambert explains: "Conditions specific to Polish politics and Poland's relationship with the Soviet Union during the Thaw after Stalin's death provided the opponents of 'Lysenkoism' in Poland with an opportunity to criticize Lysenko, and to restore Polish genetics."[73]

One of the pillars of Stalinist Plan found a special place in Polish science—reception of the so-called "new biology," known as Lysenkoism in the 1948–1956 period, was very thoroughly studied by Polish and foreign science historians.

Lysenkoism in Poland

The Stalin Plan was based on new, as we now know, erroneous, biological and farming theories developed by Trofim Lysenko (1898–1976), who triumphed in Soviet science for many years. His pseudo-scientific ideas, though irrational, prevailed for many years and found followers in other countries of the Eastern Bloc, Poland included. These pseudo-scientific theories were promoted in the 1940s and 1950s under various names, such as "the new biology," Michurinism, Lysenko-Michurinism, Soviet creative Darwinism, and "the new genetics." At present, these theories as a set are known as "Lysenkoism." Lysenko's theories were additionally supposed to validate the Stalin Plan from the scientific viewpoint.

Lysenko directed his theories primarily against the "bourgeois" formal genetics, ignoring the scientific findings collected up until then, research principles (methodology), testing, and the repetitiveness requirement. He tried to prove that organisms may change under the influence of the environment. It fitted in well with Marxism–Leninism. Lysenko mainly directed his theories at the party apparatus of the USSR. Therein, as well as in the Marxist ideology, he searched for confirmation of his theories. The party apparatus in turn expected Lysenko's theories to confirm its ideology, considered to be the only legitimate one. Public criticism of Lysenkoism led to the imprisonment in 1940 and subsequent death of the prominent Russian geneticist N. Vavilov.

The reception of Lysenko and his theories among Polish naturalists was not impressive. As P. Köhler claims, based on thorough historical research, there was a small audience in this group, and those few had a decisive say in science, even if not the most influential—they are the previously mentioned trio of Dembowski, Michajłow, and Petrusewicz.[74]

Lysenko developed the basis for his theories in the field of biology before World War II, and they were founded to a certain extent on the achievements of Ivan Michurin (1855–1935). His name, as well as Lysenko's, gave name to the new theories henceforth referred to as Lysenko-Michurinism. Michurin was a Russian horticulturist and he did not live to see his achievements serve as an instrument for the Lysenkoism propaganda. In his work Michurin focused on crossing geographically distant races and species in order to acquire new hybrids that would be more resistant—for example, to difficult climate conditions. His theories, however, including the creation of vegetative hybrids, the mentor theory, the theory of destabilized heredity in juvenile organisms, and hybridization should be considered pseudo-scientific. Michurin wanted to attain frost resistance in good varieties of fruit trees. He believed that favorable environmental conditions during crossing would make the young plant (the mix-breed) reveal and preserve the desired properties. He expected to obtain increasingly better properties—for instance, bigger and sweeter fruit in frost-resistant varieties—through such pseudo-scientific methods of tree cultivation.

Michurin is believed to be the author of the saying, used often in Poland in the 1950s: "We cannot wait for favors from nature. We must take them from it—that is our task."[75] This rule, guiding the horticulturist for many years in his work on new plant hybrids, was adopted by Lysenko and used in formulating new theories. Lysenko used it as an argument that hereditary traits are not fixed but can be changed by external conditions. There was only one step from that point to new biology slogans on the heredity of acquired traits. Details of Michurin's horticultural work were ignored. Actually, the creator of Michurin's Winter Beurre (a winter variety of pear introduced in 1903) waited for a dozen years for his results. Lysenko for his part proposed instant changes and promised spectacular results from his work on farmland plant varieties, as well as great yields.

An interesting sign of the times is a Polish postage stamp from 1954, with a picture of Michurin looking at apple trees bearing beautiful fruit and a slogan: "Let's use Michurin's experience." It was issued to commemorate the Polish–Soviet friendship month, and it was a symbol of hope for an abundant fruit harvest thanks to Soviet science.

Lysenko's methods and ideas for the improvement of nature using pseudo-biological means captured the interest of Stalin himself. Lysenko found political support for the new biology. At a session held from 31 July to 7 August 1948, the V. Lenin All-Union Academy of Agricultural Sciences adopted "the new biology" as the only legitimate theory in biological and agricultural sciences applicable in the USSR.[76] Support of the Soviet regime for Lysenko lasted for a very long time—until 1964.

Lysenko's "new biology" found its place in the Stalin Plan. It complements it well, due to the proposed fast pace of influence on nature, as well as the promised decreased costs and labor of farmers and horticulturists. As was explained at the time, including to Polish readers:

> The Stalinist Plan for the Transformation of Nature faces Michurinists with many great new challenges, which require immediate solutions. For example, it is necessary to find out which afforestation methods are the most economical, efficient, and favorable to farming. It has to be determined in what set of plants new tree stands will develop with the best results; it has to be established what fauna should inhabit them, what soil germs should be introduced in the forests, etc.[77]

In the Polish context, the following conviction provided the basis for combining Michurin-Lysenkoism with the plan:

> Implementing the Stalinist Plan for the Transformation of Nature requires such soil fertility that many farmland plants will not be able to fully use this capacity. And within one and a half decades many of them will have to be replaced with different varieties. Therefore Michurinists are faced with the challenge of creating new varieties that will be able to fully use the opportunity provided by soil fertility.[78]

The support that Stalin provided to Lysenko was described by W. Michajłow, mentioned above, in a propaganda book on the attainments of Soviet science:

> When Lysenko, the creator of the new Soviet agrobiology, while fighting for the development of science, encounters resistance from scholars who use traditional models, and when those scholars, who have dominated many research and education institutions accuse him of not meeting science standards and preach old genetics *ex cathedra*, Stalin rules in Lysenko's favor and guarantees the development of the new biology in the USSR. With reference to and based on the new agrobiology development, Stalin initiated a gigantic Plan for the Transformation of Nature for steppes and semi-deserts in the Soviet Union. The thousands of kilometers of shelterbelts, which sprout on huge areas of the Soviet Union, are the work of the Stalinist ideas.[79]

Among other things, Lysenko developed the erroneous "star-shaped method of tree planting," based on the conviction that there is no intraspecies struggle for life. According to this theory, specimens of the same plant species support each other, even if they are planted in thick clumps. S. Amsterdamski wrote: "Little oak trees, whose planting cost millions of roubles, were clearly unfamiliar with the leading biological theory, and wanted to neither cooperate nor sacrifice themselves for the sake of the collective, and almost all of them died within a few years."[80] This quote refers to the introduction of a method complying with the Stalin Plan in Central Asia. Lysenko also developed a method of wheat vernalization ("jarovization"), which changed the properties of winter wheat (so it is able to sprout in spring) and obtaining spring wheat (able to sprout in late summer). Vernalization consisted in soaking the seeds and exposing them to low temperature. This method was promoted in

Poland as a way to improve wheat harvests. It should be mentioned that nowadays some vernalization researches are conducted by plant physiologists in different countries.

As P. Köhler remarks, some botanists made a significant contribution to promoting Lysenkoism in Poland.[81] These were mainly: S. Pieniążek, who was the head of the Research Institute of Pomology in Skierniewice; A. Listowski, who was the director of the Institute of Soil Science and Plant Cultivation; and Aniela Makarewicz, who was a professor at the Agriculture University in Warsaw.[82] According to her, a positive attitude towards Lysenkoism reflected progressiveness in biology, while resistance in the new biology adoption was in her opinion proof of unfamiliarity with Lysenko's and Michurin's works, as well as ignorance of Marxist literature and the rudiments of dialectical materialism.[83] In one of her papers she wrote:

> As Soviet agrobiological achievements and ideas penetrate Poland, one can more easily recognize a scientist's attitude, based on his or her approach to the science. Interest in agrobiology, and the urge to apply it and develop it in our country, is characteristic of progressive academics. Scholars' conservatism is reflected in their negative attitude to agrobiology. Difficulty in understanding agrobiology follows from: (1) Ignorance. Not all team members read Lysenko's works, and a vast majority have not read Michurin; (2) Poor familiarity with Marxist literature; and (3) Ignorance of the basics of dialectical materialism.[84]

Lysenkoism was dealt with in Poland relatively quickly. In mid December 1955 the periodical *Po prostu* published a paper by Polish biologist Leszek Kuźnicki (born in 1928) titled "Darwinism and Lysenkoism: Let Us Break the Conspiracy of Silence,"[85] which was a sign of biologists' growing resistance to Lysenkoism. It provided the stimulus for a debate in the academia, which was still ongoing in 1956. At the end of June 1956, the debate was closed, and Lysenko's unconfirmed theories were rejected to the benefit of science based on reliable knowledge and methodology.[86]

Agriculture and Horticulture

In agricultural and horticultural practice, the Stalin Plan found followers interested in crop and yield improvement. Plant introductions and acclimatization modeled on the achievements in the USSR inspired particular interest in Poland. Such procedures have been used by humans from the beginning of time, but they require a lot of time and patience, as well as appropriate environmental conditions. In the Stalin Plan and in Lysenko's "biological theories" this was completely ignored—nature transformations were supposed to be fast, easy, and unconditional.

Experiments with different plant acclimatizations conducted in Poland in the early 1950s resulted from Lysenkoism followers' conviction of unlimited adaptation abilities of species. It was believed that it was enough to introduce tropical or subtropical plants to Poland's climate conditions, to obtain, after a period (short or long) of acclimatization, varieties adapted to a colder climate.[87] Fascination with Soviet achievements inspired Poles to introduce exotic plants to large-scale farming. Such plants as rice, citruses, corn, special varieties of wheat, soya, and oilseed crops (castor oil plant and sunflowers) were supposed to be acclimated. However, these plans were reduced to the experimental stage in ministry-controlled institutions such as the Plant Breeding and Acclimatization Institute and the Institute of Soil Science and Plant Cultivation. As mentioned earlier, one of the latter's departments was the Department of Soil Science run by M. Strzemski. It had a protective arable land laboratory which conducted experiments on new plants introduction on the slopes of the southern hills in Kazimierz Dolny. It made sense for plants such as peach trees, apricot trees, and walnut trees, due to good insolation and low rainfall in the area. Frost-resistant varieties of these fruit trees are still to be found in Polish orchards, but they have never been planted on a large scale, as some visionaries imagined would happen.

Polish Rice

The chief propagator of new plant species acclimatization in Poland was J. Lekczyńska, the Plant Breeding and Acclimatization Institute founder, and director in the 1951–1969 period.[88] The institute's success in creating new varieties of potatoes under Lekczyńska's direction is remembered even now in a positive context.

Lekczyńska played a significant role in promoting Lysenkoism and imposing dangerous cases of acclimatization—also in terms of decision-making and organizational work.[89] One of Lekczyńska's subordinates was an engineer-agriculturalist, Piotr Sobczyk (1887–1979),[90] who was involved in acclimatization and experimental cultivation of rice in Poland. He started his work before the Stalinist Plan was promoted to any extent. Between 1940 and 1946 he gained experience in rice production in Romania.[91] Later on, it turned out that Sobczyk's work fitted in perfectly well with promoting the nature transformation vision. Initially, studies on rice were conducted at the State Research Institute of Rural Husbandry under the supervision of Lucjan Kaznowski. After the restructuring of the institute in 1951, the task was taken over by the Plant Breeding and Acclimatization Institute and Lekczyńska's team.

Citing rice growing in the USSR, in Hungary, and in Romania as a model, scientists began research on the possibilities of growing the plant in Poland.[92] Sobczyk wrote in a brochure promoting rice plantations in Poland:

> The government of the People's Republic of Poland decided in 1948 to begin experimental work on rice acclimatization and growing. Firstly, seeds were brought from Romania and Bulgaria, and in 1951 seven varieties were transported from the USSR. In the spring of 1948 fields were set up in fourteen places in Poland. They were small and measured between 10 and 600 m². Rice grains were obtained from only three places, where farmers were skilful and careful. In the first year, it turned out that in order to obtain yield in our climate, new methods of cultivation would have to be invented, different from the ones used in hot climate countries in the South. For this purpose, a lot of trials and experiments have been conducted on seedling production, sowing dates, vernalization, field arrangement, protective belts, irrigation methods, etc.[93]

The basic problem was adapting the plant to Poland's climate so as to ensure large yields. The goal was to be attained with agro-engineering methods based on the "new biology," among others on the vernalization of seeds proposed by Lysenko. In addition, wind buffers were recommended. In 1950, thanks to experiments and the introduction of new rice growing methods, a fairly good yield was obtained. Sobczyk assumed that his work would give birth to a new variety of rice. He tried different varieties, including "Okute Szynriki" and "Agostano." He grew as many as fourteen varieties at the Plant Breeding and Acclimatization Institute, two of which turned out to be the best, as grain and hay ripened at the same time. By selecting a few hundred plants from these two varieties, he got plants that ripened early, had full ears of corn (panicles) and robust grains.[94] For successful rice production, Sobczyk made the following recommendations to his readers, potential rice farmers.

He advised them to take the matter seriously—firstly to select an area close to fresh water (river, canal, pond) with the right soil, of lumpy structure. An important point was to use crop rotation and to plant protective belts of sunflowers around the fields. The author advised to prepare seeds by using the vernalization method to improve their ability to germinate. He described this method the same way as its creator, Lysenko. The other significant instructions were to irrigate the rice fields and to harvest rice manually with sickles.

Yields from the Plant Breeding and Acclimatization Institute plantation in 1951 were estimated, depending on the variety—from 1 hectare it was possible to get between 9 and 40 quintals. It was not even close to the achievements in the Soviet Union (160 q per ha), or in other European countries (104 q per ha).

The State Commission for Economic Planning ordered the Plant Breeding and Acclimatization Institute workers to assist in setting up rice plantations. Sobczyk's paper—only a thousand copies were published—contained a description of rice growing on experimental fields in Puławy, and basic information on the possibilities for rice cultivation in Poland. As the publishing house said, it was addressed at the general public: farming practitioners, workers on state agricultural farms and production cooperatives, and peasants and young people interested in the subject. People were encouraged to acquire seeds for their plantations at the Plant Breeding and Acclimatization Institute, in the department of special industrial plants in Puławy. Lekczyńska and her subordinates were also known for promoting corn and other plants proposed by Soviet practitioners and scientists.[95]

Kazakh Dandelion
Biologist L. Kaznowski,[96] who used to be the State Research Institute of Rural Husbandry director and was subsequently deputy director of the Institute of Soil Science and Plant Cultivation, also worked on acclimatization. He was ascribed the task of breeding a new variety of castor oil plant (*Ricinus communis* L.), and he worked on the Kazakh dandelion (*Taraxacum kok-saghyz* Rodin)[97] and pea hybrids. He was the creator and co-creator of many varieties of tobacco, hops, medical plants, fodder plants, oil seed plants, fiber plants and other plants.[98] Using Lysenkoism, he attempted and observed attempts of the acclimatization of other farming plants: sorghum (*Sorghum bicolor* (L.) Moench),

pepper (*Capsicum* sp.), chrysanthemum (*Chrysanthemum cinerariaefolium* (Trev.) Vis.), marjoram (*Majorana hortensis* Moench), lavender (*Lavandula* sp.), sage (*Salvia* sp.), rice, cotton, and sesame.[99]

Michurinist Clubs
Plant acclimatization was also taken up by amateur clubs in Poland, usually at schools. They were not, however, regular biology clubs. In fact, they were set up to imitate young Michurinist clubs in the USSR, and their ambition was to use the theories of Michurin, Lysenko, and Viliams.[100] The aim was to recreate the new biology achievements on a small scale. The youth could get to know the new theories by experimental cultivation of rice and other exotic plants, medical plants, wheat improvement procedures, and impressive grafting between related plants, such as grafting potatoes on tomato plants. In that particular case, grafts were usually successful, as the two plants belong to the same family—nightshades (*Solanaceae* Juss.). We shall quote a paper excerpt about one club's work:

> It would seem that winter should put a stop to the school Michurinist club at the State Pedagogy Secondary School in Szczekociny (Włoszczowa district), which between the spring and autumn of last year conducted innovative experiments on growing barely known and exotic plants. This, however, is not the case: regular training meetings of young enthusiasts of experimental agriculture are still being held. Young people learn about the experiences of Soviet scientists—Michurin, Lysenko, and Viliams—by writing thorough papers, organizing talks and lectures, as well as through lively, interesting debates. It has to be added that experimental work of the Michurinists club in Szczekociny will be extended in scale this year. Apart from growing rice and conducting experiments in rice acclimatization, and apart from growing medical plants, like castor-oil-plant, chennopodium, hybridum, camomile, and valerian, a lot of attention is paid to sorghum (Chinese millet), which provides high yields and is known for its pleasing taste, but can also be used for the production of rice root brushes. The sorghum sown area will be increased threefold this year. The Michurinist youth will also put a lot of effort into growing vernalized wheat and vernalized winter wheat. On ten special fields, beautiful seeds will be sown in March. The grain is much more productive than the variety used by peasants. "The apple of the eye" of the young Michurinists in Szczekociny is the experiments on grafting

tomatoes onto potato plants, which last year bore over four hundred fruit and about seven potato bulbs on average. The culture will now increase to twenty plants.[101]

Young Polish Michurinists were encouraged to follow the example of Michurinists from the USSR, and the Ministry of Education recommended an adequate reading list for teachers. One of the brochures was the *Schoolyard Garden of Young Michurinists*,[102] proposed as an item to be included in teachers' libraries. It contained necessary guidelines for conducting botany experiments in schoolyard gardens. The book's scientific editor was W. Michajłow, and it was written by a group of experts, including a recognized authority in horticulture, S. Pieniążek. The proposed experiments were the following: grafting melons on pumpkin, grafting potatoes on tomato, vernalization of wheat and of other plants, and Viliams's method demonstration. The brochure discussed Lysenko's theories, supposedly to strengthen the dialectical materialist worldview, which, as was explained, assumed the possibility and necessity to actively transform nature, abiding by its laws.[103]

Tree Planting and Shelterbelts

Planting stands of trees and shelterbelts has had an impressive tradition in the world, dating back to the beginning of rational forestry. In Russia, and Poland too, shelterbelts were planted long before the official version of the Stalinist Plan was introduced.[104] Tree planting is a landscape-forming component of nature and has many functions. In Poland, both before and after the war, trees were planted by roads.[105]

In the case of Poland, the tradition of tree planting on land in order to improve farming dates back to the nineteenth century, to the works of Dezydery Chłapowski (1788–1879), a Polish general and economic activist. After his stay in England (1818–1819) and reading books on farming, he brought modern farming concepts to Poland[106] at a time when modern soil science had not yet developed. The English concept of modern agriculture was then based on the introduction of new metal tools which facilitated nature transformation. These features of agricultural revolution allowed English agriculture to take the lead in the late eighteenth and the early nineteenth century in Europe. This was the kind of agriculture that Chłapowski wanted to introduce on his family estate in the village of Turew near Poznań. His "new agriculture program," or as he called it "English agriculture," was a comprehensive approach to increasing farming output, cattle breeding, food processing, and transformation and protection of the environment. The program recommended keeping the same form of fields by a fixed natural fence, preferably lines of trees and shrubs. According to Chłapowski, trees should be planted in the areas remaining after fields outlining. His method consisted in introducing a network of tree stands and wind buffers, which enriched the landscape and contributed to crop increase. Tree stands in fields also change the microclimate in adjacent fields. By introducing tree stands in his fields, Chłapowski got about a dozen years ahead of the works of Russian scholar Dokuchaev.[107] This unique agricultural landscape with tree stands in fields has been retained in Turew, and since 1992 the estate has been know as the D. Chłapowski Natural Landscape Park.[108] Obviously, Chłapowski's field tree stands and wind buffers cannot be compared with shelterbelts in Stalin's concept, for the mere reason of scale. The former were several meters wide, the latter several hundred kilometers wide.

After the announcement of the Stalinist Plan, one of the first scholars to present the subject of shelterbelts was biologist and forester T. Dominik, in the *Sylwan*—a renowned scientific periodical on forestry, in the paper "The Issue of Forest Protective Belts."[109] The author, who had begun his scientific work before the war, was already a prominent botanist, mycologist and phytopathologist. After World War II, Dominik initially worked at the State Research Institute of Rural Husbandry in Puławy, then in 1949–1954 at the Wrocław University Of Life Sciences.[110] In 1955 he was appointed associate professor, and wrote many papers on field tree stands and shelterbelts,[111] and co-authored one of the first accessible Polish publications on the Stalin Plan. It was a popular scientific book published in 1950 under the title *Stalin Plan for the Transformation of Nature*[112] by St. Ehrlich[113] and T. Dominik. The book was issued in many copies (over ten thousand) by a popular natural sciences publishing house Państwowe Wydawnictwo Rolnicze i Leśne (State Publishing House for Agriculture and Forestry, PWRiL). It is worth mentioning here that in the 1950s this publishing house often released publications of this kind about nature—in line with the accepted trend in science.

The book contained flattering statements on the implementation of the Plan for the Transformation of Nature in the USSR, and more cautious opinions of the authors on the outlook of nature management in Poland through the Stalinist Plan. Its theoretical bases were presented in detail, supported by research done by V. Viliams, V. Dokuchaev, and others; the theory and methodology of Lysenkoism were also discussed. The plan itself was referred to by the authors as "Great." As to its application in Poland, the authors lacked similar enthusiasm; they limited themselves to presenting the possibilities of applying the Stalin Plan in afforestation and planting shelterbelts. Based on their own observations and results of foreign studies (with examples of afforestation in the USSR, England, Denmark, and Germany), the authors honestly presented advantages and limitations when planting shelterbelts in Poland. Above all, they believed one could not indiscriminately imitate the experiences of shelterbelt planting in the USSR: "In our country conditions are different, and exact recreation of ready-made models from a different area could lead to poor results and discourage farmers from protective belt planting."[114] Dominik was occupied with the matter of shelterbelts until 1955, then he left the subject for his chief interests of many years relating to phytopathology.

Interesting information on shelterbelts and their role according to the Stalin Plan was presented in a brochure published in 1950, written by botanist Stefan Macko (1899–1967).[115] At the time, he was already a recognized specialist in forest ecology and author of the book *Forest as a Biological Complex* (1947). He wrote: "The issue of shelterbelts as wind buffers is not new, but in our country there was hardly any interest before the last war, and after the war it has not yet been sufficiently updated."[116] The author added that the Soviet regulation dated 20 October 1948 on the plan initiation "excited the whole world. A downright flood of comments and remarks flowed, also in Poland. Thanks to the political parties' stance and government agents, the matter was solved adequately with a law—the Six-Year National Plan."[117] Stefan Macko tried to familiarize readers with the issue of favorable changes in microclimate thanks to shelterbelts, which would in turn influence the entire microclimate. In a popular science manner, he described the main objectives of the Stalinist Plan with respect to afforestation. He referred to the research of Russian scholar Kostychev on carbon dioxide assimilation by plants. The author quoted multiple achievements in planting and efficiency of belts in the USSR, but as he stated, "we cannot logically compare those conditions to ours, more modest and different in every way."[118] Discussing conditions for belt planting, he remarked on their favorable influence on reduction of wind force, change of air and soil temperature, change of air humidity and evaporation, snow retention, and quantity and quality of yields.

Stefan Macko acknowledged the future need for shelterbelts in Poland to be modeled on the ones in the USSR, recalling the problem of steppe formation in Wielkopolska: "Climate conditions on the Russian steppe areas near Poltava are almost identical with the climate conditions in Wielkopolska. Therefore the alarming statements of Professor A. Wodziczko (1887–1948) about steppe formation in Wielkopolska were not an empty threat, but a very serious warning."[119] Wodziczko's research on Wielkopolska steppe formation was often referred to, justifying the need for shelterbelts in Poland. The main aim of planting shelterbelts was the improvement of agriculture: "If we also, given the size of our country, begin planting shelterbelts for protection from winds, increasing farming output, in the future we will no longer worry about our daily bread."[120] However, this enthusiasm was followed by caution, based on solid natural science education and the common sense of the author:

> Planting shelterbelts as wind buffers has to be preceded with very thorough and detailed studies of the field conditions, mainly the climate, type of soil, its physical and chemical properties, direction and force of winds, applied crop rotation, etc. in order to avoid making a mistake, both in terms of the type of construction of tree belts in wind buffers, and in terms of selection of tree and bush species used for afforestation.[121]

Moreover, Macko remarked: "Shelterbelts are by definition an artificial addition to the landscape, they represent a specific biotope that has to fit in the given biocoenosis, and enter in its natural cycle, without disturbing its natural biological balance."[122]

Equally cautious about wind buffers was a horticulture authority of the time, S. Pieniążek, in his book *Frost-Resistant Orchard*.[123] Discussing conditions for keeping an orchard, he pondered upon the problem of protection from frost using tall trees as buffers, analogically to the field protection proposed by the Stalinist Plan. The author, however, stated that "the role of buffers has not been sufficiently clarified,"[124] and that "when planning to plant a buffer, one should always consider whether it is really necessary and whether it plays its role adequately."[125] Summarizing the conditions for setting up and keeping a frost-resistant orchard, Pieniążek ultimately did not mention protective belts.

However, at the same time the propaganda related to the Stalinist Plan was being promoted, books dealing with the shelterbelts and their role in the Stalinist Plan started to appear. An example is the book by engineer Kazimierz Pietkiewicz, *Shelterbelts*.[126] The author discussed protective shelterbelts in the USSR as an element of the Stalin Plan. The book was recommended in Poland to foresters and readers interested in nature.

At the time, despite erroneous scientific ideas, the achievements of Polish horticulturists such as S. Pieniążek[127] were not always pure propaganda, and society remembers them to this day.

Forestry

After World War II, Poland's forestation rate was very low and in 1949 it stood at approximately 20.8 percent of the country's total surface.[128] The reasons were complex, two of them being irrational management of forest resources, and losses inflicted on the environment by the two world wars.[129] Therefore, forestry faced a huge challenge after the war. It was necessary to restore and reforest the foundations of extensive war-damaged areas as well as weak land and wasteland, and to improve the effectiveness of afforestation. Developing forests in the Regained Lands was a separate issue. The territory that used to belong to Germany, and then was annexed by Poland, was highly industrialized or covered with single-species forests (monoculture forests). As monoculture forests are prone to vermin infestations and windfalls, they were to be reconstructed. For this purpose, new rules of forest management were introduced—selection cutting.[130] The system was introduced by an order from the minister of forestry of 29 September 1948, on change of rules for state-owned forest management. The enactment ordered the transformation of monoculture stands into mixed stands, and the introduction of a habitat-based selection cutting forest management. It consisted of managing woodland in combination with basic silvicultural actions (restoration, exploitation, cultivation). The selection cutting experiment was an idea of H. Minc's advisors. Minc took decisions on economic plan objectives for the whole economy, and for particular sectors. In the Six-Year Plan Law there was a provision on forestry: "The forest management system shall be transformed by the gradual relinquishing of clear-cutting, and replacing it with the selection cutting system' (article 46). However, the main objectives of the selection cutting system—the elimination of defective and redundant trees, as well as extensive farming in woodlands—were not met, as this method served as an excuse for the excessive exploitation of forests. The forecast pace of reaching goals was consistent with the rest of the Plan for the Transformation of Nature, but from a biological point of view it was unattainable.[131] Developed selection cutting forests only cover a small surface of Poland nowadays.[132]

Another method was the habitat-based tree stand method, founded on natural sciences, and introduced in March 1950, by virtue of a disposition issued by the Ministry of Forestry—disposition on changes in forest exploitation. On the whole, the planned economy system, especially in the manner implemented by the State Commission for Economic Planning in the 1949–56 period, led to the ravaging of forest resources. Despite the flaws of the methods described above, they should be deemed an important stage in the process of forest management improvement. They marked the beginning of the usage of research findings in ecology, and of complex forest ecosystems functioning for the purposes of silvicultural management.[133] Perhaps the failed experiments in forest management discouraged foresters at the time from implementing the ideas imposed by the authorities.

As for the scale of shelterbelts predicted by the Stalin Plan, they failed to find many enthusiasts in Poland because of the differences in climate conditions, for example the lack of dry winds with which Russians could not cope.

Afforestation was in fact conducted in Poland, but in different conditions and for different reasons than in the USSR. Research on afforestation in the country was carried out by the Forest Research Institute Afforestation Department in 1948. The department studied fast-growing tree species, primarily the poplar. In 1951 an experimental poplar plantation was set up, in order to select the best varieties, create new hybrids, and determine the conditions for growing them.[134] Afforestation research, commenced in the 1940s, was continued throughout the People's Republic of Poland period, and still remains in the scope of interest of natural scientists.

In the Polish law on protection of the environment dated 1949, one can find mentions of afforestation: "Due to forests and afforestation having exceptional importance for the public interests, their area shall be increased by the following measures: afforestation of quicksands, fallow lands, lands unsuitable for farming, watersides of open water, steep mountain slopes, and stream wellheads, and by introducing tall greenery belts."[135]

On 19 March 1955, the cabinet passed a resolution on further afforestation of the country. The regime obliged the State Forest Holdings to increase the output of tree seedlings for the purpose of extended afforestation in the whole country. At the same time, an intra-ministry commission was constituted to coordinate and supervise afforestation. In the 1950s and 1960s there were afforestation actions organized on a massive scale in the countryside, using mainly poplar trees, which were supposed to account for about 40 percent of afforestation, according to the plan. However, the afforestation growth in 1950–1955 was slower than expected—about 5.4 million trees and 2.2 million shrubs were planted. As encouragement, from the early 1960s, the state distributed free poplar trees to farmers.[136]

Grass-Crop Rotation System

The grass-crop rotation system (Viliams's grassland improvement system), was not promoted in Poland as intensely as Lysenkoism or forest belt introduction, and his concepts were never fully accepted. It seems likely that Polish scholars and practitioners, due to the geographical and climatic differences between Poland and the USSR, did not find advantages in this system. One of few books promoting Viliams's system, accessible to Polish readers, was *Grass-Crop Rotation System*[137] by M. Avaev, published in 1950 having been translated from Russian. Additionally, there was a translation of the Russian book *Vasily Viliams Great Reformer of Nature*[138] published in Poland in 1951.

A. Listowski, director of the newly established Institute of Soil Science and Plant Cultivation, presented possibilities for using Viliams's system in the paper "Struggle for Agrobiology in Poland,"[139] published in 1950 in the periodical *Nowe Drogi*. He wrote, albeit with certain reservations, that "Viliams's system concepts are accurate—that the guidelines regarding fertilizing and farming crop rotation are correctly perceived by Viliams as optimal in the conditions of, for example, steppe climate and soil—but may not be optimal for Belarus or our country."[140] Listowski believed that certain parts of Viliams's system could be adopted in Eastern Poland, but in Western Poland, in Lower Silesia where the climate was warmer, he was concerned about disturbing the balance of soil processes. Listowski thought that these processes should be studied, stating that "Viliams gave us a system, but never devised a recipe."[141] Such caution and the appeal for further studies are proof of broad-mindedness and a science-like attitude to a subject, for which, in the author's own words, there was a fight.

Nature Conservation

Along with the border shift, loss of the Eastern Borderlands and annexation of the Regained Lands, geographical and natural conditions changed quite notably in Poland. It was a new challenge for the society, and in particular for natural scientists. At the 19th convention of the State Council for Nature Conservation [Państwowa Rada Ochrony Przyrody, PROP] in September 1945 (and the first after the war),[142] prominent Polish botanist and nature protection activist Władysław Szafer (1886–1970)[143] explained it very well. He presented a profit and loss account of the border shift, as far as areas valuable for the environment and areas under protection were concerned, and outlined new tasks for naturalists. As he noted in his paper "General Program of Work in Nature Protection," "the most important change is of course the shift of our borders . . . Polish nature has changed. Some of its traits are vanishing, others appearing. . . . Polish nature has changed so dramatically in its entirety that as a result the whole nation needs to change as well."[144] It was supposed to be an attempt at a compensation for the lost valuable nature reserves in the East, of which Szafer enumerates: "The rich nature of Chornohora, gorges in Podolia, peculiar landscapes of the Volhynia granite plate with bushes of *Rhododendron luteum*, the Medobory reefs, table mountains on the north edge of the Podolia plate with historic fauna and flora, landscapes typical of Polesia and beautiful landscapes on the bank of Neman River—Poland of yesterday."[145] Naturalists, primarily botanists, responded to Szafer's appeal by engaging in natural sciences research and writing inventories of natural resources.[146]

One of the causes of the limited reception of the Stalin Plan in Poland may have been the country's own widespread tradition of nature conservation, dating back to the beginning of the twentieth century. It was continued after the war. In 1949, a new act on the protection of the environment was drafted. It determined in detail the scope of authority in the domain of nature protection (i.e., the competences of the Ministry of Forestry and the State Council for Nature Conservation as an advisory body). The act stipulated: "The task of the nature protection bodies is ensuring that natural resources management is in line with the rules aiming at protecting and enhancing the creative forces of nature."[147] Wildlife under protection was excluded from industrialization, including natural monuments, reserves, and national parks. It is noteworthy that, based on this law, in postwar Poland further national parks were created or reactivated: Białowieża National Park (reactivated in 1948), Świętokrzyski National Park (created in 1950), Tatra National Park (created in 1955.), Pieniny National Park (created in 1932, reactivated in 1955), Babia Góra National Park (reactivated in 1955), Ojców National Park (created in 1956), Wielkopolska National Park (reactivated in 1957), Karkonosze National Park (created in 1959), and Kampinos National Park (created in 1959). All the new national parks, the reactivated pre-war parks included, were located in the most valuable areas, well documented and well researched by naturalists. Note that by the end of the 1950s, as many as nine national parks had been established, including one that encompassed a large forest just on the outskirts of Warsaw, the Kampinos National Park.

Currently, there are twenty-three national parks in Poland, most of them set up in the People's Republic of Poland era. Undoubtedly, it was an active period for naturalists and they held a high position in society and in the academia. It should also be pointed out that in the discussed period there were many nature lovers and activists in Poland, such as scholars (mainly biologists): Bolesław Hryniewiecki (1875–1963), Walery Goetel (1889–1972), Roman Kobendza (1886–1955), and the above-mentioned W. Szafer and A. Wodziczko. At the time there were two important periodicals on nature protection. *Ochrona Przyrody* (Nature Protection) is one of the first science publications of this kind in the world. It was set up in 1920 by W. Szafer. Up to 1950, *Ochrona Przyrody* was published by the State Council for Nature Conservation, in the 1950–52 period by the Polish Academy of Arts and Sciences Committee for Nature Protection, and since 1953 by the Polish Academy of Sciences Department of Nature Protection and Natural Resources. The other periodical is *Chrońmy przyrodę ojczystą* (Let's Protect Our Indigenous Nature), a bimonthly published since 1945.

Due to the promoted publications and a widespread interest in the subject, the slogans that accompanied the Stalinist Plan (i.e., fighting nature, taming nature, and nature transformation) did not get the expected positive reception in Poland. Obviously, nature protection did not stand much of a chance should it conflict with any strategic investments, such as the construction of power plants, steelworks, and factories. Nevertheless, the nature transformation slogans did not find full understanding due to the Polish tradition of nature protection and the Polish natural resources management, which resulted from the pre-war work of prominent botanists such as A. Wodziczko. He emphasized that a balance has to be kept in nature and its resources should be managed rationally.[148] After the war, he initiated and organized the Commission for Afforestation of the Country in Poznań and Szczecin, and he was an expert in the field at the Central Office of Spatial Planning. Wodziczko drew public attention to the phenomenon of steppe formation in Wielkopolska, and he initiated team studies on the subject,[149] used on multiple occasions by proponents of the Stalinist Plan to support their claims. Wodziczko presented his concepts and views on nature protection in many papers,[150] as well as a paper under a very telling title, *Guarding Nature*, which was published four times in the People's Republic of Poland period.[151]

Stalinist Plan Propaganda

Propaganda in favor of the Stalinist Plan and the supporting theories, primarily the "new biology," was ubiquitous in the early 1950s. It was spread through the daily press, popular science publications, radio, newsreels, and numerous books and brochures. Lectures and science sessions were organized, and the youth were encouraged to participate in the so-called Michurin societies. Besides, there were no other possibilities—countries like Poland "were gradually sunk into the heavy atmosphere of European periphery."[152] Scientific knowledge and discoveries from the West did not reach the country, and novelties such as the Plan for the Transformation of Nature and Lysenko's theory got some interest in the society.

By the end of 1955, Lysenko's theories had become the obligatory scientific theories in natural science, which was eagerly highlighted in presentations of the Plan for the Transformation of Nature. The suitable background for the propagated knowledge was founded on the ideology associated with the cult of Stalin, who was the leader of the Plan for the Transformation of Nature, or as it was also referred to, "nature taming." The concept of taming the forces of nature by humans is, from today's perspective, not only erroneous but detrimental. As poet Adam Ważyk, one of the chief propagators of social realism in Poland, wrote in his poem "River":

> Stalin's wisdom, wide river,
> Decants water in heavy turbines,
> Flowing, sows wheat in tundra,
> Puts forests in steppes, raises gardens.[153]

The poem reflects the very essence of the cult—it ascribes to Stalin the capacity to transform nature and to drastically change whole plant formations. Turning tundra into farming fields, and steppes into forests, was not even meant as a metaphor. All changes seemed easy and fast, thanks to Stalin's wisdom. These words were supposed to be understandable to every citizen, and citizens could of course lack natural science knowledge, but they expected fast modernization of the country and improvement of the economic situation, especially in the most promising sectors: agriculture, forestry, and horticulture. Nowadays, such statements are perceived as obvious propaganda.

News of Stalin's death in March 1953 was received in Poland, at least officially, with despair; party activists, journalists, and scholars competed in writing statements and papers for the occasion. In the main press outlet of the Polish United Workers' Party, *Trybuna Ludu*, biologist Teodor Marchlewski (1899–1972), who was one of few academics from Krakow to promote Lysenko's theories, wrote, very much in the spirit of Stalin's cult: "Stalin's genius brought light to science, gave it wings, and showed it new paths of development. The contribution of comrade Stalin to economics, linguistics, biology, physiology, physics, and other domains of science cannot be overestimated. Thorough studies of his works will help science in tackling difficult problems."[154] Such a great overstatement coming from a famous Krakow scientist is a good example of how academics tried to flatter the Communist regime.

In the late 1940s, a lot of propaganda publications about Stalin's plan started appearing; some of them were translations of Soviet books. For instance, in 1949 the brochure "46 Quintals of Spring Wheat from a Hectare" was published. It was a collection of memoirs and musings on the work conducted by I.N. Rakitin—perceived as a Socialist hero thanks to his record high harvests.[155]

In the 1950s, a series of educational brochures for agricultural workers was published in Poland under the title "Farming Training Talks." In one of the brochures we can read enthusiastic words of an anonymous author about the USSR and the agricultural achievements of the country: "In the years of Stalin's Five-Year Plan, wheat cultivation was moved far to the East up to the Arctic Circle. At present, thanks to the irrigation of large desert-like steppes, farming areas have increased further by millions of hectares. It was possible thanks to the efforts of the entire Soviet nation, both scientists and millions of practitioners."[156] Some varieties of plants with valuable properties grown in the USSR inspired awe, such as a wheat variety with branched ears of grain.[157]

Another popular science book which ought to be noted is *Co-creators of Nature*, by D. Jarząbek, published in 1951.[158] It was written with slight irony and supplemented with funny images, describing mankind's struggle against nature, from prehistory up to the writer's time. The author explained the stages of humanity's efforts to tame nature, enumerating scientists who significantly accelerated the process with respect to plant cultivation and animal breeding: Ch. Darwin, L. Burbank, and I. Michurin. Obviously, he did not skip T. Lysenko, whose working methods were described in an easy-to-follow manner, supplementing the propaganda reasoning with numerous examples of success in Soviet agriculture. American agriculture was put in opposition, as functioning under bloodthirsty rules: "The American solution for famine, according to which the starving should be killed."[159]

Periodicals published in Poland at the time, such as the popular scientific *Problemy*[160] and the ideological *Nowe Drogi*, are also a good source of information. In these periodicals, scientists and activists presented various aspects of the Stalin Plan in the USSR and its Stalinist version in Poland: opportunities provided by shelterbelts were discussed, the achievements of Lysenkoism were admired, and Viliams's system was analyzed. The leader in this respect was the party's periodical *Nowe Drogi*, which covered the situation in Poland and the USSR in detail, presenting the main achievements of both countries in mouthful expressions characteristic of this period. As an outlet of the Polish United Workers' Party Central Committee, the periodical combined propaganda with news. As far as new theories and progress in the new biology was concerned, the most active authors were W. Michajłow and K. Petrusewicz. The list of all the titles of their papers on the new biology and related issues as well as new theories of science organizations published in *Nowe Drogi* in the 1948–1955 period show how these authors promoted new ideas.[161]

To gain a full view of the new theories propaganda voiced in *Nowe Drogi*, it is worth noting titles of papers of other scholars who wrote about natural sciences in line with the doctrine in force. Such slogans as *Fight for Agrobiology*[162] and *Stalin – Champion of Progress and Development in Science*,[163] and other papers,[164] suggests that popularization of the new ideas was of great importance for their authors, who tried to compete in the science race between the West and the East.

So many ideological papers published within a few years by renowned scholars, mostly biologists, paint the image of *Nowe Drogi* as a periodical that presented matters of natural science related with the Stalinist Plan very actively. Additionally, *Nowe Drogi* featured many propaganda papers on farming in Poland and in the USSR, referring to the Stalin Plan both in style and in subject.

People who remember the early People's Republic of Poland era have a recollection of the promotion of the Stalinist Plan reduced mostly to anecdotal expressions, like "reversing the flow of rivers," or "pears on a willow tree" (which has become a Polish idiom equivalent to "pie in the sky"). At the time it was proclaimed that "pears on a willow tree and tomatoes on a potato bush are no longer a nuisance, but reality."[165] V. Stoletov, president of the Moscow Agricultural Academy, wrote about it in his paper *Mankind Transforms Nature*, which was published in Poland, in an abbreviated version, in the popular science monthly *Problemy* (Problems), in 1948. It was one of the first popular science publications in Poland, presenting to a wider audience the possibilities of nature transformation based on the Michurin–Lysenko theory.[166]

Problemy covered the subject on many other occasions. In the 1948–1955 period it featured many papers by recognized Polish scholars and specialists propagating the Stalinist Plan and its theories. The subject of shelterbelts was raised in the journal by renowned forestry specialist and head of the Polish Forest Society, Franciszek Krzysik (1902–1980), from 1949. In the paper "Forests and Forestry in the USSR," he presented great afforestation plans. Despite the author's competence, the paper was purely informative and did not contain any commentary or opinion on whether using the Soviet experience might have a future in Poland.[167]

In the 1948–56 period, *Problemy* often published papers on biology and agriculture. Out of the 2,067 papers released in this time frame, 260 were on natural sciences and 277 on physics. There were 263 humanities papers (history, literature, philosophy, and psychology) published, and 203 in the field of science history and organization. Biology and agriculture were best covered in 1956 (when there were 44 papers published, which accounted for 16 percent of the content) in comparison to papers in other fields of science.[168] However, also in 1956, nature transformation according to the Stalinist Plan was less enthusiastically described. Interest had shifted to new subjects, such as the protection of forests from excessive exploitation, and protecting fields from the fearsome potato beetle.[169] Again in the same year, prominent Polish geneticist Wacław Gajewski (1911–1997) settled matters with Lysenko's theories in his paper "A Few Words on the State of Affairs in Biological

Sciences," published in *Problemy*. After critical opinions about Lysenko had been voiced by other biologists, for instance Kuźnicki, Gajewski finally unmasked the new biology. He also issued a warning for the future: "It is unacceptable for state bodies to deem any scientific theory official."[170]

In the early 1950s, one could see interest in the Stalinist Plan in the daily press. Planting belts in the USSR, in Poland (around Warsaw), and in Czechoslovakia (in the Poprad Valley) was recounted exhaustively.[171]

One of the issues of *Głos Koszalina* (a local newspaper) published at the beginning of 1953, before Stalin's death, provides an interesting read. The daily, which was a media outlet of the Communist Party's local committee, in February presented news on actions taken on the occasion of the 1st National Congress of Production Cooperatives. The main endeavor was supposed to be an agro-engineering deliberation, "Koszalin Region Villages Struggling for a Good Harvest," done by correspondence and covered by the newspaper. Lysenko's and Viliams's methods were to be discussed, along with such subjects as sowing actions, mechanization in agriculture, and production cooperatives organization. The paper was planning to write "about the best experiences in plant growing, primarily about learning and applying the leading methods of the Soviet agro-engineering, Michurin's and Lysenko's methods, as well as Viliams's ley farming system."[172]

P. Köhler wrote: "Apart from publishing their works and presenting papers at innumerable conferences, naturalists also promoted the new biology on Polskie Radio (the public radio station). In 1948–1952, it broadcast a series "Natural Science's Basis for a Worldview" as part of the program *Wszechnica Radiowa* [Radio University].[173]

Polish listeners were actually interested in Stalin and his far-reaching plans. During the period of mourning following Stalin's death, papers published poems eulogizing the leader. For instance *Słowo ludu* from Kielce published Mikhail Isakovsky's poem "Song about Stalin."[174] It glorifies in a pompous form the merits of the late Stalin in the transformation of Russian nature:

> Fertile steppes are rustling, mighty rivers rolling,
> Bright dawn comes with blessing to our land,
> Let us sing, my comrades, of the greatest man
> Our beloved, dearest, our great leader Stalin . . .
> His breath warmed up the coldest Arctic night,

> Steep mountains stand apart, roads lead to the sky,
> Streams spring in the sand, thus was his might,
> Orchards blossom joyful where the land was dry.

The Plan for the Transformation of Nature clearly inspired this piece of poetry.

For managers and farmers of production cooperatives to be able to apply modern science findings, they first had to get to know about them. Therefore, there were special lectures organized by regime officials for managers of production cooperatives, and they included "information on modern farming." In the lecture curriculum there were Stalin Plan–related issues, such as Viliams's method, protective forests in the USSR, human influence on soil fertility, plant resistance to negative climate factors (based on Michurin's work), and acclimatization (divided into regions in horticulture and farming).[175]

Polish Film Chronicle

A good example of ideological propaganda and the indoctrination of the masses about Soviet science's victory over nature was the Polish newsreel, *Polish Film Chronicle*. It is worthwhile to follow the voice-over comments for those short propaganda films[176]. A selection of quotes from the films presented to the public in 1949 alone shows how success in science, the new biology, and farming practices, achieved in the USSR and other countries of the Eastern Bloc, were propagated.

The USSR: The Karakum Desert (1949) was one of the first films about the implementation of the Stalin Plan. It showed to the Polish viewers the attainments of Soviet science in the struggle against the conditions in the Karakum Desert in Central Asia, in the Turkmen Soviet Socialist Republic. The voice-over comment stated:

> Soviet scientists have waged a ruthless war on the sand of the immense Karakum Desert in Central Asia. Against all odds, despite heat and drought, workers of the scientific experiment station decided to prove that there could be life in Turkmen sand. First, the sand has to be entrenched and forest barriers need to be planted around people's dwellings and wells. The haloxylon shrub serves perfectly for this purpose; its roots reach to subsoil water. Soon there will be ten million new drought-resistant shrubs flourishing in the Karakum Desert. There will be immense grazing land created. Soviet people change the face of the Earth.[177]

The Karakum Desert is located on the territory of Turkmenistan; in the North it borders with the Aral Sea, located on the territory of Kazakhstan and Uzbekistan. In the 1960s, the Aral Sea was the world's fourth biggest lake in terms of surface. Since then, it has shrunk due to water being diverted from the rivers that supply it—the Amu Darya and the Syr Darya—to irrigate the desert. The lake has almost entirely dried out because of careless nature transformation and decisions taken by the Soviet regime.

The following films illustrated various aspects of the Stalin Plan implementation and the spectacular achievements it brought to Soviet farming. A short documentary, *The USSR: Growing Exotic Fruit* (1949), starts with the acclamation:

> Never before have there been lemons and oranges growing in Uzbekistan. Ten years ago, Michurin's student, Vasily Panin, brought here the first branch of a lemon tree. These are the results of his work: a new kind of lemon—Uzbekistan—and juicy, aromatic grapefruit which may weigh over one kilogram. The fruit are adapted to Central Asian conditions. Abbas Mazurov came here for grafts. Firstly, he will learn how to handle them and then he will take them to his kolkhoz. Thanks to Michurin's teachings and findings, Uzbekistan will be covered with lemon groves.[178]

Another short documentary, *The USSR: Tea Plantations* (1949), described the tea plantations in Georgia that were cultivated on the basis of Michurin theories:

> Stalin Prize winner Xenia Bakhtadze does research on growing tropical plants in Georgia. Using Michurin's findings, Soviet scientists have obtained, through selection, new species of tea that are resistant to low temperature (up to 20 degrees below zero). Skillful cultivation has largely increased efficiency of the new varieties of tea shrubs. Thanks to the pioneer work of Soviet scientists, hot climate countries no longer have a monopoly on tea growing.[179]

The film titled *The USSR: Steppes and Deserts Turned into Fields* (1949) showed the successes of the Soviet Union in afforestation:

> Soviet science has declared war on deserts and steppes. A gigantic governmental plan was devised to create 5.5-kilometer-long

shelterbelts which will change the climate of an area of six million hectares. The impact of afforestation on yields is studied by the Dokuchaev Institute. Young scientists check instruments recording air humidity and temperature, soil temperature, and changes in climate. Grains ripen beautifully under the shelter of shelterbelts. Only in the Socialist economy conditions is it possible to implement an immense plan of climate change and make agriculture independent from the weather.[180]

Another film about the successes of the Soviet Union was titled *The USSR: Soviet Horticulturists* (1949) and showed the work of Michurinists:

Soviet people, following Michurin's example, create new better plant species. Horticulturists in Soviet Armenia have grown a new variety of tomatoes. The fruit are so large that one can hardly see the leaves. There are over one hundred fruit on one plant. Melons also yielded abundantly. Anait Anajan collects huge aubergines. Thanks to Michurinists' work, grapes no longer grow only in the South but also in Central Russia, near Saratov. Soviet science changes the face of the Earth.[181]

Polish Film Newsreels also covered the results of the implementation of new theories in Poland. They showed both agriculture and horticulture achievements, and the problems that practitioners faced, as it was shown in a film titled *July Potatoes* (1949):

These beautiful potatoes were grown in a field of a state agricultural farm in Małogoszcz, thanks to using modern methods of the great Soviet agronomist Michurin and his follower Lysenko. The potatoes were planted in the first week of July, right after winter rapes were harvested. Due to this method, the land was used twice. Potato lifting began in early October. From one hectare there were 160 quintals harvested, instead of the 110 quintals collected previously. Farming manager in Małogoszcz, Jakub Rościszewski, promotes modern Soviet agronomy. The Polish potatoes planted in line with the Michurin–Lysenko method have no fear of the farmers' greatest concern—the potato beetle.[182]

It is noteworthy that Polish farmers were really scared of potato beetles. The first occurrence was recorded in 1944 near Gorzów. In 1947, there were nine hot spots in the country. A real invasion took place in 1950, when the vermin was almost ubiquitous in western Poland. There were over eleven thousand hot spots registered then. Subsequently, the beetles moved to the east of Poland, and in 1954 there were a few hot spots noted by the border with the Soviet Union.[183] The propaganda said, and it was commonly believed, that the so-called Colorado beetle had been dropped on purpose by American imperialists in order to destroy crops.

The film titled *Michurin Orchards: Lecture on Trees* (1949) showed the effects of the acclimatization of fruits with Michurin methods:

> Kórnik Works, located near Poznań, are leading a campaign aimed at disseminating the horticultural achievements of the great Soviet biologist Michurin. Horticulturists come from all over the country to learn about new fruit tree species. The head of the Trees and Forest Research Department, Dr. Białobok, provides all necessary explanations. Acclimatization of species bred by Michurin is conducted in Kórnik. In addition, new domestic species are grown based on Michurin's method. After the lecture, horticulturists go into the orchard. These are Michurinist fruit trees. It is clear to the experts: the new species have been rendered resistant to frost and infertile soil, and they also provide abundant reliable yields. Trees travel from the nursery to all parts of the country.[184]

Newsreels also presented various issues related to wind buffers. Two films serve as examples here, from 1951 and 1952 respectively. Polish viewers could find out how shelterbelts were planted in Czechoslovakia and the USSR. One of these films was *Shelterbelts* (1951):

> Forests are not only beautiful but also useful. They muffle wind and help in maintaining soil humidity. Fields in southern Slovakia used to be penetrated by winds and sand storms. Losses amounted to millions of korunas. Taking the example of Soviet experiences, people in Slovakia started planting huge shelterbelts. Hundreds of volunteer groups from all over Slovakia helped in the tree planting. There are also motorized groups; ingenious machines produced in Czechoslovakia and in the USSR do the work of thirty to forty people. They plant eight thousand trees a day. This machine even places seedlings in the ground. Before we came, there was no forest—we will leave behind life-giving, protective forest belts.[185]

The other film was titled *News from the World. The USSR: Shelterbelts* (1952). The voice of reader said:

> The resolution of creating protective shelterbelts is an important link in the Stalin Plan for the Transformation of Nature. We are at one of the forest stations at Chkalov Oblast. Little trees require special care. Each station is equipped with special machines used for soil scarifying and seeding. Two years ago, when Alexandra Poliakova came to the station, there was not a single tree in this stretch. Now the birch trees are 2.5 meters tall. More seedlings are being prepared in nurseries. Afforestation is done in autumn and in spring. In the three years since the historic regulation of the Soviet government, shelterbelts have been grown on an area of over two million hectares.[186]

Transformation of Nature after 1956

When Stalin died in 1953, hopes of a thaw were raised and a new leader of the Soviet Union, Nikita Khrushchev (1894–1971), began a process of "de-Stalinization." In Poland, the previous dependence gradually weakened in many spheres of the economy and politics. Then, the death of first secretary of the Polish United Workers' Party Central Committee, B. Bierut, in the second half of 1956, was followed by the so-called "Gomułka's thaw" (October 1956). The return of W. Gomułka to power, his appointment to the position of first secretary of the Polish United Workers' Party Central Committee, as well as promises of reforms and democratization, provided an opportunity for at least a partial recovery of a rational, not only ideological, approach to the matters of science and economy. Important decisions taken at the time include Gomułka's criticism of the Six-Year Plan and putting a halt to agriculture collectivization. Notwithstanding, the influence of the USSR on economic issues remained significant.

Stalin's successor, Khrushchev, the secretary of the Central Committee of the Communist Party of the Soviet Union in 1953–1964 and also the USSR's prime minister in 1958–1964, turned out to be willing to continue previous plans, although perhaps on a smaller scale. He was known for his idea of growing corn in the entire Soviet Union, irrespective of the natural conditions. Archives of the Polish United Workers' Party Central Committee and protocols of Polish–Soviet talks show that a large portion of communication was dedicated to Polish economic issues. During conversations at the governmental level, Khrushchev boasted about the huge achievements of Soviet agriculture, described great crops of root vegetables and grains, and quoted results of Lysenko's methods, supporting them with specific data. Upon listening to Khrushchev's conversations with Polish politicians (W. Gomułka, and J. Cyrankiewicz), one gets the impression that he was obsessed with corn,[187] and he insisted on it being planted in Poland, unaware that since Lysenko's methods had been unmasked in 1955, Poles had become more cautious about Soviet ideas. During a conversation with Gomułka, Khrushchev said he had had a discussion with Polish biologist T. Marchlewski about introducing corn in Poland, and Khrushchev stated: "If he doesn't accept corn, he's no scientist."[188] At one meeting of the Polish delegation with the Soviet government in Moscow, Khrushchev addressed the Poles, saying:

> I have told you before and I'm telling you now absolutely seriously. Poland should start growing corn. No plant is more profitable. Come on, we will sign an agreement that corn will be sown on 50 to 100 ha of experimental fields in a few of your PGRs [state agricultural farms] or cooperatives. Corn will allow you to decrease or even do away with grain imports. Besides, I was in a PGR in Łódź and I saw great crops of corn. Why are you stopping? The Soviet Union is ready to provide you with the right seeds, I will send you my co-operator, he's a specialist in the matter and he will advise you. Of course, you will need appropriate equipment and farming methods. You have to select holdings whose managers will have the heart for growing corn. I guarantee you that it is possible to get crops ranging from 600 to 800 q/h of green mass. Corn is communism. The problems you can solve with corn in two to three years cannot be solved in ten without it.[189]

South American corn is a widespread fodder plant and one of the most productive farming plants. A few kinds of fodder can be obtained from it, with different composition, adjusted to animals' needs. It first appeared in Poland, most likely in the seventeenth or early eighteenth century, and it gained quite significant popularity in the Communist period, mostly because of Khrushchev's encouragement described above. Nevertheless, in recent years, the share of grain corn in the entire Polish grain farming area has not exceeded 10 percent—and according to the Central Statistical Office [Główny Urząd Statystyczny,

GUS], in 2009 it was only 3.2 percent. It seems that Poland's variable climate does not serve it well. Nowadays novelties such as genetically modified corn find both objections and support in Poland.[190]

The effects of the Plan for the Transformation of Nature in Poland are partially forgotten. The most memorable are those that failed, like rice introduction, and those that were hazardous for nature like the Wieprz–Krzna Canal. Another story was the case of Sosnowsky's hogweed (*Heracleum sosnowskyi* Manden.), brought in during the People's Republic of Poland period from the USSR. It was promoted as a medical, fodder, and honey-yielding plant, and thus was introduced in state agricultural farms.[191] It is a cautionary tale about playing with nature, because Sosnowsky's hogweed turned out to be useless, invasive, and dangerous to people and animals, due to its burning properties. It still causes problems in many regions of Poland.

Jokes are made to this day about the Stalin Plan and its supporting theories; for example: Why did Michurin graft pears on willow trees? Because a Soviet man can plant anything—even socialism in Korea, peace in Chechnya, and Sosnowsky's hogweed in Poland.

Almost sixty years have passed since the Stalinist Plan was promoted, but Poland still has difficulties in the management of natural resources. Issues such as afforestation, stands of trees in fields, agriculture, combating drought and vermin infestations, and environment protection have remained relevant to natural scientists, as well as agriculture and forestry practitioners in Poland.

It seems that nature transformation is inevitably related with civilization development and the influence of mankind (anthropogenic impact) on the environment. The impact is a result of using natural resources, both animate and inanimate, for economic purposes, and the rise of our species' headcount. Growing demand for food will force politicians and scientists to seek new solutions. Let us hope that they will be based on reliable, well-tested scientific foundations. In the twenty-first century we have to be free of any ideology that lacks respect for the laws of nature.

To sum up, under Stalin's rule there were three types of assumptions about nature transformation in Poland. First, the important economic factors related to the Polish nation's need to reconstruct and modernize the country in the postwar period. The nature transformation projects were meant to serve different economic purposes, the most important of which was initially the goal of providing enough food for citizens by redeveloping the nation's agriculture base. The different economic plans, for example the Six-Year Plan and agricultural collectivization, were to a large extent directed by the COMECON and under Moscow's control. However, the objectives of Stalinist Plan for the Transformation of Nature actually had little influence on farming and forestry practices for several reasons. For example, environmental conditions in the Polish countryside are specific, and obviously differ from the conditions in the USSR. So, for example, there was not much possibility for implementing large-scale shelterbelts or the various exotic plant acclimatization projects, such as the plan for the cultivation of rice. With its own priorities at that time, Poland's top concern was the reforestation of the countryside destroyed by the war, as well as recovering all of the Polish grain crops by redeveloping the farmland. Poland's own farming tradition and methods contributed to the weaker acceptance of the Stalinist Plan for the Polish farmers.

Ideology, always an important factor, was obviously presented in Stalinist processes in Central and Eastern Europe during the promotion of his plan. The ideology of communism, the cult of Stalin, and the propaganda of the Soviet Union's successes in agriculture and forestry served as the background for implementing the Stalin Plan. However, due to the process of the de-Stalinization in Poland after the Soviet leader's death, his plans had only been slightly implemented by the state on a large scale, and the projects were not given full time for development.

Finally, the Polish scientific community played a significant role, where some Polish scientists were fascinated with the new theories from the Soviet Union such as Lysenkoism, but many others chose not to be involved in these large-scale and environmentally risky projects, and chose to ignore them with their silence. The reception of the Stalin Plan in the Polish scientific community was to a large extent reduced to a debate of theoretical concepts by a minority of prominent scientists, dominated by the Communist Party's politics and for propagandistic reasons. Additionally, the plan was viewed by most of the Polish scientific community with extreme caution, because Poland already had a strong tradition of protecting its nature. Certain academic scholars arrived at the same conclusion as the farmers, practitioners, and others. Even if at first they were interested in the plan, they realized that it was either not useful in Poland or actually could be very detrimental, as was the case with Lysenkoism. After the publication in 1955 of Leszek Kuźnicki's paper exposing the shortcomings of Lysenko's radical ideas for the heavy-handed, human-driven transformation of nature, that paper helped to break the silence and sway the unfavorable opinions about some of these Soviet plans for the Polish scientific community.

Beata Wysokińska is a post-doctoral researcher working at the Institute for the History of Science of the Polish Academy of Sciences in Warsaw. She specializes in the history of biology and the history of nature protection.

Notes

1. J. Topolski, *Historia Polski*, Warsaw, 1994, 377.
2. N. Davies, *Boże Igrzysko. Historia Polski*, Vol. 2, Krakow, 1994, 400.
3. J. Eisler, *Zarys dziejów politycznych Polski*, Warsaw, 1992, 14–16.
4. J. Kaliński, *Zarys historii gospodarczej XIX i XX w.*, Warsaw, 2000, 201.
5. Demarcation line between Poland and Russia proposed by the British foreign minister, Lord G. Curzon, as described in document from 7 November 1920.
6. Davies, *Boże Igrzysko*, 404.
7. See Kaliński, *Zarys historii gospodarczej XIX i XX w.*
8. Davies, *Boże Igrzysko*, 395.
9. A. Paczkowski, *Pół wieku dziejów Polski*, Warsaw, 2005, 161.
10. Davies, *Boże Igrzysko*, 419.
11. Ibid., 411.
12. B. Bierut, "Zadania Partii w walce o czujność rewolucyjne na tle sytuacji obecnej" [The Party's Tasks in the Struggle for Revolutionary Vigilance against the Background of the Present Situation], *Nowe drogi*, no. 11 (1949), 27–28.
13. Act dated 21 July 1950 (Law Book no. 37, article 344).
14. R. Manteuffel, "Polskie rolnictwo po II wojnie światowej," *Rocznik Mazowiecki*, no. 9 (1987), 9–26.
15. "Wytyczne dla sporządzenia 6-letniego planu rozwoju i przebudowy gospodarczej (projekt)" [Guidelines for the Preparation of a Six-Year Plan of Economic Development and Redevelopment], AAN, KC PZPR, BP, v/3 (mkf 2819), 1949.
16. Ibid.
17. Kaliński, *Zarys historii gospodarczej XIX i XX w.*, 258–59.
18. Ibid., 259.
19. Ibid., 216–19.
20. J. Broda, *Historia leśnictwa w Polsce*, Poznań, 2000, 173–74.
21. Kaliński, *Zarys historii gospodarczej XIX i XX w.*, 258–59.
22. See R. Herczyński, *Spętana nauka. Opozycja intelektualna w Polsce 1945–1970*, Warsaw, 2008.

23 J. Sutyła and L. Zasztowt, "Popularyzacja nauki w Polsce w latach 1918–1951," in *Historia nauki polskiej*, vol. V: 1918–1951, cz. 1, ed. B. Suchodolski, Wrocław, 1992, 644–46.
24 L. Zasztowt, "Science for the Masses: The Political Background of Polish and Soviet Science Popularization in the Postwar Period," in *Communicating Science in 20th Century Europe: Comparative Perspectives*, ed. Arne Schirrmacher, Max Planck Institute for the History of Science Preprints, No. 385, 2009, 133–45.
25 Sutyła and Zasztowt, "Popularyzacja nauki w Polsce w latach 1918–1951," 650.
26 R. Manteuffel, "Polskie rolnictwo po II wojnie światowej," *Rocznik Mazowiecki* 9 (1987), 9–26.
27 P. Jaroszyński, *Nauka w komunizmie*, Lublin, 2008; P. Köhler, "Łysenkizm w botanice polskiej," *Kwartalnik Historii Nauki i Techniki* 53, no. 2 (2008), 86.
28 See P. Hübner, *I Kongres Nauki Polskiej jako forma realizacji założeń polityki naukowej państwa ludowego*, Warsaw, 1983.
29 "Projekt uchwały Biura Politycznego w sprawie Kongresu Nauki, załącznik do protokołu no. 17 posiedzenia Biura Politycznego KC PZPR z dnia 5.08.1949 r.," 5 August 1949, AAN, KC PZPR, BP, v/2 (mkf 2819).
30 Act issued on 30 October 1951, Law Book 1951, no. 57, article 391.
31 Ibid.
32 http://historiaiung.pulawy.pl (accessed 5 June 2013).
33 Act issued on 22 March 1951, Law Book, 1951, no. 18, article 140.
34 "Rozwój i dorobek Instytut Badawczego Leśnictwa," *Prace badawcze IBL*, no. 590, Warsaw, 1981, 283.
35 Based on: T. Majewski, "Choroby roślin w Polsce, dzieje ich zwalczania i poznania," manuscript, 2013.
36 A. Rubaszewski, *Filozoficzne podstawy nauki Miczurina*, Warsaw, 1951.
37 P. Wieczorkiewicz, "Imperium Stalina," in *Historia Rosji*, ed. L. Bazylow, Wrocław, 2005, 479.
38 P. Köhler, "Zarys historii łysenkizmu w ZSRR," in *Studia nad łysenkizmem w polskiej biologii*, ed. P. Köhler, Krakow, 2013, 16.
39 A. Czermiński, *Dzieje wielkiego planu*, Warsaw, 1951, 6.
40 W. Michajłow, *Nauka Związku Radzieckiego – Kraju Socjalizmu*, Warsaw, 1951, 20.
41 Series of small books produced in the 1950s, which presented topics such as: Transformation of Species, and Brave Wheat (translation from Russian). These pamphlets were written in order to present the achievements of science to the average citizen, in the spirit of the current ideology—in this case, the victory of man over nature.
42 Czermiński, *Dzieje wielkiego planu*, 143.
43 A. Bilnik, T. Świercz, and A. Siudy, *Zbiornik Goczałkowicki wczoraj i dziś*, Goczałkowice, 2004, www.gpw.katowice.pl/uploaded/files/goczalkowice (accessed 3 July 2013).

44 J. Kwapiszewski, *Budujemy Kanał Wieprz–Krzna*, Warsaw, 1956.
45 "Pierwszy odcinek kanału Wieprz-Krzna uruchomiony," *Dziennik Polski*, no. 190 (1958), 2.
46 S. Radwan, *Przyrodnicze podstawy ochrony i odnowy ekosystemów wodnotorfowiskowych w obszarze funkcjonalnym poleskiego parku narodowego na tle antropogenicznych przekształceń środowiska przyrodniczego*, Warsaw, 2003.
47 See S. Amsterdamski, *Życie naukowe a monopol władzy (casus Łysenko)*, Warsaw, 1981; S. Amsterdamski, "O patologii życia naukowego – casus T.D. Łysenko," *Przegląd powszechny* 4, no. 912 (1989).
48 Z. Kępa, "'Nowa biologia' w walce z religia," *Zagadnienia filozoficzne w nauce* 22 (1998), 63–78; Z. Kępa, *Marksizm i ewolucja. "Twórczy darwinizm" jako narzędzie propagandy antyreligijnej w latach 1948–1956*, Krakow and Tarnów, 1999.
49 P. Köhler, *Łysenkizm w botanice polskiej*, Krakow, 2013; P. Köhler, "Lysenkoism in the Satellite Countries of the Soviet Union: The Case of Botany in Poland," at XXII International Congress of History of Science, "Globalization and Diversity: Diffusion of Science and Technology Throughout History," Beijing, 24–30 July 2005. Book of Abstracts. Institute for the History of Natural Science, Chinese Academy of Sciences, 419; P. Köhler, "An Outline of Short History of Lysenkoism in Poland," *Folia Mendeliana* 45 (2009), 44–54; P. Köhler, "Casual Study of Lysenkoism in Polish Botany," *Folia Mendeliana* 46, no. 1/2 (2010), 41–53; P. Köhler, "Lysenko Affair and Polish Botany," *Journal of the History of Biology* 44, no. 2 (2010), 305–43 (DOI 10.1007/s10739-010-9238-4); P. Köhler, "Botany and Lysenkoism in Poland," *Studies in the History of Biology* 3, no. 2 (2011), 32–53; P. Köhler, "Łysenkizm," *Alma Mater – Miesięcznik Uniwersytetu Jagiellońskiego*, no. 141 (November 2011), 39–42.
50 W. deJong-Lambert, "Lysenkoism in Poland," *Journal of the History of Biology* 45 (2012), 499–524.
51 L. Kuźnicki, *Autobiografia. W kręgu nauki*, Warsaw, 2002.
52 Manteuffel, "Polskie rolnictwo po II wojnie światowej."
53 A. Strządała, "'Nowa biologia' a zoologia. Rola Włodzimierza Michajłowa w propagowaniu 'nowej biologii' w Polsce," in *Studia nad łysenkizmem w polskiej biologii*, ed. P. Köhler, 127–44.
54 S. Feliksiak, ed., *Słownik biologów polskich*, Warsaw, 1987, 420.
55 W. Michajłów, "Dyskusja o naukach biologicznych w ZSRR," *Głos Ludu*, 7–11 October 1948. See also Köhler, *Studia nad łysenkizmem w polskiej biologii*, 33.
56 W. Michajłów, *Co dali Miczurin i Łysienko wsi radzieckiej*, Warsaw, 1948.
57 W. Michajłów, *Na pierwszym etapie rozwoju biologii w Polsce* 3, no. 15, (1949), 123–33.
58 Ibid., 132–33.
59 Strządała, "'Nowa biologia' a zoologia," 132–33.

60 A. Pieniążek et al., *Działka szkolna młodych miczurinowców*, Warsaw, 1951, 3.
61 Ibid., 4–5.
62 *Zajęcia w przedszkolu. Program tymczasowy*, Warsaw, 1950, 4 and 62–63. See D. Jarosz, "Główne kierunki działalności państwa w zakresie stalinizacji wychowania dzieci w Polsce w latach 1948-1956," *Mazowieckie Studia Humanistyczne*, no. 2 (1998), 127.
63 Pieniążek et al., *Działka szkolna młodych miczurinowców*, 22.
64 *Materiały Konferencji Agrobiologów, Biologów i Medyków w Kuźnicach*, vol. I–II, Warsaw, 1951.
65 *Materiały Konferencji agrobiologicznej leśników*, Warsaw, 1953, 43.
66 Mycorrhizae was discovered in 1880 by Polish botanist F. Kamieński.
67 *Materiały Konferencji agrobiologicznej*, 225–26.
68 Köhler, "Łysenkizm w botanice polskiej," 113.
69 In magazines such as *Sylwan*, no. 1–2 (1949), 231–34; also *Problemy* and *Nowe Drogi*, as well as in numerous books and brochures.
70 Köhler, "Łysenkizm w Polsce na tle ówczesnej sytuacji politycznej," in *Studia nad łysenkizmem w polskiej biologii*, 53–55.
71 L. Kuźnicki, "Dyskusja nad łysenkizmem w Po prostu (18 grudnia 1955—17 kwietnia 1956) jako sprzeciw wobec politycznej indoktrynacji nauki," in *Studia nad łysenkizmem w polskiej biologii*, 144–54.
72 deJong-Lambert, "Lysenkoism in Poland," 499.
73 Ibid.
74 Köhler, "Łysenkizm w Polsce na tle ówczesnej sytuacji politycznej," 25–70.
75 "Мы не можем ждать милостей от природы. Взять их у нее - наша задача," based on: Rubaszewski, *Filozoficzne podstawy nauki Miczurina*; also J. Dembowski, "Teoria Miczurina – Łysenki," *Problemy*, no. 6 (1949), 412–14.
76 T.D. Łysenko, *O sytuacji w biologii. Referat wygłoszony na sesji Wszechzwiązkowej Akademii Nauk Rolniczych im. W.I. Lenina 31 lipca 1948 roku*, Warsaw, 1948, 54. See also Köhler, *Łysenkizm w botanice polskiej*.
77 Rubaszewski, *Filozoficzne podstawy nauki Miczurina*, 7.
78 Ibid.
79 Michajłow, *Nauka Związku Radzieckiego*, 35–36.
80 Amsterdamski, "O patologii życia naukowego," 69.
81 Köhler, *Łysenkizm w botanice polskiej*, 94.
82 Makarewicz obtained the title of associate professor in 1954 under special procedure. From 1951 until 1957 she worked in the the Warsaw University of Life Sciences, then at the Department of Genetics, PAN.
83 Köhler, *Łysenkizm w botanice polskiej*, 89.
84 A. Makarewicz, "Kurs agrobiologiczny w Kuźnicach," *Postępy Wiedzy Rolniczej* 1, no. 3–4 (1949), 177–82. See Köhler, *Łysenkizm w botanice polskiej*.

85   L. Kuźnicki, "Darwinizm a lysenkizm," *Po prostu*, no. 42–43, 18 December 1955, 4.
86   Köhler, *Łysenkizm w botanice polskiej*, 92.
87   Ibid., 148.
88   Lekczyńska was a member of the PAN as well as professor of the Warsaw University of Life Sciences, and from 1952 until 1956 he was also a member of the parliament.
89   P. Köhler, "Polska botanika łysenkowska," in *Studia nad łysenkizmem w polskiej biologii*, ed. P. Köhler, Krakow, 2013, 84–85.
90   Z. Gonet, "Sobczyk Piotr Ignacy," in *Polski Słownik Biograficzny*, vol. 39, Warsaw and Krakow, 1999–2000, 446–48.
91   P. Sobczyk, "Ryż rośnie w Polsce," *Problemy*, no. 1 (1949), 32–36.
92   P. Sobczyk, *Uprawa ryżu*, Warsaw 1952, 3.
93   Ibid., 6.
94   Ibid.
95   J. Lekczyńska, "O właściwą ocenę znaczenia upraw kukurydzy w Polsce," *Nowe drogi* no. 5 (1955), 57–60.
96   Feliksiak, *Słownik biologów polskich*, 256–57; "Lucjan Kaznowski – wybitny hodowca roślin," *Rolniczy Magazyn Elektroniczny*, no. 37, May–June 2010, http://www.cbr.edu.pl/rme-archiwum/2010/rme37/dane/2_1.html (accessed 1 July 2013).
97   L. Kaznowski, *Mniszek gumodajny koksagiz*, Warsaw, 1951.
98   M. Milczak, *Lucjan Kaznowski (1890–1955)*, http://www.pan-ol.lublin.pl/biul_2/kaznowski.html (accessed 2 July 2013).
99   Köhler, "Łysenkizm w botanice polskiej," 105.
100 *Program pracy kółek młodych miczurinowców (materiały dla szkolnych kółek zainteresowań)*, Warsaw, 1954.
101 "Z wizytą u młodych miczurinowców w Szczekocinach," *Słowo Ludu*, no. 42 (829), 18 February 1952.
102 Pieniążek et al., *Działka szkolna młodych miczurinowców*, 77.
103 Ibid., 7.
104 Poles deported in the 1940s to the USSR sent back various reports of the implementation of shelterbelts. Between 1943 and 1946, Dr. Adam Wolk, who was staying with his family near Piertropawlowsk, the region of Karaganda (Kazakhstan), witnessed the shelterbelts already planted there by Polep.
105 J. Szczuka, "Zagadnienia zadrzewień," in J. Broda, *Dzieje lasów, leśnictwa i drzewnictwa w Polsce*, Warsaw, 1965.
106 D. Chłapowski, *O rolnictwie*, Poznań, 1843.
107 L. Ryszkowski, "Koncepcje rolnicze Dezyderego Chłapowskiego a kształtowanie i ochrona środowiska," in *Studia z dziejów ochrony przyrody w Polsce*, ed. J. Babicz, W. Grębecka, and Z. Wójcik, Warsaw, 1985, 42–45.
108 http://regionwielkopolska.pl/katalog-obiektow/park-krajobrazowy-im-gen-dezyderego-chlapowskiego-w-turwi.html (accessed 2 July 2013).

109 T. Dominik, "Problematyka leśnych pasów ochronnych," *Sylwan* 93, no. 3–4 (1949), 56–75.
110 Feliksiak, *Słownik biologów polskich*, 135.
111 T. Dominik, "Zagadnienie pasów wiatrochronnych w nowej gospodarce rolniczo-leśnej," *Prz. Ogrod.* 26, no. 2 (1949), 51–54; T. Dominik, "Krajobraz – biocenoza – ekologia – leśne pasy ochronne," *Biul. Kom. Ekol. PAN* 1, no. 1 (1953), 10–28; T. Dominik, "O potrzebie badań nad zadrzewieniami ochronnymi ze szczególnym uwzględnieniem zadrzewień śródpolnych," *Kosmos biol.* 3, 1954, 755–66; T. Dominik, "Metody badania wpływu zadrzewień ochronnych na zdrowotność roślin uprawnych," *Biul. Kom. Ekol. PAN* 2, no. 3 (1954), 59–70; T. Dominik, "Na marginesie konferencji w sprawie zadrzewień śródpolnych," *Chrońmy Przyrodę Ojczystą* 11, no. 4 (1955), 30–40. http://www.zor.zut.edu.pl/Glomeromycota/Dominik.html (accessed 2 July 2013).
112 S. Ehrlich and T. Dominik, *Stalinowski plan przeobrażenia przyrody*, Warsaw, 1950.
113 Initials of one of the authors cannot be decoded. St. Ehrlich is probably the engineer Stefan Ehrlich, who translated M. Awajew into Polish, see M. Awajew, *Trawopolny system rolnictwa*, Warsaw, 1950.
114 Ehrlich and Dominik, *Stalinowski plan przeobrażenia przyrody*, 47.
115 S. Macko, *Leśne pasy wiatrochronne*, Warsaw, 1950, 86.
116 Ibid., 5
117 Ibid.
118 Ibid., 84.
119 Ibid.
120 Ibid., 87.
121 Ibid., 20.
122 Ibid., 83.
123 A. Pieniążek, *Sad, który nie wymarznie*, Warsaw, 1950, 32.
124 Ibid., 25.
125 Ibid., 26
126 K. Pietkiewicz, *Ochronne pasy leśne*, Warsaw, 1952, 92.
127 *Współcześni uczeni polscy. Słownik biograficzny*, Vol. III: M–R, Warsaw, 2000.
128 Currently about 28 percent.
129 Broda, *Historia leśnictwa w Polsce*, 168.
130 Ibid., 223.
131 Ibid., 246–48.
132 A. Jaworski and J. Paluch, "Charakterystyka cech morfologicznych jodeł w drzewostanach o strukturze przerębowej Beskidów Zachodnich," *Leśne prace badawcze*, no. 3 (2007), 7–31.
133 K. Przybylska and S. Zięba, "Siedliskowe uwarunkowania prac urządzeniowych i decyzji planistycznych," *Studia i Materiały Centrum Edukacji Przyrodniczo-Leśnej* 10, no. 3 (2008), 205–6.

134 S. Kocięcki and W. Strzelecki, "Hodowla lasu," *Rozwój i dorobek Instytut Badawczego Leśnictwa, Prace badawcze IBL*, no. 590, Warsaw, 1981, 81.
135 Act issued on 7 April 1949, Law Book 1949, no. 25, article 180.
136 J. Szczuka, "Zagadnienia zadrzewień," in Broda, *Dzieje lasów, leśnictwa i drzewnictwa w Polsce*, 462–69.
137 Awajew, *Trawopolny system rolnictwa*.
138 I. Krupenikow and L. Krupenikow, *Wasyl Viliams wielki reformator przyrody*, Warsaw, 1951.
139 A. Listowski, "Walka o agrobiologię w Polsce," *Nowe Drogi*, no. 2 (1950), 239.
140 Ibid.
141 Ibid.
142 W. Grębecka, "Problem ochrony rodzimego krajobrazu a tożsamość narodowa," *Kwartalnik Historii Nauki i Techniki* 55, no. 3–4 (2010), 255.
143 See *Polski Słownik Biograficzny*, vol. 46, Warsaw and Krakow, 2009–2010, 401–7.
144 W. Szafer, "Ogólny program pracy na polu ochrony przyrody," in Pamiętnik XIX Zjazdu Państwowej Rady Ochrony Przyrody odbytego w Krakowie 21 i 22 września 1945, Krakow, 1945, 23–26.
145 Ibid., 13–26.
146 For example: A. Wodziczko and W.A. Czubiński, "Materiały do inwentarza rezerwatów przyrody na odzyskanych Ziemiach Zachodnich", Państwowa Rada Ochrony Przyrody, no. 57 (1946), 32.
147 Act issued on 7 April 1949, Law Book 1949, no. 25, article 180.
148 A. Wodziczko, "Planowanie kraju drogą do utrzymania równowagi w przyrodzie," *Ochrona Przyrody* 17 (1937).
149 A. Wodziczko, "Wielkopolska stepowieje," *Prace Komisji Matematyczno-Przyrodniczej Pozn. Tow. Przyj. Nauk*, Seria B, Vol. X, 1947, 4.
150 See A. Wodziczko, "Ochrona przyrody to podstawowe zagadnienie państwowe," *Chrońmy przyrodę ojczystą* 3, no. 3–4 (1947).
151 J. Dembiński and J. Matysiak, "Materiały Adama Wodziczki," *Biuletyn archiwum PAN*, no. 48 (2007), 172–204.
152 Zasztowt, "Science for the Masses," 136.
153 A. Ważyk, "Rzeka," in *Bój to jest nasz ostatni*, Warsaw, 1953.
154 M. Jarosiński, *"Kochaliśmy Go jak Ojca i jak Przyjaciela." Żałoba po śmierci Stalina*, http://dzieje.pl/aktualnosci/smierc-stalina (accessed 3 July 2013).
155 "46 kwintali pszenicy jarej z hektara," Warsaw, 1949, 29.
156 "Uprawiamy pszenicę i jęczmień, Pogadanka pierwsza," in *Pogadanki dla przysposobienia rolniczego*, no. 16, Warsaw, 1950, 5.
157 Ibid., 16.
158 D. Jarząbek, *Współtwórcy przyrody*, Warsaw, 1951, 112.
159 Ibid., 108.
160 L. Zasztowt, J. Włodarczyk, and M. Jabłońska, "A Closer Look at the Popularization of Science in Poland during the Stalinist Period. The Magazine Problemy, 1945–1956," *Organon* 44 (2012).

161 W. Michajłow, "Twórcza dyskusja o darwinizmie" [Creative Discussion on Darwinism], *Nowe Drogi*, no. 7 (1948); W. Michajłow, "Zwycięstwo teorii biologicznych Miczurina i Łysienki [sic] w ZSRR" [Victory of Biology Theories of Michurin and Lysenko in the USSR], *Nowe Drogi*, no. 12 (1948); W. Michajłow, "Niektóre zagadnienia biologii współczesnej w świetle materializmu dialektycznego" [Selected Issues of Modern Biology in View of Dialectical Materialism], *Nowe Drogi*, 1948; W. Michajłow, "Na pierwszym etapie rozwoju biologii w Polsce" [At the First Stage of Biology Development in Poland], *Nowe Drogi* , no. 3 (1949); W. Michajłow, "Nowy kierunek rozwoju biologii współczesnej" [New Direction in Modern Biology Development], *Nowe Drogi*, 1949; K. Petrusewicz and W. Michajłow, "O twórczy rozwój nauk biologicznych" [In Favor of Creative Development of Biological Sciences], *Nowe Drogi*, no. 1 1951; W. Michajłow and K. Petrusewicz, "O obecnym etapie walk ideologicznych w biologii" [On the Current Stage of Ideological Struggle in Biology], *Nowe Drogi*, no. 9 1955.

162 Listowski, "Walka o agrobiologię."

163 J. Dembowski, "Stalin – bojownik o postęp i rozwój nauki," *Nowe Drogi*, 4 (46), 1953.

164 W. Krajewski, "O właściwą ocenę charakteru nauk przyrodniczych" [In Favor of Fair Evaluation of Natural Sciences], *Nowe Drogi*, 2 1951.

165 W. Stoletow, "Człowiek zmienia przyrodę," *Problemy*, no. 10 (1948), 592.

166 Ibid., 586–92.

167 F. Krzysik, "Lasy i leśnictwo w ZSRR," *Problemy*, no. 6 (1949), 394–98.

168 Zasztowt, Włodarczyk, and Jabłońska, "A Closer Look at the Popularization of Science."

169 T. Studziński, "Ratujmy lasy, o konieczności oszczania drewna," *Problemy*, no. 9 (1956), 624–26; M. Bielewicz, "Stonka ziemniaczana, Metody zwalczania pasiastego chrząszcza," *Problemy*, no. 8 (1956), 561–64.

170 W. Gajewski, "Parę słów o sytuacji w naukach biologicznych," *Problemy*, no. 10 (1956), 705.

171 L. Życki, "Człowiek walczy z posuchą. Strategia wielkich batalii," *Trybuna Ludu* 2, no. 70, 12 March 1949, 6; D. Wileński, "Ludzie radzieccy zmieniają przyrodę," *Trybuna Ludu* 2, no. 289, 20 October 1949, 3; H. Dankiewicz, "Człowiek radziecki zmienia przyrodę," *Trybuna Ludu* 4, no. 296, 25 October 1951, 3; "Pasy leśne wokół Warszawy zasłonią miasto przed wiatrami," *Trybuna Ludu* 4, no. 335, 3 December 1951, 5; "Leśne pasy wiatrochronne w dolinie Popradu," *Trybuna Ludu* 5, no. 118, 28 April 1952, 1; A. Oparin, "Nauka w służbie pokoju i budownictwa," *Trybuna Ludu* 4, no. 104, 16 April 1953, 3.

172 *Głos Koszalina*, 19 February 1953.

173 Köhler, "Łysenkizm w botanice polskiej," 94; Köhler, "Łysenkizm w Polsce na tle ówczesnej sytuacji politycznej," 45–46.

174 M. Isakowski, "Pieśń o Stalinie," *Słowo Ludu* 5, 1953, no. 59 (Kielce).

175 "Wiadomości o nowoczesnym rolnictwie," 1949, Archiwum Akt Nowych, Komitet Centralny Polskiej Zjednoczonej Partii Robotniczej, Biuro Polityczne, v/3 (mkf 2819).
176 Courtesy of the Filmoteka Narodowa, www.repozytorium.fn.org.pl
177 ZSRR. *Pustynia Kara-kum*, 1 June 1949, PKF, 23/49.
178 ZSRR. *Hodowla egzotycznych owoców*, 15 June 1949, PKF, 25/49.
179 ZSRR. *Plantacje herbaciane*, 4 August 1949, PKF, 32/49.
180 ZSRR. *Stepy i pustynie w pola uprawne*, 6 October 1949, PKF, 41/49.
181 ZSRR. *Ogrodnicy radzieccy*, 9 November 1949, PKF, 46/49.
182 *Lipcowe ziemniaki*, 3 November 1949, PKF, 45/49.
183 N. Zwolińska-Niedźwiadek, "Stonka ziemniaczana (*Leptinotarsa decemlineata*) – zagrożenie dla polskich upraw ziemniaków," *Nowości Warzywnicze* 52 (2011), 85–86. See D. Sosnowska, J. Pruszyński, and J. Lipa, "Ewolucja metod i środków w zwalczaniu stonki ziemniaczanej (Leptinotarsa decemlineata Say)," *Progress in Plant Protection/ Postępy w Ochronie Roślin* 49 (2), 565–76.
184 *Miczurinowskie sady. Wykład o drzewach*, 7 December 1949, PKF, 50/49.
185 *Ochronne pasy leśne*, 30 May 1951, PKF, 23/51.
186 *Ze świata. ZSRR. Ochronne pasy leśne*, 3 December 1952, PKF, 50/52.
187 A. Paczkowski, "Wstęp," in *Tajne dokumenty Biura Politycznego, PRL–ZSRR 1956–1970*, Warsaw, 1998, 3.
188 "Dokument no. 8, Rozmowy polsko-radzieckie w Moskwie w dn. 25.10-11.11.1958, 5.11.1958 – rozmowa trzecia," in *Tajne dokumenty Biura Politycznego, PRL–ZSRR 1956–1970*, 107.
189 Ibid., 100.
190 It is possible to plant better-suited hybrids, and also Genetically Modified Organisms, which are a contemporary attempt at nature transformation by humans. The corn issue has gained a new dimension in Poland, as the government has recently forbidden growing genetically modified corn MON 810. Even though the new law on seeds does not prohibit using genetically modified plants in the country, individual varieties may be forbidden by virtue of a regulation based on the law, as was done in the case mentioned above.
191 R. Wojtkowiak, H. Kawalec, and A. Dubowski, "Barszcz Sosnowskiego," *Journal of Research and Applications in Agricultural Engineering* 53, no. 4 (2008), http://www.pimr.poznan.pl/biul/2008_4_WKD.pdf (accessed 2 July 2013).

Bibliography

"46 kwintali pszenicy jarej z hektara," Warsaw, 1949.
Amsterdamski, S., "O patologii życia naukowego – casus T.D. Łysenko," *Przegląd powszechny* 4, no. 912 (1989).
——, *Życie naukowe a monopol władzy (casus Łysenko)*, Warsaw, 1981.

Archiwum Akt Nowych, Komitet Centralny Polskiej Zjednoczonej Partii Robotniczej.
Awajew, M., *Trawopolny system rolnictwa*, Warsaw, 1950.
Babicz, J., W. Grębecka, and Z. Wójcik, eds, *Studia z dziejów ochrony przyrody w Polsce*, Warsaw, 1985.
Bilnik, A., T. Świercz, and A. Siudy, *Zbiornik Goczałkowicki wczoraj i dziś*, Goczałkowice, 2004.
Broda, J. *Dzieje lasów, leśnictwa i drzewnictwa w Polsce*, Warsaw, 1965.
———, *Historia leśnictwa w Polsce*, Poznań, 2000.
Chłapowski, D., *O rolnictwie*, Poznań, 1843.
Czermiński, A., *Dzieje wielkiego planu*, Warsaw, 1951.
Davies, N., *Boże Igrzysko. Historia Polski*, Vol. 2, Krakow, 1994.
Dembiński, J., and J. Matysiak, "Materiały Adama Wodziczki," *Biuletyn archiwum PAN*, no. 48 (2007), 172–204.
Dembowski, J., "Teoria Miczurina – Łysenki," *Problemy*, no. 6, 1949, 412–14.
Ehrlich, S., and T. Dominik, *Stalinowski plan przeobrażenia przyrody*, Warsaw, 1950.
Eisler, J., *Zarys dziejów politycznych Polski*, Warsaw, 1992.
Feliksiak, S., ed., *Słownik biologów polskich*, Warsaw, 1987.
Grębecka, W., "Problem ochrony rodzimego krajobrazu a tożsamość narodowa," *Kwartalnik Historii Nauki i Techniki* 55, no. 3–4 (2010), 255.
Herczyński, R., *Spętana nauka. Opozycja intelektualna w Polsce 1945–1970*, Warsaw, 2008.
Hübner, P., *I Kongres Nauki Polskiej jako forma realizacji założeń polityki naukowej państwa ludowego*, Warsaw, 1983.
Jarosz, D., "Główne kierunki działalności państwa w zakresie stalinizacji wychowania dzieci w Polsce w latach 1948–1956," *Mazowieckie Studia Humanistyczne*, no. 2 (1998), 127.
Jaroszyński, P., *Nauka w komunizmie*, Lublin, 2008.
Jarząbek, D., *Współtwórcy przyrody*, Warsaw, 1951.
Jaworski, A., and J. Paluch, "Charakterystyka cech morfologicznych jodeł w drzewostanach o strukturze przerębowej Beskidów Zachodnich," *Leśne prace badawcze*, no. 3 (2007), 7–31.
deJong-Lambert, W., "Lysenkoism in Poland," *Journal of the History of Biology* 45 (2012), 499–524.
Kaliński, J., *Zarys historii gospodarczej XIX i XX w.*, Warsaw, 2000.
Kaznowski, L., *Mniszek gumodajny koksagiz*, Warsaw, 1951.
Kępa, Z., *Marksizm i ewolucja. "Twórczy darwinizm" jako narzędzie propagandy antyreligijnej w latach 1948–1956*, Krakow and Tarnów, 1999.
Kocięcki, S., and W. Strzelecki, "Hodowla lasu," *Rozwój i dorobek Instytut Badawczego Leśnictwa, Prace badawcze IBL*, no. 590, Warsaw, 1981, 81.
Köhler, P., "An Outline of Short History of Lysenkoism in Poland," *Folia Mendeliana* 45 (2009), 44–54.

———, "Botany and Lysenkoism in Poland," *Studies in the History of Biology* 3, no. 2 (2011), 32–53.
———, "Casual Study of Lysenkoism in Polish Botany," *Folia Mendeliana* 46, no. 1/2 (2010), 41–53.
———, *Łysenkizm w botanice polskiej*, Krakow, 2013.
———, "Łysenkizm w botanice polskiej," *Kwartalnik Historii Nauki i Techniki* 53, no. 2 (2008), 86 and 113.
———, "Łysenkizm," *Alma Mater – Miesięcznik Uniwersytetu Jagiellońskiego*, no. 141 (November 2011), 39–42.
———, "Lysenko Affair and Polish Botany," *Journal of the History of Biology* 44, no. 2 (2010), 305–43 (DOI 10.1007/s10739-010-9238-4).
———, "Lysenkoism in the Satellite Countries of the Soviet Union: The Case of Botany in Poland," at XXII International Congress of History of Science, "Globalization and Diversity: Diffusion of Science and Technology Throughout History," Beijing, 24–30 July 2005. Book of Abstracts. Institute for the History of Natural Science, Chinese Academy of Sciences, 419.
———, ed., *Studia nad łysenkizmem w polskiej biologii*, Krakow, 2013.
Krupenikow, I., and L. Krupenikow, *Wasyl Viliams wielki reformator przyrody*, Warsaw, 1951.
Kuźnicki, L., *Autobiografia. W kręgu nauki*, Warsaw, 2002.
Kwapiszewski, J., *Budujemy Kanał Wieprz-Krzna*, Warsaw, 1956.
Listowski, A., "Walka o agrobiologię w Polsce," *Nowe Drogi*, no. 2 (1950), 239.
Łysenko, T.D., *O sytuacji w biologii. Referat wygłoszony na sesji Wszechzwiązkowej Akademii Nauk Rolniczych im. W.I. Lenina 31 lipca 1948 roku*, Warsaw, 1948.
Macko, S. *Leśne pasy wiatrochronne*, Warsaw, 1950.
Majewski, T., "Choroby roślin w Polsce, dzieje ich zwalczania i poznania," manuscript, 2013.
Materiały Konferencji agrobiologicznej leśników, Warsaw, 1953.
Materiały Konferencji Agrobiologów, Biologów i Medyków w Kuźnicach, vol. I–II, Warsaw, 1951.
Michajłow, W., *Co dali Miczurin i Łysenko wsi radzieckiej*, Warsaw, 1948.
———, "Dyskusja o naukach biologicznych w ZSRR," *Głos Ludu*, 7–11 October 1948.
———, *Na pierwszym etapie rozwoju biologii w Polsce* 3, no. 15 (1949), 123–33.
———, *Nauka Związku Radzieckiego – Kraju Socjalizmu*, Warsaw, 1951.
Nowe drogi.
Paczkowski, A., *Pół wieku dziejów Polski*, Warsaw, 2005.
Pieniążek, A., *Sad, który nie wymarznie*, Warsaw, 1950.
Pieniążek, A., Z. Gąsiorowska, St. Kubicki, R. Łęski, and M. Mańczak, *Działka szkolna młodych miczurinowców*, Warsaw, 1951.
Pietkiewicz, K., *Ochronne pasy leśne*, Warsaw, 1952.
Polski Słownik Biograficzny, vol. 39, Warsaw and Krakow, 1999–2000.
Polski Słownik Biograficzny, vol. 46, Warsaw and Krakow, 2009–2010, 401–7.
Problemy.

Program pracy kółek młodych miczurinowców (materiały dla szkolnych kółek zainteresowań), Warsaw, 1954.
Przybylska, K., and S. Zięba, "Siedliskowe uwarunkowania prac urządzeniowych i decyzji planistycznych," *Studia i Materiały Centrum Edukacji Przyrodniczo-Leśnej* 10, no. 3 (2008), 205–6.
Radwan, S., *Przyrodnicze podstawy ochrony i odnowy ekosystemów wodno-torfowiskowych w obszarze funkcjonalnym poleskiego parku narodowego na tle antropogenicznych przekształceń środowiska przyrodniczego*, Warsaw, 2003.
Rubaszewski, A., *Filozoficzne podstawy nauki Miczurina*, Warsaw, 1951.
Schirrmacher, Arne, ed., *Communicating Science in 20th Century Europe: Comparative Perspectives*, Max Planck Institute for the History of Science Preprints, No. 385, 2009, 133–45.
Słowo Ludu.
Sobczyk, P., *Uprawa ryżu*, Warsaw 1952.
Sutyła, J., and L. Zasztowt, "Popularyzacja nauki w Polsce w latach 1918–1951," in *Historia nauki polskiej*, vol. V: 1918–1951, cz. 1, ed. B. Suchodolski, Wrocław, 1992.
Topolski, J., *Historia Polski*, Warsaw, 1994.
Trybuna Ludu.
Wieczorkiewicz, P., "Imperium Stalina," *Historia Rosji*, ed. L. Bazylow, Wrocław, 2005, 479.
Wodziczko, A., "Ochrona przyrody to podstawowe zagadnienie państwowe," *Chrońmy przyrodę ojczystą* 3, no. 3–4 (1947).
———, "Planowanie kraju drogą do utrzymania równowagi w przyrodzie," *Ochrona Przyrody* 17 (1937).
———, "Wielkopolska stepowieje," *Prace Komisji Matematyczno-Przyrodniczej Pozn. Tow. Przyj. Nauk*, Seria B, Vol. X, 1947, 4.
Zajęcia w przedszkolu. Program tymczasowy, Warsaw, 1950.
Zasztowt, L., J. Włodarczyk, and M. Jabłońska, "A Closer Look at the Popularization of Science in Poland during the Stalinist Period. The Magazine *Problemy*, 1945–1956," *Organon* 44 (2012).

Conclusion
Environmental History, East European Societies, and Totalitarian Regimes

Doubravka Olšáková

The consequences of the Stalinist Plan for the Transformation of Nature in Eastern Europe are still very visible today. Although they may not be noticeable at first sight, a well-trained eye can easily discover the impacts the Plan had on the modern landscape. Whereas large, wide fields are the direct result of forced collectivization, which, with the exception of Poland, was a process that occurred en masse across all of Eastern Europe, the reservoirs, hydroelectric plants, and new cities and their extraordinary esthetics are the direct products of the Stalinist Plan.

Today, the Great Plan for the Transformation of Nature has been forgotten, covered in the dust of collective memory, which is often extraordinarily forgiving to collective sin. The French sociologist Maurice Halbwachs (1877–1945) claims in his books on social and collective memory that the social framework of our memory is to a large degree formed on a generational basis.[1] Thus, collective memory is not a short-term affair; it is not just a mere collection of memories that makes up collective memory. It is memory itself that can offer us an image of our past that suits our society.

In contrast to the image of the history of Czechoslovakia embedded in collective memory, documents have been preserved in archives that indicate that at the beginning of the 1950s Eastern Europe was supposed to become one large, colorful, blooming Michurinist garden, in which the pseudoscientific breeding theories of Trofim Denisovich Lysenko would be applied. Lysenko was one of the main theoreticians behind the Stalin Plan for the Transformation of Nature, who was both the chairman of the Lenin All-Union Academy of Agricultural Sciences and the director of the Institute of Genetics at the USSR's Academy of Sciences.

It is therefore natural that the main question that arises from this disproportion between collective memory and archival sources is an absolutely fundamental question of continuity and discontinuity in the environmental history of totalitarian regimes. In one of the few publications to explicitly deal with the issue of science in totalitarian regimes, Paul Josephson suggests about this crucial issue in the history of science that, whereas the theories of historians and philosophers of science, such as Robert Merton, Michael Polanyi, and Karl Popper, are based on the concept of science in democratic regimes and are founded on a specific dynamic of "adversarial and cumulative processes,"[2] a completely different concept exists in totalitarian regimes. However, this is not a static concept, as one might logically deduce from the differences in democratic and totalitarian regimes; it is also a dynamic process in which many more actors and interests are involved and play a much more important role (the state, ideology, national interests, etc.) than in democratic regimes, where this dynamic is created and largely influenced by various schools of scientific thought. The existing dynamics of science and scientific disciplines in totalitarian societies is an important condition for continuity in fields of science, even under these nonstandard conditions. This facilitates, among other things, the ability to explain the continual evolution of certain scientific disciplines despite the objective ideological barriers existing in totalitarian societies.

As can be seen with the example of the implementation of the Stalin Plan in Eastern Europe, continuation in certain fields of science and scientific disciplines was disrupted, but in no way was it stopped. In the 1950s, mainly biological sciences were affected, and even more so the agricultural sciences, which are somewhat overlooked in today's research. In most countries, however, Soviet methods were not mechanically and monolithically adopted over the long-term; more often they were adopted to the extent that suited the situation of local politicians and scientists. Thus, new methods and procedures, which were often centrally directed and coordinated, to a large extent suited certain groups of technocratic planners, who had fruitlessly attempted to promote their ideas in the interwar period. Despite the fact that in the second half of the 1950s pseudoscientific theories were pushed aside and the scientific community was able to wrest its way from its ideological binds, the leading stratum of scientific managers remained in their positions, at least officially, until the 1960s. In the USSR, T.D. Lysenko embodied this phenomenon. In Eastern Europe, this fact is clearly visible in the construction of reservoirs and the regulation of water courses. Work on such projects culminated in the late 1950s, but in the 1960s this boom came to an end. The Stalinist technocratic approach naturally facilitated the rise of a technocratic group of planners. This phenomenon naturally faded away once de-Stalinization occurred.

The rise of these technocrats can easily be explained and the reasons are quite understandable: the postwar situation in Europe, the restoration of destroyed areas, and the rapid push for economic growth meant that European society was placing extraordinary demands on the environment. Once the Iron Curtain had descended across Europe, approaches to the environment were only thoroughly instrumentalized in Eastern Europe—that is to say, in the countries that followed the leadership of the Soviet Union on the path to communism. Thus, nature and natural processes were instrumentalized here in the name of Marxism–Leninism, the ideology in which all the countries of this bloc were fully indoctrinated. To the west of the Iron Curtain, where political parties—and therefore their ideological concepts—competed with each other in a democratic system, there was no similar bond in the form of a unified ideology. The monopoly of a single interpretation of Communist ideology at the very beginning of the international Communist movement, at least until the Sino-Soviet Rift of 1963 when the idea of brotherhood was definitively extinguished, facilitated the advancement of an all-around technocratic approach to the environment. Thus, in reality, despite certain differences from past scientific and environmental projects, the Stalin Plan for the Transformation of Nature can be considered in this context to be one of the first transnational, large-scale environmental projects ever, the goal of which was none other than to achieve changes on a global scale by applying a coordinated approach. When examining the implementation of this plan, the same characteristics can be found in most countries of Eastern Europe: the utilization of the same theoretical bases (Michurinism, Lysenkoism), the utilization of completely identical procedures (the Viliams system, the establishment of shelterbelts), and the introduction of identical species and varieties of plants throughout the entire Eastern Bloc.

The adoption of the Great Plan for the Transformation of Nature, however, called up doubts from the very beginning. In 1947, one year before the Stalin Plan was officially accepted in the USSR, the results of shelterbelt plantings, which had occurred as a result of a 1938 directive from the People's Commissariat of Agriculture, were known. These results were not good. In the Stalingrad region, half of all seedlings were either not successfully transplanted or died later. Entire shelterbelts disappeared, and those that remained lost more than one-third of their original volume as not all the trees planted survived. The results from the Stalingrad region, however, were not so alarming when compared to results from other regions. As a report from 1947 claims: "In the Saratov region 59 percent of shelterbelts perished, in the Kuibyshev regions it was 76 percent, in the Astrakhan region as much as 86 percent, and in the Kazakh SSR 72 percent."[3] The reasons given for such high failure rates included insufficient maintenance of shelterbelts once they were planted, insufficient supplies of seedlings from nurseries, and finally the mechanical establishment of shelterbelts in areas where they had no chance of success.[4]

This is startling because, despite these facts, the decision was made nonetheless to execute the Stalin Plan in which shelterbelts played a critical role. Soon after the resolution was adopted, the number of trees planted in 1949–1950 doubled, and in 1950, 760,000 seedlings were planted. In total, by 1951, 2,073,000 hectares of land had been covered in shelterbelts, meaning that the Fifteen-Year Plan was one-third fulfilled. The remaining two-thirds, encompassing another four million hectares of land, were to be planted with seedlings in the following years. However, as is evident from other sources, in 1953 the establishment of further shelterbelts was completely abandoned, and shelterbelts that had already been planted ceased to be maintained. Assuming the same rate of shelterbelt failure in the future, and with the halting of the establishment of further shelterbelts in 1953, it is clear that not even one-third of the plan could have been implemented in its originally planned scope, and the figure is more like a mere 10 percent. The same calculation can certainly also be applied to the countries of Eastern Europe, where the plan was mechanically implemented by the political elite and where it met with the same or similar problems. The ambitious aims of this project to change not only microclimates but also the macroclimate were therefore never achieved.[5]

A quarter of a century later a similarly, if not more, ambitious plan appeared in neighboring China. In the 1950s, the Great Leap Forward took place there, followed by the Cultural Revolution in the second half of the 1960s, during which the conflict between China and the USSR intensified. Once this conflict died out in the second half of the 1970s, one of the largest environmental projects in China, sharing certain identical characteristics with the Stalin Plan for the Transformation of Nature, was prepared: the Three-North Shelter Forest Program, to cover north, northeast, and northwest China. Work on the Three-North shelterbelt began on 25 November 1978, a quarter of a century after the establishment of further shelterbelts in the USSR had been unofficially halted. In contrast to the Stalin Plan, however, the Three-North project is planned to take seventy-three years, not fifteen, and this great Chinese project should therefore be completed in 2050. Similar plans, albeit on a smaller scale, existed earlier in China.[6] Nonetheless, the question arises of to what extent such a large-scale project as the Three-North Shelter Forest Program was inspired by the Great Stalin Plan for the Transformation of Nature. In any case, the Chinese plan completely exceeds the Stalin Plan in scope; whereas the Great Plan projected the afforestation of more than six million hectares of forest, the Chinese project involves planting nearly twenty-five million hectares of forest (specifically 24,469,000 ha). The goal of the "Great Green Wall of China" is to reduce and stop the expansion of the desert, to stabilize sand dunes (by making mobile dunes semi-fixed or fixed), to reduce the number of sandstorms, and to significantly contribute to improving soil quality. According to the data made available by Chinese offices, in some areas these goals have actually been met: for example, the area under control and protected from desertification has doubled, and the number of sandstorms has been reduced. However, some Chinese scientists have the completely opposite view. They consider the plan to be a "fairytale" and claim that in aggregate the number of sandstorms in north China has not decreased but instead has drastically increased by a factor of more than two.[7]

Although it is not very simple to estimate the full scope of the Stalin Plan for the Transformation of Nature, it can justifiably be assumed that if Stalin had not died in 1953, the plan would have been implemented in its original scope in the USSR as well as in Eastern Europe.[8] If that had been the case, it would have been one of the greatest ecological projects ever and, assuming that it had been successfully implemented in full, would have actually successfully changed not only individual microclimates but also the macroclimate of Central and Eastern Europe. In China, it truly seems as if changes in microclimate are occurring in the north, northeast, and northwest based on analysis of precipitation and less frequent sandstorms. According to criticism emanating from China, primarily from economists and scientists well versed in the agricultural sciences, the Stalin Plan and the Chinese Three-North shelterbelt share common weaknesses. Critics of the Great Green Wall of China cite the 85 percent failure rate of seedlings, meaning that only 15 percent survive.[9] These are roughly the same as figures from the USSR in the 1940s and 1950s.[10] In comparison to the failure rates of older projects, such as the Prairie States Forest Project initiated in the United States by President Franklin D. Roosevelt in July 1934, the differences are quite startling: in the United States, survival rates ranged from 71 to 74 percent at the start of this program and dropped to approximately 62 percent when maintenance was turned over to local farmers.[11]

There are doubtless many more similarities between the Soviet and Chinese projects. Here the important question of the role of continuity comes up again. Certain elements of this plan indicate a certain degree of continuity as well as Soviet inspiration, such as an emphasis on the technocratic aspects of these measures and how they were implemented.

In the case of the Soviet Union and China, are these plans just the same project implemented in a similar totalitarian society a quarter of a century apart from each other? This question cannot be answered yet, as there is a lack of information and modern research has not been conducted. On this issue, modern environmental history has indicated much greater interest in Nazism than Communism.[12] However, comparing Communism and Nazism is not a fruitless effort, as it can facilitate a better understanding of how nature conservation worked in these two types of totalitarian regime. Some of the first studies on the relationship between Nazism and nature conservation, for example, focused on a certain fascination that German conservationists had with Nazism, and purposefully sought out ideological crossover between nature conservation and Nazi ideology.[13] Findings about the implementation of the Great Plan in Eastern Europe tend to follow a second working hypothesis about totalitarian regimes that German historians have recently begun to explore: it was not ideology, but the technology of power in these regimes that is the key to understanding the interpretive framework of environmental history.

In Eastern Europe, where this type of research is just getting started, there is a community of environmental historians facing the same temptations that German historians studying nature conservation faced in the past: the temptation of viewing environmental change during Communism only through the prism of Communism. Operating in this frame of mind, it can certainly be expected that in the future we will read about how landscape changes that took place under the guise of Communist principles were merely the expression of Communist attempts at "controlling nature." The science that was the driving force behind changing approaches to the environment has its own rules and dynamics however. The continuity of a scientific community based on the continuity of different schools of thought has the remarkable ability of being able to preserve its viability during periods of totalitarian rule.

The linking of these two ideologies—Nazism and Communism—however opens up new fields of research. From the perspective of environmental history, we can search for similar characteristics between both of these systems—and most likely we will find them. Besides adopting a technocratic approach, both systems also attribute an esthetic-propaganda function to the transformation of nature. Communism drew from a new, large-scale technocratic esthetic of nature, and was deeply fascinated by the ornamentalism of shelterbelts, as in the visual arts ornamentation rhythmically enlivens surfaces.[14] Here we can find a fascination with the opportunities promised by the transformation of nature, as the ease with which new cities were built during Stalinism can only be compared with similar projects from Nazi Germany and Fascist Italy. Communist rulers did not ignore the esthetics of nature, as is often said today, but instead they attempted to control these esthetics just as they attempted to "control" nature itself.[15]

Now, let us return to the question posed in the introduction of the conclusion: the question of continuity and discontinuity of environmental history in totalitarian regimes. Janusz Groszkowski, the chairman of PAN in 1962–1971, noted in his thoughts on the role of Polish science in the new Polish state that in 1956 it was the first time that a modern concept of nature conservation had been adopted anywhere in the world. It can be debated whether or not Poland really was a world leader in this regard, nonetheless it is a fact that Poland was well on the way to creating a modern system of nature conservation. It seems that a certain amount of continuity was preserved thanks to the high-ranking position of the leading figure in the biological sciences, Kazimierz Petrusewicz, who was also the chairman of the Department of Science and Education of the KC PZPR, the Central Committee of the Polish United Workers' Party. As Janusz Groszkowski notes in his publication from 1970, after World War II eleven new national parks were created in Poland, a network of nature reserves was established that would grow by 1970 to encompass 510 such protected areas, and there were approximately 5,500 "natural monuments" in the country.[16]

The situation in Czechoslovakia was similar: in 1956 Act No. 40/1956 Coll. on State Nature Conservation was adopted. The introductory clause of this law still emphasized an instrumentalized concept of nature and the role of science in "controlling" nature, while still essentially laying the groundwork for modern nature conservation and landscape protection: "Nature conservation is based on the scientific study of natural and human-influenced occurrences in nature, and is conducted in close cooperation with leading scientific institutes." In January 1949, the first national park in Czechoslovakia was opened, the Tatra National Park, which today is located in Slovakia. In 1958, the Implementing Regulation on the Protection of Plant Species was adopted, followed seven years later by the Implementing Regulation on the Protection of Animals.[17] Between 1948 and 1960, in the Czech lands alone more than thirty national nature reserves were founded, which is nearly one-third of all current nature reserves in the Czech Republic today. In May 1963, a second national park was opened, the Krkonoše National Park (located in the Czech Republic today).

Slightly behind these events, the first national parks in Hungary were established in the 1970s: the Hortobágy National Park (1972), the Kiskunság National Park (1975), and the Bükk National Park (1976). At that time, the International Biological Program was running on both sides of the Iron Curtain, which was, after the International Quiet Sun Year (1964–1965) and the International Geophysical Year (1957–1958), the third largest international project under the aegis of UNESCO (more specifically the International Council of Scientific Unions, ICSU), in which countries from the Eastern Bloc were also involved. This was an important program, a fundamental milestone in environmental history for understanding how we currently view and conceive of the biological sciences and their role in the interaction between man, nature, and other scientific disciplines.

The mingling and crossing of various stimuli, factors, and priorities in the Eastern Bloc led to interactions that set further stimuli into action. The Stalin Plan for the Transformation of Nature was implemented in Eastern Europe as part of the larger program of the Sovietization of Central and Eastern Europe as a space—that is, as a buffer zone and transit zone between two political blocs. The failure of this plan was a catalyst for the development of scientific fields and disciplines that paradoxically strayed from Stalinist dogma. This includes the biological and agricultural sciences, which played a key role in the 1950s in the implementation of the Stalinist Plan.

It can be said without exaggeration that never in the modern history of Central and Eastern Europe has there been greater interest in these sciences than in the 1950s during the Stalinist Plan for the Transformation of Nature. The greater the pomp and glory that Lysenko's theories were welcomed with at the onset of this plan, the quieter the results of the plan were discussed at meetings later, and the quieter failed plantings of rice and other crops were cleaned up. The fundamental failure of this plan in the biological and agricultural sciences meant that the later return to using proven methods and a "normal" approach to the environment could take place without any significant break. The failure of the implementation of this plan in the biological and agricultural sciences, however, had two fundamental consequences. First and foremost, the parts of the plan that involved building a "country of steel and iron" were bolstered; in places where nature had failed, man and technology would have to step in. The second consequence was that the development of biological and agricultural sciences was pushed aside to the edges of societal interest. This paradoxically allowed them to develop quietly on the sidelines, outside of the close ideological monitoring they had been subjected to in the 1950s.

These two phenomena, prioritizing industrial landscapes and pushing environmental sciences to the background, started to become significantly apparent a generation later. In the 1980s, growth in environmental activities and activism can be observed that fought against the technocratic approach to the environment, and attempted to "rediscover" an original, complex approach to the environment—one that valued nature in and of itself. It is no great surprise that most of these institutions and organizations that focused on actively struggling for nature conservation were born in the academies of sciences or were directly linked to them.

Despite this, the generational gap between the generation active in the 1950s and later generations is quite clear in this era. People born at the height of the Stalinist Plan for the Transformation of Nature frequently joined these environmental organizations. The experience of the "transformation of nature," which significantly influenced the social framework of collective memory of the generation involved in it, was completely foreign to later generations. Their memory was influenced by other factors, mainly of an ecological character—very often a consequence of the emphasis on technical progress at the expense of environmental protection. In the 1970s, when the report "The Limits to Growth," initiated and funded by the Club of Rome, had already made its way to the Eastern Bloc, a teleological concept of the progress of science and technology was officially promoted, primarily by ruling political leaders. František Toman, the vice-chairman of the Czech National Council, reacted to the conclusions made in "The Limits to Growth" by stating: "The main course the fight against environmental pollution shall follow is, therefore, not the 'freezing' of technological progress. . . . Instead it shall be dealt with by speeding up technological progress and the application of technology for satisfying the material and immaterial needs of mankind." In 1974, one of the first environmental organizations to emerge in Czechoslovakia was founded—Brontosauraus—the name and motto of which referenced the Club of Rome's publication: "The brontosaurus didn't survive because he outgrew his limits." Most of its founding members were born at the time of the Stalinist Plan for the Transformation of Nature.[18]

The Stalin Plan for the Transformation of Nature had significant, complex impacts on both the landscape and society. It was to a large extent a generational experience that is very narrowly connected to the first active postwar generation. Generational issues, however, pose greater challenges to collective memory than do day-to-day occurrences. It is an enormous challenge to collective memory to find a shared framework for grasping this issue that, despite potential problems, could facilitate the further successful development of society.

Therefore, it can be expected that in the near future a new type of tourism will most likely develop; all of the new cities constructed in Eastern Europe according to the plans of Stalinist architects will become welcome places of interest for future generations—just as the Nazi-designed Prora, the Stalinist architecture in the former states of the Soviet Union, and Chinese projects such as the Three-North shelterbelt and the Three Gorges Dam all have. According to official Chinese statistics, the number of visitors to the Three-North shelterbelts increased from 2.3 million in 1977 to a full 90 million in 2007.[19] And thus, despite the fact that this Great Plan for the Transformation of Nature failed in the Eastern Bloc, there is certainly no more important subject for environmental history research in totalitarian regimes, and their consequences in this area. Failed plans—a vast yet ignored topic of research!

Doubravka Olšáková is a senior researcher at the Institute for Contemporary History of the Academy of Sciences of the Czech Republic in Prague, where she leads a working group on contemporary environmental history. She has published a book titled *Science Goes to the People!* (2014, in Czech), which examines mass indoctrination and the dissemination of science in Communist Czechoslovakia.

Notes

1. M. Halbwachs, *Les cadres sociaux de la mémoire*, Paris, 1925; M. Halbwachs, *La mémoire collective*, Paris, 1968; M. Halbwachs and L.A. Coser, eds, *On Collective Memory*, Chicago, 1992.
2. P. Josephson, *Totalitarian Science and Technology*, Amherst, 2005, 10.
3. F. Šrom, "Překonávání následků války v zemědělství a některé problémy rozvoje obilnářství SSSR v poválečných letech," *Slovanský přehled* 70, 1984, 43.
4. Ibid.
5. Ibid.
6. I would like to thank Nicholas Orsillo for calling this fact to my attention. I am also grateful to him for his valuable comments and suggestions for changes to this chapter, as well as to its overall conceptualization. In this regard, my thanks also go to Paul Josephson, David Moon, and Dolly Jørgensen.
7. M.-M. Li et al., "An Overview of the 'Three-North' Shelterbelt Project in China," *Forestry Studies in China* 14, no. 1 (March 2012), 70–72. S. Cao, "Why Large-Scale Afforestation Efforts in China Have Failed To Solve the Desertification Problem," *Environmental Science & Technology* (March 2008), 1828.
8. J.V. Stalin was of course a key figure; however, it must be realized that the sciences in Eastern and Central Europe were not emancipated from Stalinist ideology immediately after his death or as a result of changes in the political climate brought about by Nikita Khruschev's "secret speech" at the Twentieth Congress of the Communist Party of the Soviet Union held in Moscow in February 1956. This point came much earlier: at the Geneva Summit in July 1955. It is strange how significantly the traditional periodization of the Cold War and the periodization of the global history of science differ from each other. That is, here 1955 was the key date, not 1956.
9. Cao, "Why Large-Scale Afforestation Efforts in China Have Failed," 1830.
10. Li et al., "An Overview of the 'Three-North' Shelterbelt Project in China," 75.
11. "Shelterbelts," *Encyclopedia of Oklahoma History and Culture*, http://digital.library.okstate.edu/encyclopedia/entries/S/SH022.html (accessed 5 January 2014).
12. *How Green Were the Nazis?: Nature, Environment, and Nation in the Third Reich* [online], Athens, 2005, http://site.ebrary.com/lib/natl/Doc?id=10141108 (accessed 15 December 2013).
13. Cf. F. Uekötter, "Green Nazis? Reassessing the Environmental History of Nazi Germany," *German Studies Review* 30 (2007), 267–87.

14 Cf. volume 1953 of the journal *Výtvarné umění*, published by the Association of Czechoslovak Visual Artists, which contains many papers focusing on this topic (pp. 94, 134, 275).
15 K. Stibral, O. Dadejík, and V. Zuska, *Česká estetika přírody ve středoevropském kontextu*, Prague, 2009, 45.
16 J. Groszkowski, "Nauka i jej rola w okresie 25-lecia Polski ludowej," *Z problemów polityki naukowej*, Warsaw, 1970, 33.
17 F. Urban, "Nature Conservation in the Czech Republic with Respect to Species Protection," *Workshop on Nature Conservation in Central and Eastern Europe – Present Situation, Needs and the Role of the Bern Convention (Environmental Encounters, no. 18)*, Council of Europe, Strasbourg, 1994, 15.
18 A. Štanzel, "Meze růstu a Československo," *Literární noviny*, no. 30, 25 July 2013, 23.
19 Li et al., "An Overview of the 'Three-North' Shelterbelt Project in China," 73.

Bibliography

Cao, S., "Why Large-Scale Afforestation Efforts in China Have Failed To Solve the Desertification Problem," *Environmental Science & Technology* (March 2008), 1828.
Carson, R., *Silent Spring*, London, 1963.
Čeřovský, J., *Jak jsme zachraňovali svět, aneb, Půl století ve službách mezinárodní ochrany přírody*, Prague, 2014.
David, J., *Smrdov, Brežněves a Rychlonožkova ulice: kapitoly z moderní české toponymie: místní jména, uliční názvy, literární toponyma*, Prague, 2011.
Encyclopedia of Oklahoma History and Culture, http://digital.library.okstate.edu/encyclopedia/entries/S/SH022.html (accessed 5 January 2014).
Förster, H., J. Herzberg, and M. Zückert, eds, *Umweltgeschichte(n): Ostmitteleuropa von der Industrialisierung bis zum Postsozialismus: Vorträge der Tagung des Collegium Carolinum in Bad Wiessee vom 4. bis 7. November 2010*, Göttingen, 2013.
Gestwa, K., *Die Stalinschen Großbauten des Kommunismus: sowjetische Technik- und Umweltgeschichte, 1948–1967*, Munich, 2010.
Groszkowski, J., "Nauka i jej rola w okresie 25-lecia polski ludowej," *Z problemów polityki naukowej*, Warsaw, 1970.
Halbwachs, M., *La mémoire collective*, Paris, 1968.
———, *Les cadres sociaux de la mémoire*, Paris, 1925.
Halbwachs, M., and L.A. Coser, eds, *On collective memory*, Chicago, 1992.
Holdgate, M.W., *The Green Web: A Union for World Conservation*, London, 1999.
How Green Were the Nazis?: Nature, Environment, and Nation in the Third Reich [online], Athens, 2005, http://site.ebrary.com/lib/natl/Doc?id=10141108 (accessed 15 December 2013).
Janáč, J. *European Coasts of Bohemia: Negotiating the Danube–Oder–Elbe Canal in a Troubled Twentieth Century*, Amsterdam, 2012.

Josephson, P.R., *Industrialized Nature: Brute Force Technology and the Transformation of the Natural World*, Washington, 2002.
———, *Totalitarian Science and Technology*, Amherst, 2005.
Kremencov, N.L., *Stalinist Science*, Princeton, NJ, 1997.
Krige, J., *American Hegemony and the Postwar Reconstruction of Science in Europe*, Cambridge, MA, 2006.
Li, M.-M., A.-T. Liu, Ch.-J. Zou, W.-D. Xu, H. Shimizu, and K.-Y. Wang, "An Overview of the 'Three-North' Shelterbelt Project in China," *Forestry Studies in China* 14, no. 1 (March 2012), 70–71.
Stibral, K., O. Dadejík, and V. Zuska, *Česká estetika přírody ve středoevropském kontextu*, Prague, 2009.
Strakoš, M., *Nová Ostrava a její satelity. Kapitoly z dějin architektury 30.–50. let 20. století*, Ostrava, 2010.
Šrom, F., "Překonávání následků války v zemědělství a některé problémy rozvoje obilnářství SSSR v poválečných letech," *Slovanský přehled* 70, 1984, 43.
Štanzel, A., "Meze růstu a Československo," *Literární noviny*, no. 30, 25 July 2013, 23.
Uekötter, F., 'Green Nazis? Reassessing the Environmental History of Nazi Germany," *German Studies Review* 30 (2007), 267–87.
———, *The Green and the Brown: A History of Conservation in Nazi Germany*, Cambridge, 2006.
Urban, F., "Nature Conservation in the Czech Republic with Respect to Species Protection," *Workshop on Nature Conservation in Central and Eastern Europe – Present Situation, Needs and the Role of the Bern Convention (Environmental Encounters, no. 18)*, Council of Europe, Strasbourg, 1994, 15.
Vaněk, M., *Nedalo se tady dýchat: ekologie v českých zemích v letech 1968 až 1989*, Prague, 1996.
Weiner, D.R., *A Little Corner of Freedom: Russian Nature Protection from Stalin to Gorbachëv*, Berkeley, 1999.

Index

afforestation, 10, 13, 18–20, 28, 31, 51, 64–68, 111, 114, 177, 180–81, 194, 197, 231, 236, 238–41, 246, 249, 258–62, 265, 268, 271–72, 274, 276, 294
Africa, 166
agricultural engineering (agro-engineering), 4, 253, 269
Agricultural Science Center (Agricultural Experiment Center; Hungary), 145, 147–49, 152, 163, 168, 171, 178, 180, 183, 208n115
agricultural sciences, 49, 53, 57–59, 63, 143, 235–36, 249, 291, 294, 297–98
agrobiology, 17, 20, 26, 50, 53–54, 57–59, 63, 73, 141–42, 144–45, 147, 162, 245–46, 250–51, 262, 267
Akali (Balatonakali), 168, 170
Albania, 88
Alföld (Great Hungarian Plain), 27, 29, 170, 172, 176, 180–84, 193–94
Algeria, 88
All-Union Communist Party (Bolsheviks). *See* Communist Party of the USSR
Alsógalla, 135
Amsterdamski, Stefan, 243, 250
Anajan, Anait, 272
Anzherodzhensk, 36
Armenia, 14, 272
Astrakhan, 7, 293
Austria, 88
Avaev, M., 262
Azerbaijan, 14

Babos, Imre, 185
Bacílek, Karol, 87
Bakhtadze, Xenia, 271
Bakonszeg, 187
Baláže, 89
Balázs, Ferenc, 164
Bálint, Andor, 142
Bánhida, 135
Banská Štiavnica, 88
Baranov, Pavel Aleksandrovich, 169, 213n191
Barcika, 135
barley production, 12, 31, 231
Baťa, Jan Antonín, 79–80
Baťa, Tomáš, 79
Békés, 152, 157
Belarus, 262
Belgium, 64, 199n2
Bencsik, István, 167
Beneš, Edvard, 45, 48
Bereczkei, Tamás, 140
Berente, 135, 202n31
Beria, Lavrentiy Pavlovich, 129, 132
Berlin, 45–46, 59–60
Bezruč, Petr, 85
Białobok, Stefan, 246, 273
Bielik, Paľo, 94
Bierut, Bolesław, 30, 32, 227, 230, 274
biology, 17, 26, 31–32, 35, 52–53, 61, 112, 140–41, 143, 147–48, 150–51, 195–97, 203n45, 206n95, 235, 237–38, 243–55, 257, 264–70, 273, 275, 291, 296–98
Błędów, 241
Bludovice (Havířov), 84
Bobrowski, Czesław, 230
Bodnăraș, Emil, 131–32
Bohumín, 86
Borsod, 25, 135, 157, 190

Boyko, Vasily Romanovich, 132–33
Bozsó, Sándor, 159
Brain, Stephen, 18
Braşov (Oraşul-Stalin), 138
Bratsk, 105, 134
Brazil, 4
Brezhnev, Leonid Ilyich, 26, 36, 108
Brno, 52–53, 86
Brussels, 98
Budapest, 29, 131, 152, 161, 163, 170, 200n6
Budzko, Igor Alexandrovich, 61
Bükk, 297
Bulgaria, 3, 5, 88, 110, 132–33, 167, 253
Burbank, Luther, 267
Bydgoszcz, 236

Caillois, Roger, 77
California, 33–34
canals, 2, 4, 6, 8–10, 12–13, 15–16, 31–33, 92, 100–101, 106, 193, 197, 226, 236, 240–41, 254
  Danube–Oder(–Elbe) Canal, 100, 110, 241
  Danube–Tisza Canal, 28, 183–84, 218n266
  Eastern Main Canal (Hungary), 172, 183, 185–87
  Fergana Canal, 14–15, 19
  Moscow–Volga Canal, 8, 19
  Panama Canal, 4, 16, 101
  Suez Canal, 4, 16
  Turkmen Canal, 16, 100, 104
  Volga–Don Canal, 3, 103, 105, 240
  White Sea–Baltic Canal, 8, 19
  Wieprz–Krzna Canal, 32, 242–43, 276
Celldömölk, 149
Central Asia, 2, 8, 10–13, 15–16, 26–27, 79, 239–40, 250, 270–71
Čepička, Alexej, 131–32
Chaikin, P. I., 56
Charles IV, 98
Chervenkov, Valko, 132–33
China, 4, 34–35, 110, 167, 293–95
Chkalov, 274
Chłapowski, Dezydery, 256–57

Chornohora, 263
Churchill, Winston, 30, 227
Čierná nad Tisou, 86
citruses production, 2, 4, 26–27, 32, 36, 149, 151, 167–71, 174, 177, 191, 197, 207n98, 252, 271
collectivization, 3, 6, 10–11, 13–14, 19–21, 25, 27, 30–31, 33, 44, 49–52, 65–66, 128, 134, 184, 231–32, 274, 277, 290
Colorado potato beetle, 76–79, 268, 272–73
Colorado (state), 34
Communist Party of Czechoslovakia (KSČ), 20, 45–49, 51, 58–59, 61–62, 69, 81, 83, 89, 94, 101, 111, 131–32
Communist Party of Hungary. *See* Hungarian Workers' Party
Communist Party of Poland. *See* Polish United Workers' Party
Communist Party of Slovakia (KSS), 101, 115n7
Communist Party of the USSR, 1–2, 6, 23, 25, 51, 54–56, 61, 102, 104, 108, 141, 196, 238, 274, 300n8
corn production, 15, 23, 32–33, 36, 55, 138, 155, 157, 195, 197, 214n191, 252–54, 274–76, 286n190
cotton production, 2, 4, 10, 12, 15, 19, 26–29, 36, 71, 74, 148–64, 166–67, 173–74, 176–77, 186, 190–92, 195–97, 207n107, 208nn115–16, 208n120, 209n126, 209n129, 210n140, 210n143, 214n191, 255
Council of Mutual Economic Assistance (COMECON), 27, 29–30, 62, 153, 196, 229, 277
Csepel, 25, 136
Csongrád, 157
Cyrankiewicz, Józef, 275
Czchów, 242
Czech Republic, 22, 297
Czechoslovak Academy of Agricultural Sciences (CAAS), 20, 43, 52–63, 70, 74, 108–109, 112–13, 116n35
Czechoslovak Academy of Sciences, 53, 57–63, 111–13, 122n201

Czechoslovakia, 3, 5, 19–23, 29, 32, 43–52, 54–58, 62–65, 67–83, 85–86, 88–89, 91–92, 94, 96–108, 111–13, 132, 137, 210n140, 228, 247, 269, 273, 290, 297, 299
Czermiński, Adam, 240–41
Czetwertyński, Edward, 242

dams and reservoirs, 1–2, 4, 6, 8–10, 12–13, 16, 20, 22–23, 32, 34–36, 44, 50–51, 65, 68, 84–86, 89–90, 92–93, 95–100, 102–107, 111, 177, 181, 185, 193, 197, 226, 239, 241–42, 290, 292
   Czorsztyn Reservoir, 32, 242
   Dnieperstroi Dam, 22, 90
   Gabčíkovo-Nagymaros Reservoir, 22, 90
   Goczałkowice Reservoir, 32, 242
   Kružberk Reservoir, 83, 99
   Lipno Reservoir, 23, 96–98, 101, 107, 121n169
   Orava Reservoir, 22–23, 90, 93–95, 101
   Orlík Reservoir, 23, 92, 95–96, 98–101, 105, 108, 122n201
   Rożnów Reservoir, 32, 242
   Slapy Reservoir, 23, 96–98, 101, 104–105
   Three Gorges Dam, 4, 34–35, 299
   Tiszalök Dam, 28, 172–73, 181–84, 186–87, 190, 217n258
   Váh Cascade, 23, 90, 93, 95–96, 98, 101
   Vír Reservoir, 99
   Žermanice Reservoir, 99
Danzig, 228
Darwin, Charles, 267
Davies, Norman, 230
Davydov, Mitrofan Mikhailovich, 100, 181
deJong-Lambert, William, 243, 247
Dembowski, Jan, 235, 244, 248
Denmark, 64, 258
Derera, Miklós, 154
de-Stalinization, 32, 47, 72, 107, 113, 247, 274, 277, 292
Diósgyőr, 25, 136

Dmitrovgrad, 3
Dnieprostroi, 19
Dokuchaev, Vasily Vasilievich, 19, 37n4, 177, 180, 182, 236, 257–58, 272
Dominik, Tadeusz, 243, 246–47, 257–58
drought, 1, 3, 10–11, 13–14, 18–19, 34, 68, 91, 109, 182, 193, 195, 197, 236, 238, 270, 276
Dubnica, 95
Dunaújváros (Sztálinváros, Dunapentele), 3, 23–25, 29, 128, 135–39, 190, 193
Ďuriš, Julius, 49, 65
Dushanbe (Stalinabad), 134
Dyemchenkova, Maria, 75

ecology, 7, 18, 36, 112, 238, 258, 261, 294, 298. *See also* nature conservation and protection
Ehrlich, St. (Stefan), 257, 283n113
Eisenhüttenstadt (Stalinstadt), 138
England, 64, 166, 256, 258
Erdei, Ferenc, 185, 207n107, 208n116, 218n263
Eszterházy, Pál, 190

Falenty, 237
famine, 3, 13, 238, 267
Farkas, Mihály, 132, 201n24
Federal Republic of Germany, 77
Felsőgalla, 135
Fertőd, 168–71, 191
fiber crop production, 27, 166–67, 254
France, 64, 88, 199n2
Frank, Josef, 83
Frank, Melanie, 184–85
Fuchs, Vítězslav, 83

Gajewski, Wacław, 268–69
Ganazhenkova, Olga, 75
genetics, 17–18, 26, 29, 31, 139–43, 196, 203n45, 239, 247–48, 250, 268, 291. *See also* Michurinism
geoengineering, 3–4, 6, 9, 12, 23, 31–35
Georgia, 271
German Democratic Republic (GDR, East Germany), 5, 62, 77, 110, 247

Germany, 45, 88, 200n6, 202n32, 228, 258, 260, 296
Germuska, Pál, 134
Gerő, Ernő, 128, 136, 144–45, 149, 199n2, 201n24
Gheorghiu-Dej, Gheorghe, 132
Gluschenko, Ivan, 142, 145
Goetel, Walery, 264
Gomułka, Władysław, 32, 229, 274–75
Gorbachev, Mikhail Sergeyevich, 48
Gorky (city), 9
Gorky, Maxim, 9
Gorzów, 272
Gottwald, Klement, 19, 45, 47, 49, 54, 57, 65, 79–80, 108, 111
Gottwaldov. *See* Zlín
grass-crop rotation system, 54, 58, 65, 69–71, 73, 148, 177–81, 185, 196, 239–40, 262. *See also* Viliams's ley farming system
Great Works of Socialism, 44, 50, 53, 85, 88–89, 98, 101–105, 107, 111, 114
Greshnevo, 6
Groszkowski, Janusz, 296
Guevara, Ernesto "Che", 108
Gusev, Nikolai Ivanovich, 132
Gyula (city), 157

Halbwachs, Maurice, 290
Havířov, 84–85
Hegedűs, András, 185, 195
Hódmezővásárhely, 184
Hofman, Jaroslav, 64
Holý, Ivan, 80
Horáková, Milada, 47
horticulture, 138, 148, 151, 237, 248–49, 251, 256, 259, 260, 266, 270, 272–73
Hortobágy, 23, 27, 172–73, 175, 184, 197, 297
Hronská Dúbrava, 88
Hryniewiecki, Bolesław, 264
Hungarian Academy of Sciences, 24, 29, 143–49, 169, 178, 185, 195, 197, 204n65, 212n164, 214n201
Hungarian Science Council, 144–46, 148, 205n74

Hungarian Workers' Party (Hungarian Communist Party), 23–26, 29, 126–30, 132, 136–37, 141, 143–44, 147, 149–50, 163, 166–68, 171, 174, 184–85, 188–90, 192, 195, 199n2, 199–200n6, 201n24, 206n95, 208n116, 210n144, 213n173, 217n258, 218n263
Hungary, 3, 5, 19–20, 22–27, 29, 32, 59, 71, 78, 88, 110, 126–28, 130–32, 135–39, 142–43, 145–46, 149–55, 160–63, 166–69, 172–74, 176–81, 184, 186–87, 189–92, 194, 196–97, 198n2, 200n7, 202n32, 207n98, 208n115, 208n117, 209n129, 210n140, 210n145, 212n166, 214n192, 217n258, 218n261, 222n337, 247, 253, 297
hydroelectric power, 1–3, 10–14, 16, 22–23, 34, 36, 44, 85, 90, 95–99, 101, 103–105, 241, 290
hydro-engineering, 8, 51, 100, 106, 239–42. *See also* canals; dams and reservoirs

Igali, Sándor, 142
Ilava, 95
India, 4, 88, 166, 191
Indonesia, 4
industrialization, 1, 3, 5–6, 10, 19–20, 24–25, 29, 36, 62, 80–86, 89, 91, 99, 101, 107, 128–31, 133–37, 139, 144, 156, 187, 193, 210n145, 230–33, 260, 263
Iregszemcse, 161, 166–67, 207n106, 212n164
irrigation, 1–2, 7–8, 10–14, 17, 20, 27–28, 32–34, 90, 92, 95, 106, 145, 151, 153, 160, 170, 172–75, 179, 181–87, 190, 193–95, 197, 218n266, 226, 236, 240–42, 253–54, 266, 271
Isakovsky, Mikhail, 269
Italy, 64, 167–68, 170, 296
Iváncsa, 156
Ivanitstky, N. A., 73

Jakoby, Július, 87
Jantokhova, Darikha, 75

jarovization. *See* vernalization
Jarząbek, Dawid, 267
Jaworzno, 31, 231
Jeszenszky, Árpád, 168
Jindra, Jan, 64
Jordán, Mikuláš, 87
Josephson, Paul, 291

Kafka, Franz, 68, 114
Kansas (state), 34
Kapitsa, Pyotr Leonidovich, 61
Kaplan, Karel, 131
Karakalpakiia, 15
Karakum, 15, 270–71
Karviná, 81–82, 84, 99
Kasimovski, Evgeni, 12
Katowice (Stalinogród), 138
Kazakhstan, 13, 14, 160, 239, 271, 282n104
Kazimierz Dolny, 237, 252
Kazincbarcika, 24–25, 128, 135, 202n31
Kaznowski, Lucjan, 237, 253–54
Kemerovo, 36
Kępa, Zbigniew, 243
Keszthely, 168, 170, 206n85
Kharkov, 16
Khrushchev, Nikita Sergeyevich, 2, 23, 32–33, 36, 47, 54–55, 129, 196, 274–75
Kielce, 269
Kirichenko, Fedor Grigoryevich, 61
Kirovsk, 10
Kisbattyán, 135
Kiskunság, 297
Kisújszállás, 170, 184, 207n106
Klečka, Antonín, 43, 54, 57, 63, 112–13
Kliment, Gustav, 83
Kobendza, Roman, 264
Kodaj, Dušan, 23, 94
Kodály, Zoltán, 143, 204n65
Köhler, Piotr, 243, 248, 251, 269
Komárom, 135, 157
Komló, 24–25, 128, 135–36
Korea, 78, 192, 276
Korean War, 24, 78, 131–32, 192, 231
Kórnik, 238, 273
Kortowo, 247

Košice, 45–46, 86–87, 110
Kostychev, Pavel, 37n4, 236, 258
Kőtelek, 159
Kovács, Béla, 126
Kovátsits, László, 161, 211n155
Krajski, Wacław, 246
Krakow, 22, 31, 33, 231, 234, 238, 242, 266
Krasnoyarsk, 73
Krivoj Rog, 137
Křižanovice, 99
Kroměříž, 76
Krzhizhanovskii, Gleb, 15–16
Krzysik, Franciszek, 268
Kuibyshev, 3, 181, 293
Kunčice (Ostrava), 81–82
Kunhegyes, 184
Kurnik, Ernő, 162, 167, 212n164, 212n170
Kuznetsk, 36
Kuźnice, 246
Kuźnicki, Leszek, 243, 251, 269, 277
Kyselovice, 76

Ladce, 95
Le Corbusier, 79
Lekczyńska, Jadwiga, 237, 252–54, 282n88
Lenin All-Union Academy of Agricultural Sciences of the Soviet Union (VASKhNIL), 18, 52–54, 56, 59, 61, 249, 290
Lenin, Vladimir Ilyich, 7, 14–15, 188, 231
Leningrad, 9, 16–17, 192
Lidice, 89
Listowski, Anatol, 237, 244, 251, 262
Lithuania, 31
Litvínov, 89
livestock production, 12, 14, 31, 44, 72, 74, 196, 231
Lobanov, Pavel Pavlovich, 61
Łódź, 233, 275
London, 45–46, 228
Los Angeles, 34
Lublin, 31, 231, 242–43
Lukashev, A. M., 73
Lviv, 31, 228, 233, 238

Lysenko, Trofim Denisovich, 4, 17–18, 20–21, 25–26, 29, 33, 52–54, 60–61, 140–42, 144, 175–76, 190, 195–97, 203n45, 243, 245–56, 265, 267–69, 272, 275, 277, 290–91, 298. *See also* Lysenkoism

Lysenkoism, 4, 12, 17–20, 25–26, 28–29, 31–32, 52, 54, 141–43, 176, 190, 195–96, 243, 245–52, 254, 256, 258, 262, 265–69, 272, 275, 277, 292, 298. *See also* Lysenko, Trofim Denisovich

Lysenkova, Maryia, 75

Macko, Stefan, 258–59
Macura, Vladimír, 76
Magnitogorsk, 10, 25, 73, 134
Majkut, Jozef, 87
Makarewicz, Aniela, 246, 251, 281n82
Málek, Ivan, 53
Malenkov, Georgy Maxilianovich, 129–30, 132
Małogoszcz, 272
Malstev, Terenty Simonovich, 70
Mánfa, 135
Manteuffel, Ryszard, 234, 243
Mao Tse-tung, 34
Marchlewski, Teodor, 266, 275
Marshak, Ilya (pseudonym M. Ilin), 9
Marx, Karl, 188
Marxism-Leninism, 44, 248, 292
Mazlumov, Avedikt Lukyanovich, 61
Mazurov, Abbas, 271
Mecsekfalu, 135
Mecsekjánosi, 135
Mělník, 96
Mendel, Gregor Johann, 63, 141, 150, 196
Merton, Robert King, 291
Mezőcsát, 184
Mezőtúr, 170
Michajłow, Włodzimierz, 244–46, 248, 250, 256, 267
Michurin, Ivan Vladimirovich, 17, 21, 26, 29, 71, 74, 140–41, 144, 188, 190, 196, 245–46, 248–49, 251, 255, 267–73, 276. *See also* Michurinism

Michurinism, 17, 19–21, 29, 31–32, 35, 50, 54, 57–59, 61, 63, 71–74, 76, 100, 140–42, 144, 151, 190, 195–97, 243, 245, 248–51, 255–56, 265, 268–73, 290, 292. *See also* Michurin, Ivan Vladimirovich

Mikołajczyk, Stanisław, 227, 229
Milly, Dezider, 94
Minc, Hilary, 230–31, 260
Mindszenty, József, 190
Minsk, 16
Mňačko, Ladislav, 87
Mohács, 136–37, 160, 206n85
Molotov, Vyacheslav Mikhailovich, 10, 132
Molotovsk, 10
Morgan, Thomas Hunt, 141, 150, 195
Moscow, 5, 9, 16, 21, 23, 27, 30, 45–48, 59, 61–62, 70, 72, 108, 129, 131–33, 145, 227, 229, 233, 268, 275, 277, 300n8
Mosonmagyaróvár, 161, 164, 206n85, 207n106, 208n115
Mosonyi, Emil, 182, 185, 217n258
Most (city), 89
Mrška, Jan, 71
Munich, 46, 54

Nagy, Ferenc, 126
Nagy, Imre, 129–30, 186, 195
Nagytétény, 184, 206n85
nationalization, 7, 20, 23, 30, 46, 49, 79, 81, 87, 126, 161, 231–33
nature conservation and protection, 7–8, 21, 32, 64–67, 102, 109, 112–13, 130, 194, 226, 257, 261, 263–65, 268, 276, 295–98. *See also* ecology
Nejedlý, Zdeněk, 59
Nekrasov, Nikolay Alexeyevich, 6
Nepomuk, John of, 97
New Mexico, 34
New Orleans, 34
Nógrád, 157, 210n140
Norilsk, 10
Norway, 88
Nosice, 95

Novosibirsk, 72
Nowa Huta, 3, 21, 242
Nowe Tychy, 84
nuclear energy, 240

oats production, 15, 31, 231
Obermayer, Ernő, 152, 172, 185
Ochab, Edward, 131–32
Oklahoma, 34
Olomouc, 86
Olonichenko, A. I., 73
Olsztyn, 247
Oregon, 34
Örkény, István, 138
Ostrava, 21, 80–85, 89, 99, 110
Oświęcim, 31, 231
Ózd, 25, 136

Pálos, László, 168
Panchevski, Petar, 132
Panin, Vasily, 271
Paris, 45
Páter, Károly,185
Pávkovits, Katalin, 191
Pechora, 10, 134
Penek, Bohumil, 75–76
Perner, Jan, 86
Peschel, Rudolf, 83
Petrozavodsk, 16
Petrusewicz, Kazimierz, 244, 247–48, 267, 296
Pieniążek, Szczepan Aleksander, 237, 251, 256, 259–60
Pietkiewicz, Kazimierz, 260
Pittaway, Mark, 25
Podolia, 263
Poland, 3, 5, 19–20, 22, 29–33, 45–46, 62–63, 80, 86, 88, 110, 132–33, 137, 226–30, 232–36, 238–49, 251–69, 272–77, 278n5, 286n190, 290, 296
Polanyi, Michael, 291
Polesia, 243, 263
Poliakova, Alexandra, 274
Polish Academy of Sciences, 31, 235, 237–38, 244, 264, 281n82, 282n88, 296

Polish United Workers' Party (Polish Communist Party), 30, 132, 227, 229–30, 235, 244, 266–67, 269, 274, 277, 296
Poltava, 259
Popper, Karl Raimund, 291
Poprad, 269
popularization, 3, 6, 21, 29, 32, 59, 64, 99, 151, 188–90, 193, 234, 240–41, 247, 257–58, 265, 267–68
Porpáczy, Aladár, 168–70, 214n201
Poruba (Ostrava), 84
Potsdam, 228
Poznań, 256, 265, 273
Prague, 45–48, 53, 59–60, 69, 74, 77, 86, 96, 102, 109
Prezent, Isaak, 17
Prokop'evsk, 36
Prończuk, Józef, 237
propaganda, 1, 4, 12, 15, 17, 28, 51, 61, 64, 67, 76–77, 85–86, 92, 94–97, 101–108, 111, 134, 140, 151, 172, 175, 178, 181, 186, 188–91, 194, 228, 232, 248, 250, 259–60, 265–68, 270, 273, 277, 296
Prosenice, 74–76
Przewóz, 242
Puławy (Nowa Aleksandria), 236–37, 254, 257
Püsk, 161, 163, 165, 208n115
Putin, Vladimir Vladimirovich, 8

Radzików, 237
railways, 10, 22, 28–29, 36, 81–82, 86–89, 173, 180, 189
Rajk, László, 127, 199n6
Rakitin, I. N., 266
Rákosi, Mátyás, 23–24, 28, 126–33, 149, 152, 160, 181, 188, 198–99n2, 200n7, 201n24
Ratkó, Anna, 191
Rázompuszta, 183
reservoirs. See dams and reservoirs
rice production, 2, 15, 21, 27–28, 32, 70–72, 74, 151–52, 154–55, 172–75, 184, 186, 194, 197, 222n337, 252–55, 276–77, 298

Rogów, 246
Rokossovsky, Konstantin Konstantinovich, 132–33
Romania, 5, 70, 78, 101, 110, 132, 253
Roosevelt, Franklin Delano, 227, 295
Rościszewski, Jakub, 272
Rożnów, 32, 242
rubber production, 18, 27, 149, 150–51, 160–66, 197, 211nn155–56, 212n162, 212n165, 212n170, 213n173, 254
Rudakov, V. I., 56
Rudnev, Lev Vladimirovich, 30
Růžička, Karel, 109
Rybinsk, 16, 239
rye production, 12, 31, 176, 231

Sacramento, 34
Sajókazinc, 135
Salgótarján, 25, 136
Samara, 7
Sándor, Ottó, 159
Saratov, 7, 104, 272, 293
Sarlai, Károly, 164, 213n173
Schabanov, Alexander, 142, 145
Schüller, Ferenc, 152, 207n107
Semipalatinsk, 72
shelterbelts (protection forests), 1–3, 10–15, 18, 20–21, 28, 31, 35, 44, 50–51, 54, 63–66, 68, 79, 91, 106, 177, 180, 185, 194, 197, 226, 236, 246, 250, 253, 256–61, 267–68, 272–74, 277, 282n104, 292–96, 299
Shtemenko, Sergei Matveevich, 132
Siberia, 2–3, 8, 10–11, 14, 21, 36, 70–71, 73, 107, 134, 179, 181, 192, 240–41
Sihanouk, Norodom, 108
Silesia, 84–86, 231, 241–42, 262
Skierniewice, 237, 251
Skoblykov, A., 153, 156, 159
Skryabin, Konstantin Ivanovich, 61
Slánský, Rudolf, 47, 51, 69, 83, 111, 132
Slovakia, 22, 45–46, 71, 86–90, 92–93, 95–96, 98, 100, 102, 104, 113, 273, 297
Smetana, Bedřich, 107–108
Smrkovský, Josef, 69, 111

Sobczyk, Piotr, 252–54
Sochi, 73
socialist cities, 3, 10, 20, 25, 27, 79–81, 83–85, 89, 128, 130, 134–39, 194, 197, 290, 296, 299
Sólyom, Barna Zoltán, 161, 212n162
Somos, András, 185
Soviet Academy of Sciences, 142, 145, 169, 203n45, 213n191, 235, 291
Soviet Union (USSR), 1–5, 7, 10–11, 13, 15–16, 18–23, 26, 28–30, 32–33, 35–36, 43–47, 51–61, 64–65, 67, 70–73, 75–78, 88, 92–94, 98, 100–104, 106, 110, 112, 126–27, 130, 132–34, 137, 140–42, 150–51, 153, 160, 162, 164, 166–68, 170–71, 175–76, 178–81, 188, 192, 196, 199n2, 202n32, 214n191, 226, 228, 230, 232, 235, 237–40, 243, 245–51, 253–62, 266–77, 282n104, 291–95, 299
Sovietization, 44, 50, 52, 54, 56, 65, 108, 113, 143, 145, 147, 227, 230, 297
Spirin, V. V., 73
Sromowce Wyżne, 242
Stalin, Joseph Vissarionovich, 1, 3–10, 12–20, 23, 28–32, 36, 43, 47, 52, 68, 100, 103, 106–108, 111, 127, 129, 131–33, 139, 186, 192–94, 196, 227, 229–30, 234, 238–40, 246–47, 249–50, 257, 265–67, 269, 274, 276–77, 294
Stalingrad (Volgograd), 12, 80, 138, 181, 293
Stalinization, 30, 43, 47, 144
Stoletov, Vsevolod Nikolaevich, 268
Štrougal, Lubomír, 62
Strzemski, Michał, 237, 252
Šturdík, Jozef, 95
sugar beet production, 12, 15, 21, 60, 72, 74–76
Šumbark (Havířov), 84
Sus, Nikolai Ivanovich, 64
Svirsk, 19
Szabadszállás, 184
Szabó, Irén Klára, 141
Szabolcs-Szatmár, 157, 217n257

Szafer, Władysław, 263–64
Szarvas, 168, 184, 206n85
Szczecin, 265
Szczekociny, 255
Szeged, 153, 166, 172, 209n126
Székkutas, 153–54, 207n106
Szelevény, 155
Szentes, 154, 207n106
Szolnok, 152, 154, 184, 208n120
Sztálinváros. *See* Dunaújváros

Tadzhikistan, 14
Taraxacum kok-saghyz. *See* rubber production
Tatabánya, 25, 135
Teheran, 228
Tengelic, 167
Texas, 34
Tihany, 168, 170
Tiszaföldvár, 159
Tiszántúl, 173–74, 181, 183–84, 186–87
Tito, Jozip Broz, 136, 156, 192
Titov, P. I., 56
Tolna, 161, 208n120
Toman, František, 299
Tomsk, 36, 73
Toruń, 233
Trefulka, Jan, 68, 114
Turew, 256–57
Túrkeve, 170
Turkmenistan, 12, 271

Ukraine, 3, 31, 75, 78, 88, 191
United States of America (USA), 2, 15–16, 35, 77–78, 121n153, 132, 160, 166, 192, 211n158, 228, 295
Ústí nad Labem, 92
Uzbekistan, 12, 14, 16, 160, 192, 239, 271

Varna (Stalin), 138
Vas, Zoltán, 185
Vasilevsky, Aleksandr Mikhailovich, 132
Vavilov, Nikolai Ivanovich, 239, 248
vernalization (jarovization), 17, 26, 53, 151, 175, 177, 179, 196, 250–51, 253–56
Veselý, Josef, 74

Vienna, 86, 198–99n2
Viliams's ley farming system, 54, 69, 73, 177, 236, 243, 262, 267, 269–70, 292. *See also* grass-crop rotation system; Viliams, Vasily Robertovich
Viliams, Vasily Robertovich, 177, 180, 182, 236, 243, 246, 255–56, 258, 262, 267, 269–70, 292
Villány, 160, 168, 170
Vilnius, 31, 228, 233
Vítkovice (Ostrava), 21, 80–81
Vlasov, Andrej Andrejevich, 161
Volgograd. *See* Stalingrad
Volhynia, 263
Volkhovsk, 19
Vorkuta, 134
Vraný nad Vltavou, 96

Warsaw, 30–31, 227, 231, 236–37, 251, 264, 269
Washington (state), 34
water management, 62, 65–66, 89–92, 95–96, 99–102, 104, 110–11, 122n201, 173, 181–84, 193, 217n258, 238, 242. *See also* canals; dams and reservoirs; hydroelectric power; hydro-engineering; irrigation
wheat production, 26, 32, 34, 55, 69, 149, 160, 175–77, 179, 186, 195, 231, 250–52, 255–56, 265–66
Wielkopolska, 259, 264–65
Wittfolgel, Karl, 101
Włoszczowa, 255
Wodziczko, Adam, 259, 264–65
Wrocław, 233, 257

Yalta, 228–29
Yugoslavia, 131, 136, 155

Záhony, 164
Zbenické Zlákovice, 98
Žďákov, 98
Zhigulyovsk, 134
Zhuk, Sergei Yakovlevich, 8
Zlín (Gottwaldov), 79–80, 84
Zorin, F. M., 73
Żuławy Gdańskie, 241

www.ingramcontent.com/pod-product-compliance
Lightning Source LLC
Chambersburg PA
CBHW072145100526
44589CB00015B/2092